ESSAYS ON
MORAL REALISM

ESSAYS ON
MORAL REALISM

EDITED BY

Geoffrey Sayre-McCord

Cornell University Press

ITHACA AND LONDON

First published 1988 by Cornell University Press.

International Standard Book Number (cloth) 0–8014–2240–X
International Standard Book Number (paper) 0–8014–9541–5
Library of Congress Catalog Card Number 88–47753

Printed in the United States of America
Librarians: Library of Congress cataloging information
appears on the last page of the book.

The paper in this book is acid-free and meets the guidelines for
permanence and durability of the Committee on Production Guidelines
for Book Longevity of the Council on Library Resources.

Contents

I   MORAL ANTI-REALISM

II   MORAL REALISM

[v]

Contents

Contributors

A. J. Ayer, Emeritus, New College, Oxford University
Simon Blackburn, Pembroke College, Oxford University
Richard Boyd, Cornell University
Gilbert Harman, Princeton University
Jonathan Lear, Yale University
J. L. Mackie, formerly of University College, Oxford University
John McDowell, University of Pittsburgh
Mark Platts, National Autonomous University of Mexico
Geoffrey Sayre-McCord, University of North Carolina/Chapel Hill
Nicholas Sturgeon, Cornell University
David Wiggins, University College, Oxford University
Bernard Williams, University of California/Berkley

Preface

People commonly think both that some actions, practices, and institutions really are good or bad, right or wrong, just or unjust, and that we often know that they are so. Few doubt, for instance, that kindness is good and honesty valuable, or that murder is wrong and slavery unjust. These seem to be obvious truths, easily known. Yet these common convictions are notoriously hard to justify. Indeed, for the greater part of this century, the dominant view in philosophy has been that these convictions cannot be justified. Moral facts, it is said, have no place within a suitably scientific picture of the world. On this dominant view, moral utterances are simply cognitively empty expressions of our parochial interests, personal attitudes, or narrow prejudices; they never report moral facts. Whatever the precise account given, the shared assumption has been that there are no moral facts for our moral claims to be about.

The standard arguments against moral facts, however, rely largely on the outmoded and discredited assumptions of logical positivism, assumptions that live on in moral theory long after they have died everywhere else. Moreover, recent developments in moral theory, philosophy of science, and philosophy of language suggest several positive arguments, over and above common conviction, for believing in moral facts. As a result, moral realism—the view that there are moral facts which we can discover—now enjoys renewed vitality and new intellectual credentials. Many people, of course, still reject moral realism. Of necessity, though, their arguments have become both more sophisticated and more penetrating, if also more tenuous.

The controversy between moral realists and their opponents turns on the realists' view that the aim of moral deliberation is to discover moral facts. Anti-realists contend, on various grounds, that moral facts are unavailable to moral inquiry. The anti-realists fall into two groups: some think the very notion of a moral fact is incoherent, while others think the search for moral facts, though intelligible, is always futile. The first group, the non-cognitivists, hold that moral claims are cognitively empty and so can never report moral facts. If the point of moral deliberation is to discover such facts, they argue, then moral deliberation is pointless. The second group of anti-realists, in contrast, grant that our everyday moral concepts do presuppose moral facts; still, they maintain, we have reason to think either that all moral claims are false (for instance, because they make some false presupposition) or that our beliefs concerning moral facts are all completely unfounded. Against anti-realists, moral realists are committed to defending both cognitivism and some sort of 'success theory' (according to which at least some cognitively packed moral claims are true).

In the introductory essay, ''The Many Moral Realisms,'' I set out in detail the picture of the moral realism debate that has structured this collection. The rest of the papers are clustered, in order, according to whether they attack or defend moral realism.

A. J. Ayer's now classic ''Critique of Ethics and Theology'' represents traditional non-cognitivism and gives a vivid and intuitively forceful idea of what is worrisome about moral realism. Like Ayer, Bernard Williams advances a form of non-cognitivism. Yet he comes to his non-cognitivism not, as Ayer does, by embracing an overarching and grand criterion of meaningfulness. Instead, Williams offers much more localized arguments that stress, in particular, the analogy between conflicting moral commitments and conflicting desires. In ''Supervenience Revisited'' Simon Blackburn emphasizes the ontological and epistemological problems that arise for any realist view that treats moral properties as supervenient upon, but not reducible to, natural properties. Ultimately, he maintains, the relation of supervenience between moral and natural properties has been introduced to explain what is better explained in terms of conventions we've adopted for the projection of our moral attitudes. Concentrating on the apparent analogies between mathematics and ethics, Jonathan Lear (in ''Ethics, Mathematics, and Relativism'') argues that any attempt to defend cognitivism in ethics, and so moral realism, by drawing parallels with mathematics will founder on the fundamental relativism that plagues ethics.

J. L. Mackie and Gilbert Harman break stride with other anti-realists by acknowledging that moral claims are perfectly meaningful. The problem with moral thoery, each argues, is that the meaningful claims it makes are, at least as far as we can tell, all false. Mackie's position is bolstered by modernized versions of arguments offered by Hume, each unabashedly based in common sense (he calls them ''the Argument from Queerness'' and ''the Argument from Relativity''). In contrast, Harman's arguments are tightly linked to recent developments in epistemology, and specifically to the view that we should believe only in what we appeal to in our best explanations of our experiences. His fundamental complaint against moral realism is that moral facts play no role in such explanations. Despite their differences, Mackie and Harman share two honors: each salvages (part of) what is appealing about the empiricists' attack on ethics, and together they bear the lion's share of responsibility for reviving anti-realism in ethics.

David Wiggins's "Truth, Invention, and the Meaning of Life" serves as a bridge between the papers attacking moral realism and those defending it. On the one hand, Wiggins espouses a sort of 'cognitive underdetermination' with respect to moral claims that is congenial to anti-realists. On the other hand, he argues that objectivity and discovery can be combined with invention to support a realist interpretation of moral values. Both the subtlety of Wiggins's arguments for realism and the reasons he gives for limiting its scope have earned his paper a central place in discussions of realism.

Against anti-realists, John McDowell defends cognitivism about moral utterances (in "Values and Secondary Qualities"), recommending that moral properties be thought of as analogous to secondary qualities like color and that such secondary qualities are genuine features of objects (and not mere projections onto objects). Taking advantage of new arguments for realism in the philosophy of science, Richard N. Boyd's "How to Be a Moral Realist" canvasses the many strategies open to realists for defending a consequentialist naturalism. Likewise concerned to defend a naturalism in ethics, Nicholas L. Sturgeon (in "Moral Explanations") argues, against Harman, that moral facts do figure in our best explanations of our experiences. In "Moral Theory and Explanatory Impotence" I explore and criticize the assumption, shared by Harman, Boyd, and Sturgeon, that the legitimacy of moral facts would be secured by, and only by, their having a place within scientific explanations. Finally, Mark Platts's "Moral Reality" draws primarily on the philosophy of language to argue for a version of intuitionism that gives moral claims an 'austerely realistic' interpretation and straightforward truth-conditions, while avoiding traditional intuitionism's reliance on peculiar mental faculties of moral intuition.

This collection illustrates the range, depth, and importance of moral realism, the fundamental issues it raises, and the problems it faces. Care has been taken to make the collection accessible, rigorous, and thought provoking. In addition, a balance has been sought between wide-ranging papers (such as those by Mackie, Wiggins, and Boyd) which advance a barrage of arguments and more focused papers (such as those by Blackburn, Lear, and Sturgeon) which develop a few arguments in great detail. What emerges, I hope, is an introduction to moral realism that exhibits the scope of the debate as well as the intricacies of the arguments marshaled on all sides. Even so, this is only a sampling of the excellent work that has been, and is being, done on moral realism. In the all-too-abbreviated bibliography are listed some of the many pieces those interested in the moral realism debate will want to read. May this collection spark such an interest.

Putting together an anthology, I've discovered, is time-consuming and sometimes exasperating. Throughout the whole affair, John Ackerman's extraordinary patience, encouragement, and good humor have been invaluable. Invaluable too

Preface

has been the support given to me by Harriet Sayre McCord and Joan McCord. The University of North Carolina at Chapel Hill (in the person of Gillian Cell) and especially Karann Durland have also given much needed, and greatly appreciated, help.

GEOFFREY SAYRE-MCCORD

Chapel Hill, North Carolina

ESSAYS ON
MORAL REALISM

Introduction
The Many Moral Realisms

Geoffrey Sayre-McCord

1. Introduction

Recognizing the startling resurgence in realism, Don Philahue (of *The Don Philahue Show*) invited a member of Realists Anonymous to bare his soul on television. After a brief introduction documenting the spread of realism, Philahue turned to his guest:

DP: What kinds of realism were you into, Hilary?

H: The whole bag, Don. I was a realist about logical terms, abstract entities, theoretical postulates—you name it.

DP: And causality, what about causality?

H: That too, Don. (Audience gasps.)

DP: I'm going to press you here, Hilary. Did you at any time accept moral realism?

H: (staring at feet): Yes.

DP: What effect did all this realism have on your life?

H: I would spend hours aimlessly wandering the streets, kicking large stones and shouting, "I refute you thus!" It's embarrassing to recall.

DP: There was worse, wasn't there Hilary?

H: I can't deny it, Don. (Audience gasps.) Instead of going to work I would sit at home fondling ashtrays and reading voraciously about converging scientific theories. I kept a copy of "Hitler: A Study in Tyranny" hidden in the icebox, and when no one was around I would take it out and chant "The Nazis were bad. The Nazis were *really* bad."[1]

This is a slightly revised version of my essay "The Many Moral Realisms," *Spindel Conference: Moral Realism, Southern Journal of Philosophy* 24 (1986), Supplement, pp. 1–22. I thank Murray Spindel for his generous support of philosophy, William Lycan for many helpful discussions, and John Ackerman, Neera Badhwar, and Douglas Butler for their extensive comments on this paper.

1. This dialogue finds its source at the end of Philip Gasper's pen. I am grateful for his permission to use it.

Realists and anti-realists alike can, of course, say that the Nazis were bad and even that the Nazis were *really* bad. What distinguishes realists from anti-realists about morals has disappointingly little to do with the particular moral claims each willingly endorses. In fact, moral anti-realists expend a great deal of energy trying to show that their views about the *status* of moral claims are perfectly compatible with saying all the same things any decent, wholesome, respectable person would.[2] In a similar way, what distinguishes realists from anti-realists about science has little to do with there being some difference in the particular scientific claims each willingly endorses. So, too, with the difference between realists and anti-realists about mathematics or about causality or about the external world. Indeed, the plausibility of anti-realism, in any of these areas, depends largely on preserving our normal ways of speaking even while challenging the natural (though perhaps naïve) realistic interpretation of what is being said.

In spite of the anti-realists' conciliatory attitude toward much of what we say, the debate between realists and anti-realists is deep and important. It affects our ontology, epistemology, and semantics. And in ethics, it makes a significant difference to our understanding of what (if anything) is valuable, to our account of moral disagreement, and to the importance we place on moral reflection. It may even make a difference to our happiness, if we desire things because of the value we take them to have and not vice versa.[3]

The realism/anti-realism debate in ethics has been around ever since people began thinking critically about their moral convictions. The problem has always been to make sense of these convictions in a way that does justice to morality's apparent importance without engaging in outrageous metaphysical flights of fancy. Some have thought it can't be done; they've held that the apparent importance of morality is mere appearance. Others have thought it can be done; they've held that whatever metaphysics is necessary is neither outrageous nor fanciful.

The debate came into especially sharp focus at the beginning of this century,

2. Simon Blackburn, for instance, takes the major challenge facing anti-realism to be that of showing that our normal (apparently realistic) way of speaking about morals is neither 'fraudulent' nor 'diseased', even though it is to be given an anti-realist construal. See *Spreading the Word* (New York, 1984), esp. pp. 189–223, and "Supervenience Revisited," in *Exercises in Analysis,* ed. Ian Hacking (Cambridge, 1985), pp. 47–67, reprinted in this volume. See also C. L. Stevenson, "Retrospective Comments," in *Facts and Values* (New Haven, Conn., 1963), pp. 186–232, and "Ethical Fallibility," in *Ethics and Society,* ed. Richard DeGeorge (New York, 1966), pp. 197–217.

3. David Wiggins argues that we can see our lives as meaningful only if we attach a value to our ends that transcends the mere fact that we have adopted them ("Truth, Invention, and the Meaning of Life," *Proceedings of the British Academy* [1976], 331–78, reprinted in this volume.) In a similar vein, Mark Platts maintains that "desires frequently require an appropriate belief about the independent desirability of the object of desire." Where the belief is absent so too is the desire and thus the satisfaction that might have come from getting what is desired. So rejecting moral realism "can be the end for a reflective being like us by being the beginning of a life that is empty, brutish, and long" ("Morality and the End of Desire," in *Reference, Truth and Reality,* ed. Mark Platts [London, 1980], p. 80). And Robert Bellah et al. suggest that accepting moral realism may be essential if we are to enjoy the psychological benefits of commitment. See *Habits of the Heart* (Berkeley, 1985). See Nicholas Sturgeon, "What Difference Does It Make Whether Moral Realism Is True?" *Spindel Conference: Moral Realism, Southern Journal of Philosophy* 24 (1986), Supplement, pp. 115–41, for a discussion of what is at stake in the moral realism debate.

when G. E. Moore identified what he called the "Naturalistic Fallacy." All at once, it seemed as if the most metaphysically attractive theories of morality, those that reconciled morality with naturalism, turned on committing the fallacy. According to Moore, one commits this fallacy by trying to define moral terms that are simple and unanalyzable. A clear example of the fallacy, he argues, is the hedonist's proposal that the goodness of an action consists simply in its pleasantness. Although many good things are pleasant, and many (perhaps even all) pleasant things are good, he maintains it is a mistake to think goodness and pleasantness are one and the same property. To see why, consider that it makes perfect sense both to say of some action that it is pleasant and to ask whether it is also good. That the question makes sense, and is not immediately answered by noticing that the action is pleasant, shows (Moore argues) that "x is pleasant" can't mean the same thing as "x is good". Moreover, the fact that the question makes sense (the fact that it remains an 'open question') shows as well, he thinks, that 'good' does have a meaning, that

> whoever will attentively consider with himself what is actually before his mind when he asks the question "Is pleasure (or whatever it may be) after all good?" can easily satisfy himself that he is not merely wondering whether pleasure is pleasant. And if he will try this experiment with each suggested definition in succession, he may become expert enough to recognise that in every case he has before his mind a unique object, with regard to the connection of which with any other object, a distinct question may be asked. Every one does in fact understand the question "Is this good?". When he thinks of it, his state of mind is different from what it would be, were he asked "Is this pleasant, or desired, or approved?". It has a distinct meaning for him, even though he may not recognise in what respect it is distinct.[4]

Thus Moore concludes that to say of something that it is good is to attribute to it a simple, unanalyzable, property, a property wholly distinct from other properties of everyday experience. And he holds that the simple property referred to by 'good' is not the object of sensory experience at all. "It is not *goodness*," he says, "but only the things or qualities which are good, which can exist in time—can have duration and cease to exist—can be objects of *perception*."[5] Goodness itself cannot be perceived, yet it nonetheless can be, and is, a property of those things we do perceive.[6]

The resulting view treats moral properties as real, but also as simple, non-natural, and seemingly mysterious. And because moral properties are taken to be

---

4. *Principia Ethica* (New York, 1971), pp. 16–17. Despite its name, the fallacy Moore is concerned with is not limited to identifying mistakenly moral properties with natural properties. The same fallacy is committed, he argues, when any simple, unanalyzable property is identified with some other, even if the properties involved are both natural (p. 13) or both nonnatural (p. 125).

5. *Principia Ethica*, pp. 110–11.

6. For a discussion of Moore's argument see, for instance, W. K. Frankena, "The Naturalistic Fallacy," *Mind* 48 (1939), 464–77; D. H. Munro, *Empiricism and Ethics* (Cambridge, 1967); W. D. Falk, "Fact, Value, and Nonnatural Predication," in his *Ought, Reasons and Morality* (Ithaca, N.Y., 1986); and A. N. Prior, *Logic and the Basis of Ethics* (New York, 1949).

[3]

nonnatural and not discoverable by normal empirical investigation, this account falls back on 'moral intuition' (which is no less mysterious) to explain how it is we come to know about morality. No other view, Moore argues, can avoid committing the Naturalistic Fallacy, if it is to respect our sense that we do mean something when we call one action good, another bad, or one person virtuous, another vicious, or one practice right, another wrong. Convinced by Moore's arguments, many people have accepted the mystery and some have even strived to make it appear commonsensical.[7]

In reaction, and in the name of removing mystery, others have questioned the idea that our moral claims are meaningful in the way Moore supposes. We might steer clear of the Naturalistic Fallacy and avoid a commitment to strange non-natural properties, they point out, by rejecting the assumption that in calling something good we are describing it. By distinguishing descriptive language from prescriptive language—the language of fact from that of value—we can acknowledge that no proposed definition of moral terms by descriptive ones will be adequate and still deny that prescriptive language involves attributing any properties whatsoever (let alone peculiar nonnatural ones) to actions, people, or practices. On this theory, to say of something that it is good is (roughly) to recommend it, and to show one's approval of it, but not to report any fact about it.[8] This argument thus neatly eliminates the need to introduce nonnatural properties into our ontology, but it does so by abandoning the notion of moral facts.

Consequently, many have found this theory just as unsatisfying as Moore's. For it seems that by doing without the notion of moral facts, the theory robs morality of its claim to importance and renders unintelligible the idea that we might make fundamental moral mistakes. At the same time, the theory has trouble giving a compelling account of the phenomenology of moral experience, both because it construes morality as a reflection of our attitudes rather than a standard for them and because it denies that, when it comes to morals, there is some fact of the matter.[9] Yet whether there is some alternative that can salvage moral facts without introducing Moorean metaphysical excesses (if indeed they are excesses) remains unclear. According to some moral realists, there is some alternative; according to other moral realists (the Mooreans) and all anti-realists, there is not.

---

7. The same general view of moral properties, and our knowledge of them, was advanced in 1757 by Richard Price in *A Review of the Principal Questions in Morals,* ed. D. D. Raphael, (Oxford, 1948), and in 1788 by Thomas Reid in *Essays on the Active Powers of Man,* ed. R. E. Beanblossom and K. Lehrer (1983). For more recent examples, see Prichard, *Moral Obligation* (1949); W. D. Ross, *The Right and the Good* (New York, 1930) and *The Foundations of Ethics* (Oxford, 1939); and A. C. Ewing, *Ethics* (London, 1953).

8. Classic examples of this view can be found in A. J. Ayer, *Language, Truth and Logic* (New York, 1952) (selection reprinted in this volume); C. L. Stevenson, *Ethics and Language* (New Haven, Conn., 1944); and R. M. Hare, *The Language of Morals* (New York, 1952).

9. See, just as examples, Kurt Baier, *The Moral Point of View* (Ithaca, N.Y., 1958); Carl Wellman, "Emotivism and Ethical Objectivity," *American Philosophical Quarterly* 5 (1968), 90–99; J. O. Urmson, *The Emotive Theory of Ethics* (Oxford, 1968); Philippa Foot, "Moral Beliefs," *Proceedings of the Aristotelian Society* 59 (1958–1959); Sabina Lovibond, *Realism and Imagination in Ethics* (Minneapolis, 1983); and Wiggins, "Truth, Invention and the Meaning of Life."

In this introductory chapter I mean to establish coordinates for locating various versions of moral realism that have been proposed (both Moorean and otherwise) and for distinguishing these positions from anti-realism. To sort out the many moral realisms from the equally plentiful anti-realisms, what's needed is some 'conceptual cartography'—a mapping out of the moral realism debate. I assume that, to be adequate, a map of the moral realism debate must do at least three things: first, since there is a *debate* about moral realism, the map should leave each side with defensible territory (neither realism nor anti-realism ought to be trivially true); second, since the debate is about *moral* realism, the map should locate positions that are recognizably positions people have taken about morality; and third, since the debate is about moral *realism,* the map should extend in some natural way to other debates about realism.

The map I offer is, I believe, one that works not just for the moral realism debate, *but also for all other debates concerning realism.* Moved from realist debate to realist debate, the same map will serve to demarcate in each the issues that separate realists from anti-realists. Unlike most maps, which elaborately detail one area but are useless in the next, this map serves as a guide to realism wherever it occurs.

Flexible as the map is, it bends always to accommodate a fixed view of the nature of realism. In every case, what marks off some particular terrain as the realist's remains the same; over and over, it is the view that some of the disputed claims literally construed are literally true. Wherever it is found, I'll argue, realism involves embracing just two theses: (1) the claims in question, when literally construed, are literally true or false (cognitivism), and (2) some are literally true. Nothing more. (Of course, a great deal is built into these two theses.)

Correspondingly, there are two ways to be an anti-realist: embrace a non-cognitivist analysis of the claims in question or hold that the claims of the disputed class, despite their being truth-valued, are none of them true (say, because they all share a false presupposition).

This characterization of realism differs in significant ways from many common definitions. For instance, some characterizations of realism give pride of place to objectivity, others to independence from the mental, and still others treat realism as a semantic thesis about the nature of truth and its transcendence of our recognitional capacities. Some even combine several of these elements. Thus, according to Michael Dummett, realism is "the belief that statements of the disputed class possess an objective truth-value, independently of our means of knowing it: they are true or false in virtue of a reality existing independently of us."[10] Yet, in the account I offer, there is no mention of objectivity or existence,

---

10. *Truth and Other Enigmas* (Cambridge, Mass., 1978), p. 146. In a later article Dummett says that realism is the view that "statements in the given class relate to some reality that exists independently of our knowledge of it, in such a way that reality renders each statement in the class determinately true or false, again independently of whether we know, or are even able to discover, its truth-value" ("Realism," *Synthèse* 52 [1982], 55).

no mention of recognition transcendence or independence, no mention of reference, bivalence, or correspondence. And this is a virtue. Independence from the mental may be a plausible requirement for realism when we're talking about macrophysical objects but not when it comes to realism in psychology (psychological facts won't be independent of the mental);[11] bivalence might go hand in hand with realism in mathematics, but realism in other areas seems perfectly compatible with acknowledging that some of our predicates are vague and have indeterminate extensions; and existence may be crucial to realism about scientific entities (since claims concerning such entities are true only if the entities exist) but not to realism about scientific laws (that makes no existence claims).

By abstracting from these (often contentious) notions, the account I offer is in a position to explain why they are so often central to the realist's position *even though none of them is always central*. The governing idea is that independence, bivalence, existence, etc., come into play when, *but only when,* they are relevant to whether the disputed claims, literally construed, are literally true.

The notions of 'literal construal' and 'literal truth' will obviously carry a lot of weight in this account of realism. Significantly, though, most realism debates are not especially sensitive to how these notions are understood. As long as we're working with theories of meaning and truth that allow the distinctions between cognitive and non-cognitive discourse, and between claims that are literally true and those that aren't, all the positions relevant to the realism debate in ethics can be reproduced.[12] An important constraint, however, is that the semantics offered must be *seamless;* whichever theories of meaning and truth are offered for the disputed claims must be extended as well to apply to all claims. Thus, one won't have defended realism in ethics by showing moral claims to be meaningful and true in a sense, if they are not meaningful and true in the same sense that other, noncontroversial, claims are. It doesn't much matter which particular theory of meaning and which particular theory of truth are being relied on.[13] What our choice of semantic theories will influence is our understanding of what it is to say that some of the disputed claims, literally construed, are literally true. The choice won't settle whether any of the claims actually is true. This insensitivity to the

11. Similarly, requiring epistemic independence is implausible once we grant (1) the possibility of a realist position according to which a necessary condition for being in pain is knowing (or at least believing) one is, or (2) the possibility of a realist position in ethics according to which one can't have moral obligations of which one is incapable of knowing. Elliott Sober explores the problems facing all attempts to build independence into the general characterization of realism in "Realism and Independence," *Noûs* 16 (1982), 369–85.

12. If these distinctions aren't preserved, then there will be no way to generate any special problem about moral realism. In any case, theories of meaning and truth that don't preserve these distinctions in one way or another don't have much credibility as accounts of meaning and truth.

13. For instance, the positions available to the anti-realist remain virtually unaffected if we accept a Dummett-style verificationist theory of meaning. An anti-realist will hold either that moral claims are not verifiable and so are meaningless or that moral claims are verifiable but that none of them are verified. In other words, an anti-realist will defend either non-cognitivism or an error theory. In the same way, the boundaries set by the map of realism debates are unaffected by whether we give an account of literal truth in terms of correspondence or in terms of warranted assertibility.

differences between various semantic theories reflects the fact that, for the most part, realism is a matter of metaphysics, not semantics.

To defend this general account of realism, it will help to have the map of the moral realism debate set out. This is best done by surveying the positions one might take concerning the status of moral claims.

2. Realism, instrumentalism, and idealism

Depending on the domain in question, one or the other of two time-honored contrasts has dominated talk of realism. One is the contrast between realism and *instrumentalism;* the other is the contrast between realism and *idealism.* Which contrast is relied on in characterizing realism will, of course, significantly color what is counted as moral realism.

When realism (moral or otherwise) is contrasted with instrumentalism the central issue is whether the claims of the disputed class should be interpreted as having truth-values. In other words, the issue is whether the claims should be given a cognitivist interpretation. When realism is contrasted with idealism, however, the issue is not whether the claims have truth-values. Realists and idealists agree that they do. Instead, the issue is whether minds (or their contents) figure expressly in the truth-conditions for the claims in question. Realists (in this sense) hold that they don't, while idealists hold that these claims are literally true or false, but that they have whatever truth-value they do in virtue of someone or other's mind: your's, mine, or God's.

Clearly the realism/idealism debate presupposes a cognitivist interpretation of the disputed claims. For if we accept instrumentalism, and so non-cognitivism, about claims involving unobservable entities or moral properties or scientific laws, then no worry arises as to the nature of their truth-conditions—they haven't any.[14]

Those who are instrumentalists, whether about scientific claims concerning unobservable entities or moral claims about what actions are right, hold that such claims are not literally true or false but are instead merely useful devices for controlling either our own experiences or others' behavior. A standard instrumentalist view of scientific theories, for example, makes them out to be uninterpreted formal systems (cognitive "black boxes") constructed to control and predict the observable world, not to describe an unobservable one. So viewed, scientific theories are thought of not in terms of truth but in terms of empirical adequacy, conceptual utility, practical fruitfulness, and formal elegance.

Along the same lines, 'moral instrumentalists' argue that in using moral language we don't ascribe moral properties to people, actions, or institutions. As Ayer puts it, "The presence of an ethical symbol in a proposition adds nothing to

---

14. Conversely, if instrumentalism is rejected, we face the challenge of figuring out what the truth-conditions are for the disputed claims. And figuring this out is often problematic. Indeed, much of the motivation for embracing instrumentalism lies in thinking that if the disputed claims were meaningful, they would have to be about very peculiar things.

its factual content. Thus if I say to someone, 'You acted wrongly in stealing that money,' I am not stating anything more than if I had simply said, 'You stole that money.' In adding that this action is wrong I am not making any further statement about it.''[15]

Our aim, moral instrumentalists say, is not to describe the world but to change it.[16] And moral language is especially well suited to this job, they point out, because people have been conditioned to use and respond to moral utterances as commands and exhortations. The point of making moral claims, Stevenson argues, "is not to indicate facts but to *create an influence*. Instead of merely describing people's interests they *change* or *intensify* them. They *recommend* an interest in an object, rather than state that the interest already exists.''[17]

Moral instrumentalists come in several stripes, the differences among them turning on the particular account each offers of how moral language works. Emotivists, for instance, hold that moral language comes by its usefulness thanks to its expressing the emotions, tastes, feelings, and other affective states of the speaker. Moral language, they maintain, is "used to express feeling about certain objects, but not to make any assertion about them.''[18] Prescriptivists are also moral instrumentalists. However, they reject as too tight the link the emotivists assume to hold between moral language and emotions. They argue that moral claims are useful not because they express emotions but because they serve as universalizable commands that a certain course of action be taken by anyone (including the speaker) under the circumstances.[19] Whichever particular account is offered, all moral instrumentalists emphasize the practical uses of moral language and think it is a mistake to treat moral claims as assertions about the world that might be literally true or false.[20]

Moral realists, like moral instrumentalists, recognize moral language as a useful instrument for controlling the world around us. What distinguishes the realist is, at least in part, the conviction that this use depends for its effectiveness on moral language expressing moral beliefs. Moral language exerts its influence, realists argue, because people can hold moral beliefs (and not just have moral reactions), and we can alter other people's attitudes and behavior through discussion and moral argument only because people's moral beliefs can change their

15. Ayer, *Language, Truth and Logic,* p. 107 (see p. 30 this volume).

16. Not surprisingly, moral instrumentalism goes hand in hand with scientific instrumentalism, and both found strong support in the verificationism of logical positivism. See my "Logical Positivism and the Demise of 'Moral Science'," in *The Heritage of Logical Positivism,* ed. Nicholas Rescher, University of Pittsburgh Philosophy of Science Series (Lanham, Md., 1985), pp. 83–92.

17. C. L. Stevenson, "The Emotive Meaning of Ethical Terms," *Facts and Values,* p. 16.

18. Ayer, *Language, Truth and Logic,* p. 108 (see p. 31 this volume).

19. R. M. Hare, *The Language of Morals,* is a classic articulation of prescriptivism. It is worth emphasizing that prescriptivists don't hold that moral claims are devoid of all cognitive content. Indeed, their account of how moral claims work requires that the claims have some content. It is just that, according to prescriptivists, the content they have is like the content of commands—it suits them for prescribing, not for describing.

20. Some instrumentalists are prepared to allow the propriety of saying things like "It's true that gratuitous cruelty is wrong." But then they emphasize that 'true' in this context plays a role significantly different from the role it plays in the context of cognitive discourse. While they grant that moral claims may be 'true' in some sense, they deny that they may be 'literally true.'

[8]

moral reactions. In explaining how moral attitudes affect inference and argument, action and interaction, we'll find ourselves appealing not just to moral reactions but also to moral beliefs (that, in turn, are expressed by moral language).[21]

Moreover, realists point out, moral language exhibits all the telltale signs of cognitive discourse: we seem to hold moral beliefs, have moral disagreements, seek evidence for our opinions; we act as if there were something to discover, as if we could be mistaken, as if there is a fact of the matter; and we even talk of moral claims being true or false, and of people knowing the better (even while doing the worse).[22] Indeed, moral claims are apparently indistinguishable in logical form and within inferential contexts from claims that are recognized as cognitive.[23] These characteristics of moral discourse are not merely flukes of the language, they reflect the phenomenology of moral experience—a phenomenology that represents obligation as a constraint on, and value as giving direction to, our actions independently of what we might happen to desire. All told, realists argue, cognitivism offers the best psychological and linguistic account of moral language.

Of course none of these considerations is conclusive. It's open to moral instrumentalists to explain away these practices and convictions as misleading indicators of cognitive content. But then they must offer some principled way of drawing the distinction between cognitive and non-cognitive discourse, and they must show that moral claims fall on the non-cognitive side of the distinction.[24] In doing so, the moral instrumentalists, not the moral realists, will be the ones denying appearances in the name of a deeper reality.[25]

3. Error and success theories

Anti-realists might simply grant that moral claims are cognitively packed.

21. Moral instrumentalists, particularly emotivists, can easily explain the intimate connection between first-person moral judgments and action. If moral language simply reflects the emotions, feelings, or conative attitudes of the speaker, it is not surprising that people are moved to pursue what they call good and to avoid what they call bad. Moral instrumentalism doesn't fare so well, though, when it comes to explaining other aspects of how moral language works. Specifically, it has difficulty giving a plausible explanation of the interpersonal impact of moral discussion and argument.

22. See Jonathan Harrison, *Our Knowledge of Right and Wrong* (New York, 1971), p. 258. The presence of similar telltale signs has played a significant role in undermining the instrumentalist's non-cognitive account of the theoretical discourse of science.

23. P. T. Geach rejects as inadequate all expressive theories of meaning on the grounds that they can't account for the fact that "a proposition may occur in discourse now asserted, now unasserted, and yet be recognizably the same proposition" (p. 449). See "Assertion," *Philosophical Review* 74 (1965), 449–65.

24. John McDowell suggests that there is no such principled distinction available that will serve the moral instrumentalists' purpose. See "Non-cognitivism and Rule-Following," in *Wittgenstein: To Follow a Rule,* ed. Steven Holtzman and Christopher Leich (London, 1981), pp. 141–62.

25. Simon Blackburn argues that 'quasi-realism' can accommodate all the appearances, and even explain the legitimacy of our practice, without indulging in what he takes to be realist excesses. In addition to *Spreading the Word,* see "Rule-Following and Moral Realism," in *Wittgenstein: To Follow a Rule,* pp. 163–87, and "Errors and the Phenomenology of Value," in *Morality and Objectivity,* ed. Ted Honderich (London, 1985), pp. 1–22.

Realism, after all, does not find salvation simply in the discovery that the disputed claims purport to report facts—that is compatible with their failing miserably. To get realism, the disputed claims must not only have truth-values, *some of them must have the truth-value true*. Thus it is not enough to say, as Putnam does, that "a realist (with respect to a given theory of discourse) holds that (1) the sentences of that theory are true or false; and (2) that what makes them true or false is something external—that is to say, it is not (in general) our sense data, actual or potential, or the structure of our minds, or our language, etc."[26] An anti-realist can perfectly well acknowledge that the disputed claims have a truth-value, and even that these truth-values depend on something external, while going on to say that none of the claims is true.[27] An anti-realist might, in other words, advance an error theory.

To take account of this possibility the realism/instrumentalism dichotomy must be replaced with two distinctions: first, the distinction between non-cognitivism and cognitivism, and second, the distinction, within cognitivism, between error theories (that maintain that none of the disputed claims are true)[28] and success theories (that hold that some of the claims literally construed are literally true). Together, these two distinctions suggest the following (partial) map of the positions one might take concerning the claims in question:

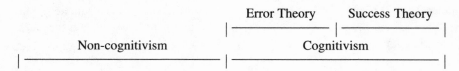

Error theorists and success theorists share the view that the disputed claims should be given a cognitivist interpretation. That is, they agree that the claims are about putative facts. What they disagree on is whether the claims live up to their pretensions.

Combining cognitivism with anti-realism in the way error theories do is neither new nor unusual. Early defenders of the error theory in ethics include Hume (on some plausible interpretations) and Spinoza, who argues that good and evil "are nothing but modes in which the imagination is affected in different ways, and, nevertheless, they are regarded by the ignorant as being specially attributes of things."[29] Similarly, in the philosophy of psychology eliminative materialists

26. "What Is Mathematical Truth?" *Mathematics, Matter and Method,* 2d ed. (Cambridge, 1979), pp. 69–70. See also Dummett's characterizations of realism on page 146 of *Truth and Other Enigmas,* and on page 55 of "Realism."

27. Mackie emphasizes this point in "Anti-Realisms," in *Logic and Knowledge* (Oxford, 1985), pp. 225–45.

28. Depending on their views of presupposition, some error theorists will hold that all the disputed claims are false, while others will simply hold that none of them are true.

29. Benedict Spinoza, *Ethics* (New York 1949), pt. 1, app., p. 77. W. D. Ross considers an error theory as an alternative to his own and then rejects it on the grounds that suspecting reason to have made such an egregious error "is in principle to distrust its power of ever knowing reality" (*The Right and the Good,* p. 82).

combine the two, maintaining that our attributions of mental states, though meaningful, are all mistaken.[30] Constructive empiricists in the philosophy of science take a similar line about all claims concerning unobservable entities, arguing that while such claims may have truth-values, we have no good reason for thinking of any of them that they have the truth-value true.[31] Likewise, the cornerstone of atheism is not the view that talk of God's existence, his Will, or his Kingdom, is meaningless but instead the view that such claims are all quite literally false. Each of these anti-realist positions moves from a cognitivist interpretation of the disputed claims to arguments against thinking any of the claims true (and so ultimately to an error theory). What all error theorists recognize is that granting cognitivism to a disputed discourse doesn't by itself secure legitimacy for its claims.

Error theories (wherever they show up) typically involve more than just the view that none of the disputed claims are true. Since such theories almost always go against 'common sense' by denying assumptions built into the disputed language, their acceptability depends on having some account of why people have gone wrong. Needless to say, the explanations offered and their plausibility will vary as the error being explained varies. In ethics, the source of the error is often traced partly to our projecting reactions onto the world ("gilding or staining all natural objects with the colours borrowed from internal sentiment," as Hume puts it),[32] and partly to the workings of effective socialization. The plausibility of social/psychological explanations of moral opinion has itself played a crucial role in making the error theory attractive. The plausibility of these explanations suggests that the error is a natural one, and the more natural the error seems, the more our conviction that no error has been will dissolve. So even if moral claims purport to report facts (as cognitivists maintain), moral realists face the serious challenge of showing that moral discourse doesn't simply embody a large-scale and systematic (albeit, natural) error. Many think the challenge can't be met.

J. L. Mackie, for instance, willingly acknowledges that moral claims pass any reasonable test for cognitive content.[33] But he then relies on his cognitivist

---

30. Paul Churchland argues, for instance, that "our common-sense conception of psychological phenomena constitutes a radically false theory, a theory so fundamentally defective that both the principles and the ontology of that theory will eventually be displaced" in "Eliminative Materialism and the Propositional Attitudes," *Journal of Philosophy* 78 (1981), 67–90. See also Stephen Stich, *From Folk Psychology to Cognitive Science* (Cambridge, Mass., 1983).

31. Bas van Fraassen, *The Scientific Image* (Oxford, 1980), p. 11. Constructive empiricism serves a bit uneasily as an example of an error theory because it reserves judgment about the falsity of claims about unobservable entities (saying only that such claims are completely unjustified). It's more like agnosticism than atheism. See Nancy Cartwright, *How the Laws of Physics Lie* (Oxford, 1983), for a clear example of an error theory concerning the laws of physics (which is combined with a success theory concerning the unobservable entities postulated by physics).

32. *An Enquiry concerning the Principles of Morals* (Indianapolis, Ind., 1983), p. 88. Hume offers a similar explanation of our view that a necessary connection holds between events in the world when he attributes it to our projecting the "customary transition of the imagination from one object to its usual attendant." *An Enquiry concerning Human Understanding* (Indianapolis, Ind., 1977), p. 52.

33. "If second order ethics were confined, then, to linguistic and conceptual analysis, it ought to conclude that moral values at least are objective; that they are so is part of what our ordinary moral

analysis of moral claims to discredit them and motivate an error theory: "the denial of objective values will have to be put forward . . . as an 'error theory', a theory that although most people in making moral judgements implicitly claim, among other things, to be pointing to something objectively prescriptive, these claims are all false."[34] Mackie offers three arguments for thinking our moral claims are all false.

First, he points to the diversity of moral opinion, arguing that in the face of such radical disagreement it is implausible to think of people's moral judgments as responses to objective moral truths. The problem, though, isn't simply that people disagree on moral questions. Disagreement is perfectly compatible with there being a fact of the matter. The problem is that "the actual variations in the moral codes are more readily explained by the hypothesis that they reflect ways of life than by the hypothesis that they express perceptions, most of them seriously inadequate and badly distorted, of objective values."[35]

Second, Mackie maintains that the motivational force and practical relevance of moral judgments undermines the plausibility of believing in moral facts. Moral beliefs, he argues, necessarily motivate, and moral facts (if there were such) would have to be necessarily action-guiding. "Plato's Forms," he suggests, "give a dramatic picture of what objective values would have to be. The Form of the Good is such that knowledge of it provides the knower with both a direction and an overriding motive; something's being good both tells the person who knows this to pursue it and makes him pursue it."[36] Yet if objective values have to be like this, allowing the existence of moral facts would require embracing a fantastic ontology and positing mysterious powers of moral perception. The best course, Mackie argues, is to deny the existence of objective values, especially since we can explain the motivational force of moral judgments without recourse to any corresponding (and mysterious) facts that necessarily motivate.

Third, Mackie raises worries about the connection that is supposed to hold between moral properties and natural properties.[37] If there are moral properties, it seems they must have two characteristics: they must be supervenient upon natural

---

statements mean; the traditional moral concepts of the ordinary man as well as of the main line of western philosophers are concepts of objective value" (Mackie, *Ethics: Inventing Right and Wrong* [New York, 1977], p. 35 [p. 109 in this volume]).

34. Mackie, *Ethics: Inventing Right and Wrong*, p. 35 (p. 109 below). "The assertion that there are objective values or intrinsically prescriptive entities or features of some kind, which ordinary moral judgments presuppose, is, I hold, not meaningless but false" (p. 40). See also Francis Snare, "The Empirical Bases of Moral Scepticism," *American Philosophical Quarterly* 21 (1984), 215–25.

35. Mackie, *Ethics: Inventing Right and Wrong*, p. 37 (p. 110 in this volume).

36. Mackie, *Ethics: Inventing Right and Wrong*, p. 40 (p. 112 in this volume). Whether Mackie is right about what ordinary moral judgments presuppose is of course crucial to evaluating his attack on moral realism. For my part, I seriously doubt that Plato's Forms give an accurate picture of what objective values would have to be according to a realist. See David Brink, "Externalist Moral Realism," *Spindel Conference: Moral Realism, Southern Journal of Philosophy* 24 (1986), Supplement, pp. 23–41.

37. The same problem is pressed with great care by Simon Blackburn in "Moral Realism," in *Morality and Moral Reasoning*, ed. J. Casey et al. (London, 1971), pp. 101–24, in *Spreading the Word*, and in "Supervenience Revisited."

properties (so that setting the natural facts sets the moral ones), and they must be such that the truth of a moral proposition is not entailed by the truth of any naturalistic, or set of naturalistic, propositions (since moral terms cannot be definitionally reduced to naturalistic terms). But these two characteristics, taken together, apparently generate insuperable difficulties: if it is a logical possibility that a thing could have had all the natural properties it has and still not have had the moral properties, why is it not possible for two things with exactly the same natural properties to have different moral properties? The realist must answer this question by appeal to some special connection between moral facts and natural facts and must rely on an extraordinary epistemology in explaining how it is we discover the connection. In contrast, an anti-realist can explain the apparent supervenience of moral properties on natural properties by rejecting moral properties altogether and substituting an account of how we form our moral beliefs as a response to the natural features of the world with which we interact.[38]

A common strategy runs through all three of these arguments. The strategy is to highlight the ontological and epistemological price of postulating moral facts and to show that we needn't incur these costs. The upshot of the arguments is that we don't need moral facts to explain moral judgments or to explain anything else; moral facts, if there were any, would be explanatorily impotent. So we do better not to believe in them. By eschewing the moral facts presupposed by moral discourse, we get a better overall account of the world, a less extravagant ontology, and a neater epistemology.[39]

Of course these metaphysical and epistemological worries are not new with Mackie. (Although in earlier incarnations they've usually been advanced as grounds for non-cognitivism rather than for an error theory.) In one form or another they go back at least to Hume. And ever since Moore offered the Open Question Argument against (definitional) naturalism, philosophers have by and large assumed that moral realism faces insurmountable ontological and epistemological difficulties. Indeed, the common (mistaken) assumption is that the only realist positions available in ethics are those that embrace supernatural properties and special powers of moral intuition.[40]

---

38. For a discussion of these arguments, see William Tolhurst, "Supervenience, Externalism, and Moral Knowledge," *Spindel Conference: Moral Realism, Southern Journal of Philosophy* 24 (1986), Supplement, pp. 43–55; and James Klagge, "An Alleged Difficulty concerning Moral Properties" *Mind* 93 (1984), 370–80.

39. This general strategy has been forcefully defended by Gilbert Harman in *The Nature of Morality* (New York, 1977), selections of which are reprinted in this volume, and "Moral Explanations of Natural Facts," *Spindel Conference: Moral Realism, Southern Journal of Philosophy* 24 (1986), Supplement, pp. 57–68. For criticisms of both the success and the legitimacy of this strategy, see my "Moral Theory and Explanatory Impotence," *Midwest Studies* 12, 433–57, reprinted in this volume; Warren Quinn, "Truth and Explanation in Ethics," *Ethics* (1986), 524–44; William Lycan, "Moral Facts and Moral Knowledge," *Spindel Conference: Moral Realism, Southern Journal of Philosophy* 24 (1986), Supplement, pp. 79–94; Nicholas Sturgeon, "Moral Explanations," in *Morality, Reason and Truth*, ed. D. Copp and D. Zimmerman (Totowa, N.J., 1984), pp. 49–78, reprinted in this volume, and "Harman on Moral Explanations of Natural Facts," *Spindel Conference: Moral Realism, Southern Journal of Philosophy* 24 (1986), Supplement, pp. 69–78.

40. Recent arguments against this assumption can be found in Richard Boyd, "How to Be a Moral

Significantly, error theorists don't just embrace cognitivism as a gallant concession to a limping foe; the particular metaphysical and epistemological misgivings they advance turn crucially on particular views about what the world would have to be like for moral claims to be true. (In Mackie's case, he thinks the truth of moral claims presupposes the existence of 'objectively prescriptive' facts—of nonnatural facts that supervene upon natural facts and that, when recognized, necessarily compel action regardless of the agent's affective states.) If error theorists are wrong about what ordinary moral judgments presuppose, then their arguments against moral realism will miss their mark. Moreover, if they rejected cognitivism, they'd have no grounds for holding that moral claims carry any presuppositions whatever, let alone the ones they identify and attack.

4. Objectivism, intersubjectivism, and subjectivism

If we suppose that cognitivism offers the best account of the claims in question, then regardless of whether we are defending an error or a success theory, the question arises: What are the truth-conditions for the claims? Error theorists no less than success theorists are committed to giving some answer since the issue between them turns on whether the appropriate truth-conditions are ever satisfied. While both error theorists and success theorists must give an account of the truth-conditions for the disputed claims, they might not agree, of course, as to how the disputed claims should be construed. Often, in fact, a major part of their disagreement centers on just what counts as the proper literal construal, and so on what would have to be the case for the claims to be true. Even so, each must give some account in order to go on to argue for either an error or a success theory.

It is in worrying about the truth-conditions for the disputed claims that the issues surrounding idealism come into play. Idealists disagree with their opponents not about whether the disputed claims, literally construed, can be literally true, but about whether the proper literal construal makes the truth of those claims depend on a mind or minds. As with the realism/instrumentalism distinction, however, the realism/idealism dichotomy is insufficiently rich. In its place we need to substitute a three-part distinction among (what I'll call) objectivism, intersubjectivism, and subjectivism.

What separates objectivist, intersubjectivist, and subjectivist accounts of the disputed claims is whether and how people figure in the truth-conditions for the claims. Truth-conditions are 'subjectivist' (as I use the term) if they make essen-

---

Realist'' (this volume); Peter Railton, ''Moral Realism,'' *Philosophical Review* (1986), 163–207; David Brink, ''Moral Realism and the Sceptical Arguments from Disagreement and Queerness,'' *Australasian Journal of Philosophy* 62 (1984), 111–25; Nicholas Sturgeon, ''Moral Explanations''; William Lycan, ''Moral Facts and Moral Knowledge'' John Post, *The Faces of Existence* (Ithaca, N.Y., 1986); and my ''Coherence and Models for Moral Theorizing,'' *Pacific Philosophical Quarterly* (1985), 179–90, and ''Moral Theory and Explanatory Impotence.''

tial reference to an individual; 'intersubjectivist' if they make essential reference to the capacities, conventions, or practices of groups of people; and 'objectivist' if they need make no reference at all to people, their capacities, practices, or their conventions.

Each of these views constitutes a position one might take regarding the nature of the relevant truth-conditions for some cognitively interpreted claims. So we can locate objectivism, intersubjectivism, and subjectivism on the cognitivist side of the map, as positions available to both error and success theorists:

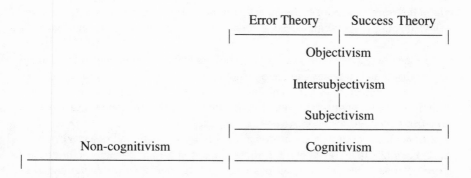

For many claims (e.g., about personal value or about pain or beliefs or mental states in general), the most plausible positions are those that make the truth of such claims dependent on mind (and so have subjectivist truth-conditions). For other claims (e.g., about the permissibility of some move in a game or the legality of some action), the most plausible positions are those that make the truth of such claims turn on social practices or conventions (and so have intersubjectivist truth-conditions). For still other claims (e.g., about the existence of electrons or other solar systems or mountains), the most plausible positions are those that make their truth independent of people's thoughts, practices, and conventions (and so have objective truth-conditions).

Each of these positions, whether subjectivist, intersubjectivist, or objectivist, is quite clearly a *realist* position as long as it is combined with the view that some of the relevant truth-conditions are actually satisfied (i.e., that some of the relevant claims are true).

For instance, a perfectly respectable and unquestionably realist position in philosophy of psychology is that the truth-conditions for reports of pain make explicit reference to an individual's mental state and that sometimes those truth-conditions are satisfied (i.e., that people do sometimes feel pain). No general account of realism that rules out subjectivism, so no account that requires 'independence from the mental,' will be acceptable. As R. B. Perry emphatically points out: "There is not the slightest ground for imputing to realism the grotesque notion that there are no such things as acts or states of mind, or that such things

[15]

cannot be known. . . . Because he [the realist] seeks to avoid a philosophical psycho-mania, there is no reason to accuse him of psycho-phobia.''[41]

In the same way, a perfectly respectable realist position in philosophy of law is that the truth-conditions for legal assertions make essential reference to the conventions and practices of particular societies and that sometimes those conditions are satisfied.[42] Of course, this isn't the only realist position. One who offers objectivist truth-conditions for legal claims (as natural law theorists do) and who thinks the claims are sometimes true will also be a realist about the law. Yet in the case of legal claims, the most plausible account of what we're saying when we claim that some behavior is illegal is that it's forbidden by the laws in force in the relevant society. Anything stronger seems to build into legal claims more than is intended.

These views, in the philosophy of psychology and the philosophy of law, are no less realist than is the view, in the philosophy of science, that the truth-conditions for claims concerning theoretical entities make no reference to people or their practices and conventions and that those truth-conditions are sometimes satisfied. *Realism is not solely the prerogative of objectivists.*

At one time or another, subjectivism, intersubjectivism, and objectivism have each been advanced as providing the right account of the truth-conditions for moral claims. Any one of these positions, if it can successfully be defended as giving the proper account of moral claims literally construed, will constitute a realist position, as long as it is combined with a success theory. The variety of positions available within subjectivism, intersubjectivism, and objectivism, and the differences among the three, will become clearer with a few examples.

Subjectivism

Moral subjectivists who accept cognitivism hold that the truth of moral claims depends on the subjective states of individuals. They maintain that moral claims are, in a perfectly straightforward sense, either true or false—it is just that their truth or falsity is mind-dependent; whether something is good or bad (or right or wrong) depends on someone's attitude toward it.[43] This sort of subjectivism, which rejects all varieties of non-cognitivism, including those that treat moral claims as reflections of an individual's subjective states, is a view about the truth-conditions of moral claims.

According to one version of moral subjectivism, judgments of value make sense only relative to the desires, preferences, and goals of the judger, so that the

41. R. B. Perry, *General Theory of Value* (Cambridge, Mass., 1926), p. 139.

42. Ronald Dworkin defends this sort of realist position in the philosophy of law. See *Taking Rights Seriously* (Cambridge, Mass., 1977).

43. "Apart from ourselves, and our human bias," Santayana argues, "we can see in such a mechanical world no element of value whatever. In removing consciousness, we have removed the possibility of worth." (*The Sense of Beauty* [New York, 1899], pp. 17–19.)

claim that 'x is good' should be treated as elliptical for 'x is good-for-me'; and this claim in turn is true or false depending on the desires, preferences, and goals of the person doing the judging. Hobbes endorses this view, arguing that "these words of good, evil, and contemptible are ever used with relation to the person that uses them, there being nothing simply and absolutely so, nor any common rule of good and evil to be taken from the nature of the objects themselves."[44] More recently, essentially the same subjectivist theory has found support in theories of rationality that tie the value something has for someone to that person's preferences.[45]

Alternatively, a subjectivist about value might hold that judgments of value are not relative to *judgers,* even though their truth is dependent on the subjective states of individuals. Such a subjectivist might say value claims are about what would satisfy someone's (but not necessarily the judger's) desires, preferences, or goals. R. B. Perry explicitly defends this account when he maintains that "a thing—any thing—has value, or is valuable, in the original and generic sense when it is the object of an interest—any interest."[46] In the same spirit, a preference satisfaction utilitarian will say that value depends on preferences, and so long as a thing satisfies someone's preferences (it doesn't matter whose), it is to that extent good.[47]

Or again, a subjectivist about value might hold that judgments of value are not relative to judgers but are still true, when they are true, only because of the desires, preferences, or goals of some particular person (rather than anyone). For instance, according to Ideal Observer theories, whether something has value (or at least whether something is right) depends on whether it would be approved of by some observer ideally constituted and situated.[48]

Importantly, subjectivism about value gives perfect sense to there being a fact of the matter (you might even say 'an objective fact of the matter') about what is good or valuable, and to that fact being reportable by any number of people. The

---

44. Thomas Hobbes, *Leviathan,* pt. 1, chap. 6 (Bobbs-Merrill, 1958), p. 53. Spinoza defends a similar view, arguing that "we neither strive for, wish, seek, nor desire anything because we think it to be good, but, on the contrary, we adjudge a thing good because we strive for, wish, seek, or desire it." *Ethics,* pt. III, prop. IX, note, pp. 136–37.

45. David Gauthier defends a preference-based subjective theory of value in *Morals by Agreement* (Oxford, 1986), chap. 2. See also Richard Brandt, *A Theory of the Good and the Right* (New York, 1979).

46. R. B. Perry, *Realms of Value* (Cambridge, Mass., 1954), pp. 2–3.

47. Of course, preference satisfaction utilitarians might combine subjectivism about value with objectivism about the right; they might hold that the utilitarian principle itself (that an action is right if and only if it maximizes preference satisfaction) is true independently of what anyone may happen to think, or feel, or want. Mill seems to defend a view along these lines. John Stuart Mill, *Utilitarianism* (Indianapolis, 1957), see esp. chap. 4: "Of What Sort of Proof the Principle of Utility is Susceptible."

48. See Adam Smith, *The Theory of Moral Sentiments,* ed. D. D. Raphael and A. L. Macfie (Oxford 1976); and Roderick Firth, "Ethical Absolutism and the Ideal Observer," *Philosophy and Phenomenological Research* 12 (1952), 317–45. Resisting idealization (in his characteristic way), Hobbes defends a version of subjectivism along these lines, arguing that, *after a society has been established,* claims concerning what is good report the commands of the Leviathan or the Leviathan's appointed surrogates. (*Leviathan,* pt. I, chap. 6, p. 53.)

(cognitivist) subjectivist's position is not that value judgments are 'true-for-someone' and not literally true. Instead their view is that value judgments are literally true, when true, but only because of the subjective states of someone (e.g., the desires, preferences, and goals of the relevant person).[49]

Of course, on the subjectivist view that treats 'x is good' as elliptical for 'x is good-for-me', two people will not be reporting the same thing when each utters 'x is good'; rather, they will each be making claims about x's relation to their own desires, preferences, and goals. But (contrary to common criticisms of subjectivism) this doesn't mean the two can't report the same facts or that they can't disagree; it just means that to do either they will have to introduce some other locution, e.g., 'x is good-for-you'. Even so, one of the many strong arguments against such a subjectivist account of moral judgments is that two people can agree (it seems) on the moral value of something, both saying simply that it is good (or bad), regardless of their respective desires, preferences, and goals. And this suggests that this version of subjectivism is implausible as an account of what we're saying when making moral claims. Thus, while this subjectivist account is realist in virtue of holding that some value claims literally construed are literally true, the implausibility of its proposed literal construal may nonetheless make it implausible as a realist position.

Intersubjectivism

Just as subjectivism is a position one might take concerning the truth-conditions of moral claims, so too is intersubjectivism. By spelling out the truth-conditions of moral claims in terms of the conventions or practices of groups of people, intersubjectivism grants (with subjectivism) that people figure in the truth-conditions, but it holds (with objectivism) that the truth of moral claims doesn't turn on facts about particular individuals.

Straightforward conventionalism is one version of intersubjectivism in ethics. It treats moral claims as being about (and not merely reflecting) the conventions and practices actually in force in the relevant society. This sort of view has figured prominently in defenses of cultural relativism. Ruth Benedict, for instance, defends an intersubjectivist account along these lines, maintaining, "Morality differs in every society, and is a convenient term for socially approved habits. Mankind has always preferred to say, 'It is morally good,' rather than 'It is habitual,' . . . but historically the two phrases are synonymous."[50]

---

49. Although the most common versions of subjectivism make the truth of moral claims depend on the *affective* states of an individual, not all do. For example, H. A. Prichard maintains that our having an obligation depends "on some fact about ourselves." Yet the fact about ourselves on which the obligatoriness of some act depends is not our having some affective state but "our having certain thoughts about the situation." The relevant thoughts, according to Prichard, are our thoughts about the consequences of the action in question. (H. A. Prichard, *Moral Obligation*, p. 37.)

50. Ruth Benedict, "Anthropology and the Abnormal," *Journal of General Psychology* 10 (1934), 73. For another version of straightforward conventionalism, see Edvard Westermarck's *Ethical Relativity* (New York, 1932).

Benedict's particular version of conventionalism runs into several problems as an account of the literal meaning of moral claims. It has trouble explaining how we can intelligibly raise moral objections to the habits approved by our society. And it suggests a peculiar view of how we are to go about discovering what is good; it recommends sociological investigation as decisive. Yet Benedict's conventionalism is not the only one available.

There are various ways of altering intersubjectivism so as to get a more plausible account of moral language.[51] One way is to reject the view that moral claims are directly about the conventions of a society and to maintain instead merely that they are the product of convention. With this in mind, one might argue that the correctness of a moral principle (and so the truth of moral claims) depends on its falling within the best coherent justificatory theory available (in principle) for the practices, conventions, and principles we happen to embrace. This view recognizes a standard against which particular practices and actions are to be measured, so it acknowledges the intelligibility of saying that there are immoral conventions and morally unacceptable practices. At the same time, however, it treats this standard as itself reflecting the institutions and practices of the moral community in question.[52]

Another way to improve on 'Benedictine' conventionalism is to abstract from actual practices and people and treat the truth of moral claims as being determined in some way by the hypothetical conventions or practices of hypothetical people. Moving away from the actual practices and conventions of real people allows sense to be made of moral criticism of these actual practices and conventions and shows why simple sociological investigation isn't sufficient for settling moral questions. Many contractarian views of morality take advantage of these benefits of abstraction by holding that the truth of moral claims turn on what appropriately idealized agents would agree to under certain specified conditions. Of course, different contractarian theories identify different conditions as relevant for determining the principles of morality, but they share the view that the appropriate truth-conditions for moral claims are intersubjectivist.[53]

Objectivism

Like subjectivism and intersubjectivism, objectivism is a view about what the truth-conditions are for moral claims as literally construed. The differences among the three lies in what each takes to be the proper literal construal. Objectivists hold that the appropriate truth-conditions make no reference to anyone's subjective states or to the capacities, conventions, or practices of any group of

---

51. More sophisticated versions of conventionalism are defended in Gilbert Harman, "Moral Relativism Defended," *Philosophical Review* 84 (1975), 3–22; and Kurt Baier, *The Moral Point of View.*

52. See my "Coherence and Models for Moral Theorizing."

53. See, for instance, John Rawls, *A Theory of Justice* (Cambridge, Mass., 1971); David Gauthier, *Morals by Agreement;* Stephen Darwall, *Impartial Reason* (Ithaca, N.Y., 1985).

people.[54] Underlying objectivism is the sense, well articulated by Ross, that "it is surely a strange reversal of the natural order of thought to say that our admiring an action either is, or is what necessitates, its being good. We think of its goodness as what we admire in it, and as something it would have even if no one admired it, something that it has in itself."[55]

In this spirit, Moore advances an objectivist theory of value when he argues that it is better that a beautiful world should exist rather than an ugly one even if no "human being ever has or ever, by any possibility, *can,* live in either, can ever see and enjoy the beauty of one or hate the foulness of the other."[56] The value of beauty, he holds, is independent of both the subjective states of individuals and the practices and conventions of groups of individuals. Elsewhere Moore does argue that the value of "personal affection and the appreciation of what is beautiful"[57] far outshines the value of beauty taken alone. This may seem to bollix up the distinction between subjective and objective accounts of the truth-conditions for moral claims, since 'personal affection and the appreciation of what is beautiful' are valuable subjective states, but it doesn't. What puts Moore squarely in the objectivist's camp is that he holds that the *value* of these subjective states, like the value of the beautiful (but unpopulated) world, is independent of the subjective states of individuals as well as of the practices of groups of people.

Objectivism in ethics has suffered from the mistaken assumption that objectivists must hold that moral properties are nonnatural and that nothing informative can be said about the truth-conditions of moral claims. For this reason, it is worth emphasizing several things about the relations among objectivism, nonnaturalism, and various ways in which the truth-conditions for moral claims might be specified.

Both Ross and Moore argue that moral terms refer to nonnatural properties. So each, in giving the truth-conditions of moral claims, will resist specifying the truth-conditions of, say, 'Murder is wrong,' in anything but moral terms. And Ross, unlike Moore, will even resist (when giving the semantics of English in English) anything more than:

(T) 'Murder is wrong' is true if and only if murder is wrong.

The difference between Moore and Ross in this respect turns on Ross's thinking 'wrong' is altogether unanalyzable whereas Moore thinks the truth-conditions of claims containing 'right' and 'wrong' can informatively be spelled out in terms of

---

54. Of course, even objectivists will grant that some of our duties (for instance, the duty to give money to the local soup kitchen) will turn on the subjective states or the practices of individuals. However, these will all be derivative duties (according to the objectivist) that have their grounding in something independent of subjective and social factors.

55. W. D. Ross, *The Right and the Good*, p. 89.

56. G. E. Moore, *Principia Ethica*, p. 84.

57. *Principia Ethica*, p. 188.

'good' and 'bad'. Since Moore believes 'good' and 'bad' are unanalyzable, though, he will follow Ross, when it comes to claims like 'pain is bad', in offering only a disquotational specification of truth-conditions.[58]

Three things, in particular, bear notice. First, a disquotational specification of truth-conditions such as the one Ross would offer for 'murder is wrong' is not trivial, despite appearances. It uses English to report a contingent yet important fact about the relation between the English language and the world. If some language other than English were used to report the same fact, even the appearance of triviality would disappear.[59]

Second, objectivists needn't always rely on, nor are they the only ones who might rely on, a disquotational specification of truth-conditions. What form the truth-conditions take depends on the resources of the metalanguage being used. In cases where we use a language in giving the semantics for that same language, we'll find that the truth-conditions have no more than a disquotational specification whenever the language contains only one way to report the relevant fact about the world. This specificational poverty might plague us regardless of whether the truth-conditions being offered are objectivist, intersubjectivist, or subjectivist. On the other hand, given a sufficiently rich metalanguage, objectivists (and not just intersubjectivists and subjectivists) might be able to offer more than merely disquotational specifications of truth-conditions.

Third, refusing to give truth-conditions in nonmoral terms for moral claims, as Moore and Ross do, is not an essential feature of objectivism; nor does it commit one to nonnaturalism. Objectivists of a naturalist bent might well identify moral properties with (perhaps very complex) physical properties and then, with the identification in hand, use the natural predicates when giving the truth-conditions for moral claims.[60] The only difficulty facing such a naturalized objectivism in ethics is finding plausible candidates for the identification. The major attraction of subjectivism and intersubjectivism to naturalists is that the most plausible candidates for an identification seem to involve reference either to the desires, preferences, and goals of individuals or to the practices and conventions of groups of

---

58. Relying on a disquotational specification of truth-conditions, in this way, is not tantamount to holding a disquotational theory of truth. The first is simply the form the specification of truth-conditions will sometimes take when the metalanguage includes the object language. The latter is a theory about the function of 'is true' in the English language (according to which, to say of some quoted sentence that it is true is to say nothing over and above what could be said by disquoting and using the sentence itself). While giving disquotationally specified truth-conditions is compatible with the disquotational theory of truth, it is also compatible with correspondence theories of truth and others. The core idea of the disquotational theory of truth can be found in F. P. Ramsey, "Facts and Propositions," in *The Foundations of Mathematics* (London, 1931). For a sophisticated version of the theory see Dorothy Grover, Joseph Camp, Jr., and Nuel Belnap, Jr., "The Prosentential Theory of Truth" *Philosophical Studies* 27 (1975), 73–125.

59. See the introduction to *Truth and Meaning,* ed. Gareth Evans and John McDowell (Oxford, 1976), pp. x–xi.

60. Richard Boyd defends this sort of objective naturalism in "How to Be a Moral Realist," as does Nicholas Sturgeon in "Moral Explanations," and Peter Railton in "Moral Realism."

individuals. In any case, even confirmed naturalists (whether they are objectivists, intersubjectivists, or subjectivists) can, alternatively, maintain that moral facts are not identifiable in nonmoral terminology; even if moral facts are just natural facts, we may not have a nonmoral vocabulary suited to reporting those facts.[61]

5. Conclusion

Subjectivism, intersubjectivism, and objectivism are each positions one might take about what the truth-conditions are for some disputed claims. So each might be proposed in support of realism in some domain or other. Yet they clearly aren't all equally plausible in each domain. While subjectivism is a plausible realist position in the philosophy of psychology, and intersubjectivism is a plausible realist position in the philosophy of law, both subjectivism and intersubjectivism are usually anti-realist positions in ethics. One of the advantages of my account of realism is that it explains why. The reason is that most versions of subjectivism and intersubjectivism give radically implausible accounts of the truth-conditions for moral claims as *literally construed*.

No simple subjectivist or intersubjectivist interpretation of the claim that 'x is wrong' does justice to the meaning of moral language (although more sophisticated versions might). And unless a proposal can be defended as offering the right truth-conditions for moral claims literally construed, it can form no part of a defense of moral realism. Realism requires that the moral claims, literally construed, be literally true. So when faced with an implausible account of the truth-conditions for moral claims, the charitable interpretation of the proposal is that it rests on the view that moral claims, as literally construed, should be jettisoned in favor of some new language. Such a proposal, if it is to be taken seriously, will have to find its motivation in a prior acceptance of anti-realism for moral claims as literally construed.

I've maintained that the heart of realism, no matter what the domain, lies in defending a success theory against both non-cognitivism and error theories. Such a defense requires two things: (1) showing that the disputed claims, when *literally construed*, have a truth value, and (2) showing that the disputed claims, when so construed, are sometimes *literally true*. If this is right, then moral realism is defensible if, but only if, there is some plausible account of the truth-conditions of moral claims that captures their literal meaning and that makes sense of some of them being literally true. It won't be enough, of course, to propose just any old account for the truth-conditions of moral claims; if the account offered doesn't capture what we mean when we make moral claims, then it will be no help in

61. It is worth noting as well that the realist, even the objectivist realist, isn't committed to introducing any extra *entities* as long as the supposition that there are moral *properties* is all that is needed to account for the truth of the claims that are held to be true.

defending *moral* realism. Nonliteral construals will be irrelevant. It also won't be enough to defend the view that some moral claims when literally construed are 'true in a sense'; if the defense doesn't show that the claims are literally true, then it will be no help in defending moral *realism*. Nonliteral truth will be insufficient. Yet, if the account does capture what we mean, and does make it reasonable to believe some of our moral claims are literally true, then moral realism will have found all the defense it needs.

I

MORAL
ANTI-REALISM

Critique of Ethics
and Theology

A. J. Ayer

There is still one objection to be met before we can claim to have justified our view that all synthetic propositions are empirical hypotheses. This objection is based on the common supposition that our speculative knowledge is of two distinct kinds—that which relates to questions of empirical fact, and that which relates to questions of value. It will be said that "statements of value" are genuine synthetic propositions, but that they cannot with any show of justice be represented as hypotheses, which are used to predict the course of our sensations; and, accordingly, that the existence of ethics and aesthetics as branches of speculative knowledge presents an insuperable objection to our radical empiricist thesis.

In face of this objection, it is our business to give an account of "judgements of value" which is both satisfactory in itself and consistent with our general empiricist principles. We shall set ourselves to show that in so far as statements of value are significant, they are ordinary "scientific" statements; and that in so far as they are not scientific, they are not in the literal sense significant, but are simply expressions of emotion which can be neither true nor false. In maintaining this view, we may confine ourselves for the present to the case of ethical statements. What is said about them will be found to apply, *mutatis mutandis*, to the case of aesthetic statements also.

The ordinary system of ethics, as elaborated in the works of ethical philosophers, is very far from being a homogeneous whole. Not only is it apt to contain pieces of metaphysics, and analyses of non-ethical concepts: its actual ethical contents are themselves of very different kinds. We may divide them, indeed, into four main classes. There are, first of all, propositions which express definitions of ethical terms, or judgements about the legitimacy or possibility of certain definitions. Secondly, there are propositions describing the phenomena of moral experi-

This essay was originally published as chapter 6 of A. J. Ayer, *Language, Truth, and Logic*, Dover Publications, Inc., New York, 1952. Used by permission of the publisher.

ence, and their causes. Thirdly, there are exhortations to moral virtue. And, lastly, there are actual ethical judgements. It is unfortunately the case that the distinction between these four classes, plain as it is, is commonly ignored by ethical philosophers; with the result that it is often very difficult to tell from their works what it is that they are seeking to discover or prove.

In fact, it is easy to see that only the first of our four classes, namely that which comprises the propositions relating to the definitions of ethical terms, can be said to constitute ethical philosophy. The propositions which describe the phenomena of moral experience, and their causes, must be assigned to the science of psychology, or sociology. The exhortations to moral virtue are not propositions at all, but ejaculations or commands which are designed to provoke the reader to action of a certain sort. Accordingly, they do not belong to any branch of philosophy or science. As for the expressions of ethical judgements, we have not yet determined how they should be classified. But inasmuch as they are certainly neither definitions nor comments upon definitions, nor quotations, we may say decisively that they do not belong to ethical philosophy. A strictly philosophical treatise on ethics should therefore make no ethical pronouncements. But it should, by giving an analysis of ethical terms, show what is the category to which all such pronouncements belong. And this is what we are now about to do.

A question which is often discussed by ethical philosophers is whether it is possible to find definitions which would reduce all ethical terms to one or two fundamental terms. But this question, though it undeniably belongs to ethical philosophy, is not relevant to our present enquiry. We are now concerned to discover which term, within the sphere of ethical terms, is to be taken as fundamental; whether, for example, "good" can be defined in terms of "right" or "right" in terms of "good," or both in terms of "value." What we are interested in is the possibility of reducing the whole sphere of ethical terms to non-ethical terms. We are enquiring whether statements of ethical value can be translated into statements of empirical fact.

That they can be so translated is the contention of those ethical philosophers who are commonly called subjectivists, and of those who are known as utilitarians. For the utilitarian defines the rightness of actions, and the goodness of ends, in terms of the pleasure, or happiness, or satisfaction, to which they give rise; the subjectivist, in terms of the feelings of approval which a certain person, or group of people, has towards them. Each of these types of definition makes moral judgements into a sub-class of psychological or sociological judgements; and for this reason they are very attractive to us. For, if either was correct, it would follow that ethical assertions were not generically different from the factual assertions which are ordinarily contrasted with them; and the account which we have already given of empirical hypotheses would apply to them also.

Nevertheless we shall not adopt either a subjectivist or a utilitarian analysis of ethical terms. We reject the subjectivist view that to call an action right, or a thing good, is to say that it is generally approved of, because it is not self-contradictory to assert that some actions which are generally approved of are not right, or that some things which are generally approved of are not good. And we reject the alternative subjectivist view that a man who asserts that a certain action is right, or that a certain thing is good, is saying that he himself approves of it, on the ground that a man who confessed that he sometimes approved of what was bad or wrong would not be contradicting himself. And a similar argument is fatal to utilitarianism. We cannot agree that to call an action right is to say that of all the actions possible in the circumstances it would cause, or be likely to cause, the greatest happiness, or the greatest balance of pleasure over pain, or the greatest balance of satisfied over unsatisfied desire, because we find that it is not self-contradictory to say that it is sometimes wrong to perform the action which would actually or probably cause the greatest happiness, or the greatest balance of pleasure over pain, or of satisfied over unsatisfied desire. And since it is not self-contradictory to say that some pleasant things are not good, or that some bad things are desired, it cannot be the case that the sentence "*x* is good" is equivalent to "*x* is pleasant," or to "*x* is desired." And to every other variant of utilitarianism with which I am acquainted the same objection can be made. And therefore we should, I think, conclude that the validity of ethical judgements is not determined by the felicific tendencies of actions, any more than by the nature of people's feelings; but that it must be regarded as "absolute" or "intrinsic," and not empirically calculable.

If we say this, we are not, of course, denying that it is possible to invent a language in which all ethical symbols are definable in non-ethical terms, or even that it is desirable to invent such a language and adopt it in place of our own; what we are denying is that the suggested reduction of ethical to non-ethical statements is consistent with the conventions of our actual language. That is, we reject utilitarianism and subjectivism, not as proposals to replace our existing ethical notions by new ones, but as analyses of our existing ethical notions. Our contention is simply that, in our language, sentences which contain normative ethical symbols are not equivalent to sentences which express psychological propositions, or indeed empirical propositions of any kind.

It is advisable here to make it plain that it is only normative ethical symbols, and not descriptive ethical symbols, that are held by us to be indefinable in factual terms. There is a danger of confusing these two types of symbols, because they are commonly constituted by signs of the same sensible form. Thus a complex sign of the form "*x* is wrong" may constitute a sentence which expresses a moral judgement concerning a certain type of conduct, or it may constitute a sentence which states that a certain type of conduct is repugnant to the moral sense of a particular society. In the latter case, the symbol "wrong" is a descriptive ethical symbol, and the sentence in which it occurs expresses an ordinary sociological proposition; in the former case, the symbol "wrong" is a normative ethical symbol, and the sentence in which it occurs does not, we maintain, express an empirical proposition at all. It is only with normative ethics that we are at present concerned; so that whenever ethical symbols are used in the course of this argument without qualification, they are always to be interpreted as symbols of the normative type.

In admitting that normative ethical concepts are irreducible to empirical concepts, we seem to be leaving the way clear for the "absolutist" view of ethics— that is, the view that statements of value are not controlled by observation, as ordinary empirical propositions are, but only by a mysterious "intellectual intuition." A feature of this theory, which is seldom recognized by its advocates, is that it makes statements of value unverifiable. For it is notorious that what seems intuitively certain to one person may seem doubtful, or even false, to another. So that unless it is possible to provide some criterion by which one may decide between conflicting intuitions, a mere appeal to intuition is worthless as a test of a proposition's validity. But in the case of moral judgements, no such criterion can be given. Some moralists claim to settle the matter by saying that they "know" that their own moral judgements are correct. But such an assertion is of purely psychological interest, and has not the slightest tendency to prove the validity of any moral judgement. For dissentient moralists may equally well "know" that their ethical views are correct. And, as far as subjective certainty goes, there will be nothing to choose between them. When such differences of opinion arise in connection with an ordinary empirical proposition, one may attempt to resolve them by referring to, or actually carrying out, some relevant empirical test. But with regard to ethical statements, there is, on the "absolutist" or "intuitionist" theory, no relevant empirical test. We are therefore justified in saying that on this theory ethical statements are held to be unverifiable. They are, of course, also held to be genuine synthetic propositions.

Considering the use which we have made of the principle that a synthetic proposition is significant only if it is empirically verifiable, it is clear that the acceptance of an "absolutist" theory of ethics would undermine the whole of our main argument. And as we have already rejected the "naturalistic" theories which are commonly supposed to provide the only alternative to "absolutism" in ethics, we seem to have reached a difficult position. We shall meet the difficulty by showing that the correct treatment of ethical statements is afforded by a third theory, which is wholly compatible with our radical empiricism.

We begin by admitting that the fundamental ethical concepts are unanalysable, inasmuch as there is no criterion by which one can test the validity of the judgements in which they occur. So far we are in agreement with the absolutists. But, unlike the absolutists, we are able to give an explanation of this fact about ethical concepts. We say that the reason why they are unanalysable is that they are mere pseudo-concepts. The presence of an ethical symbol in a proposition adds nothing to its factual content. Thus if I say to someone, "You acted wrongly in stealing that money," I am not stating anything more than if I had simply said, "You stole that money." In adding that this action is wrong I am not making any further statement about it. I am simply evincing my moral disapproval of it. It is as if I had said, "You stole that money," in a peculiar tone of horror, or written it with the addition of some special exclamation marks. The tone, or the exclamation marks, adds nothing to the literal meaning of the sentence. It merely serves to show that the expression of it is attended by certain feelings in the speaker.

If now I generalise my previous statement and say, "Stealing money is wrong," I produce a sentence which has no factual meaning—that is, expresses no proposition which can be either true or false. It is as if I had written "Stealing money!!'"—where the shape and thickness of the exclamation marks show, by a suitable convention, that a special sort of moral disapproval is the feeling which is being expressed. It is clear that there is nothing said here which can be true or false. Another man may disagree with me about the wrongness of stealing, in the sense that he may not have the same feelings about stealing as I have, and he may quarrel with me on account of my moral sentiments. But he cannot, strictly speaking, contradict me. For in saying that a certain type of action is right or wrong, I am not making any factual statement, not even a statement about my own state of mind. I am merely expressing certain moral sentiments. And the man who is ostensibly contradicting me is merely expressing his moral sentiments. So that there is plainly no sense in asking which of us is in the right. For neither of us is asserting a genuine proposition.

What we have just been saying about the symbol "wrong" applies to all normative ethical symbols. Sometimes they occur in sentences which record ordinary empirical facts besides expressing ethical feeling about those facts: sometimes they occur in sentences which simply express ethical feeling about a certain type of action, or situation, without making any statement of fact. But in every case in which one would commonly be said to be making an ethical judgement, the function of the relevant ethical word is purely "emotive." It is used to express feeling about certain objects, but not to make any assertion about them.

It is worth mentioning that ethical terms do not serve only to express feeling. They are calculated also to arouse feeling, and so to stimulate action. Indeed some of them are used in such a way as to give the sentences in which they occur the effect of commands. Thus the sentence "It is your duty to tell the truth" may be regarded both as the expression of a certain sort of ethical feeling about truthfulness and as the expression of the command "Tell the truth." The sentence "You ought to tell the truth" also involves the command "Tell the truth," but here the tone of the command is less emphatic. In the sentence "It is good to tell the truth" the command has become little more than a suggestion. And thus the "meaning" of the word "good,' in its ethical usage, is differentiated from that of the word "duty" or the word "ought." In fact we may define the meaning of the various ethical words in terms both of the different feelings they are ordinarily taken to express, and also the different responses which they are calculated to provoke.

We can now see why it is impossible to find a criterion for determining the validity of ethical judgements. It is not because they have an "absolute" validity which is mysteriously independent of ordinary sense-experience, but because they have no objective validity whatsoever. If a sentence makes no statement at all, there is obviously no sense in asking whether what it says is true or false. And we have seen that sentences which simply express moral judgements do not say anything. They are pure expressions of feeling and as such do not come under the category of truth and falsehood. They are unverifiable for the same reason as a cry of pain or a word of command is unverifiable—because they do not express genuine propositions.

Thus, although our theory of ethics might fairly be said to be radically subjectivist, it differs in a very important respect from the orthodox subjectivist theory. For the orthodox subjectivist does not deny, as we do, that the sentences of a moralizer express genuine propositions. All he denies is that they express propositions of a unique non-empirical character. His own view is that they express propositions about the speaker's feelings. If this were so, ethical judgements clearly would be capable of being true or false. They would be true if the speaker had the relevant feelings, and false if he had not. And this is a matter which is, in principle, empirically verifiable. Furthermore they could be significantly contradicted. For if I say, "Tolerance is a virtue," and someone answers, "You don't approve of it," he would, on the ordinary subjectivist theory, be contradicting me. On our theory, he would not be contradicting me, because, in saying that tolerance was a virtue, I should not be making any statement about my own feelings or about anything else. I should simply be evincing my feelings, which is not at all the same thing as saying that I have them.

The distinction between the expression of feeling and the assertion of feeling is complicated by the fact that the assertion that one has a certain feeling often accompanies the expression of that feeling, and is then, indeed, a factor in the expression of that feeling. Thus I may simultaneously express boredom and say that I am bored, and in that case my utterance of the words, "I am bored," is one of the circumstances which make it true to say that I am expressing or evincing boredom. But I can express boredom without actually saying that I am bored. I can express it by my tone and gestures, while making a statement about something wholly unconnected with it, or by an ejaculation, or without uttering any words at all. So that even if the assertion that one has a certain feeling always involves the expression of that feeling, the expression of a feeling assuredly does not always involve the assertion that one has it. And this is the important point to grasp in considering the distinction between our theory and the ordinary subjectivist theory. For whereas the subjectivist holds that ethical statements actually assert the existence of certain feelings, we hold that ethical statements are expressions and excitants of feeling which do not necessarily involve any assertions.

We have already remarked that the main objection to the ordinary subjectivist theory is that the validity of ethical judgements is not determined by the nature of their author's feelings. And this is an objection which our theory escapes. For it does not imply that the existence of any feelings is a necessary and sufficient condition of the validity of an ethical judgement. It implies, on the contrary, that ethical judgements have no validity.

[32]

There is, however, a celebrated argument against subjectivist theories which our theory does not escape. It has been pointed out by Moore that if ethical statements were simply statements about the speaker's feelings, it would be impossible to argue about questions of value.[1] To take a typical example: if a man said that thrift was a virtue, and another replied that it was a vice, they would not, on this theory, be disputing with one another. One would be saying that he approved of thrift, and the other that *he* didn't; and there is no reason why both these statements should not be true. Now Moore held it to be obvious that we do dispute about questions of value, and accordingly concluded that the particular form of subjectivism which he was discussing was false.

It is plain that the conclusion that it is impossible to dispute about questions of value follows from our theory also. For as we hold that such sentences as "Thrift is a virtue" and "Thrift is a vice" do not express propositions at all, we clearly cannot hold that they express incompatible propositions. We must therefore admit that if Moore's argument really refutes the ordinary subjectivist theory, it also refutes ours. But, in fact, we deny that it does refute even the ordinary subjectivist theory. For we hold that one really never does dispute about questions of value.

This may seem, at first sight, to be a very paradoxical assertion. For we certainly do engage in disputes which are ordinarily regarded as disputes about questions of value. But, in all such cases, we find, if we consider the matter closely, that the dispute is not really about a question of value, but about a question of fact. When someone disagrees with us about the moral value of a certain action or type of action, we do admittedly resort to argument in order to win him over to our way of thinking. But we do not attempt to show by our arguments that he has the "wrong" ethical feeling towards a situation whose nature he has correctly apprehended. What we attempt to show is that he is mistaken about the facts of the case. We argue that he has misconceived the agent's motive: or that he has misjudged the effects of the action, or its probable effects in view of the agent's knowledge; or that he has failed to take into account the special circumstances in which the agent was placed. Or else we employ more general arguments about the effects which actions of a certain type tend to produce, or the qualities which are usually manifested in their performance. We do this in the hope that we have only to get our opponent to agree with us about the nature of the empirical facts for him to adopt the same moral attitude towards them as we do. And as the people with whom we argue have generally received the same moral education as ourselves, and live in the same social order, our expectation is usually justified. But if our opponent happens to have undergone a different process of moral "conditioning" from ourselves, so that, even when he acknowledges all the facts, he still disagrees with us about the moral value of the actions under discussion, then we abandon the attempt to convince him by argument. We say that it is impossible to argue with him because he has a distorted or undeveloped moral sense; which signifies merely that he employs a different set of

1. Cf. *Philosophical Studies*, "The Nature of Moral Philosophy."

values from our own. We feel that our own system of values is superior, and therefore speak in such derogatory terms of his. But we cannot bring forward any arguments to show that our system is superior. For our judgement that it is so is itself a judgement of value, and accordingly outside the scope of argument. It is because argument fails us when we come to deal with pure questions of value, as distinct from questions of fact, that we finally resort to mere abuse.

In short, we find that argument is possible on moral questions only if some system of values is presupposed. If our opponent concurs with us in expressing moral disapproval of all actions of a given type *t,* then we may get him to condemn a particular action A, by bringing forward arguments to show that A is of type *t.* For the question whether A does or does not belong to that type is a plain question of fact. Given that a man has certain moral principles, we argue that he must, in order to be consistent, react morally to certain things in a certain way. What we do not and cannot argue about is the validity of these moral principles. We merely praise or condemn them in the light of our own feelings.

If anyone doubts the accuracy of this account of moral disputes, let him try to construct even an imaginary argument on a question of value which does not reduce itself to an argument about a question of logic or about an empirical matter of fact. I am confident that he will not succeed in producing a single example. And if that is the case, he must allow that its involving the impossibility of purely ethical arguments is not, as Moore thought, a ground of objection to our theory, but rather a point in favour of it.

Having upheld our theory against the only criticism which appeared to threaten it, we may now use it to define the nature of all ethical enquiries. We find that ethical philosophy consists simply in saying that ethical concepts are pseudo-concepts and therefore unanalysable. The further task of describing the different feelings that the different ethical terms are used to express, and the different reactions that they customarily provoke, is a task for the psychologist. There cannot be such a thing as ethical science, if by ethical science one means the elaboration of a "true" system of morals. For we have seen that, as ethical judgements are mere expressions of feeling, there can be no way of determining the validity of any ethical system, and, indeed, no sense in asking whether any such system is true. All that one may legitimately enquire in this connection is, What are the moral habits of a given person or group of people, and what causes them to have precisely those habits and feelings? And this enquiry falls wholly within the scope of the existing social sciences.

It appears, then, that ethics, as a branch of knowledge, is nothing more than a department of psychology and sociology. And in case anyone thinks that we are overlooking the existence of casuistry, we may remark that casuistry is not a science, but is a purely analytical investigation of the structure of a given moral system. In other words, it is an exercise in formal logic.

When one comes to pursue the psychological enquiries which constitute ethical science, one is immediately enabled to account for the Kantian and hedonistic theories of morals. For one finds that one of the chief causes of moral behaviour is fear, both conscious and unconscious, of a god's displeasure, and fear of the enmity of society. And this, indeed, is the reason why moral precepts present themselves to some people as "categorical" commands. And one finds, also, that the moral code of a society is partly determined by the beliefs of that society concerning the conditions of its own happiness—or, in other words, that a society tends to encourage or discourage a given type of conduct by the use of moral sanctions according as it appears to promote or detract from the contentment of the society as a whole. And this is the reason why altruism is recommended in most moral codes and egotism condemned. It is from the observation of this connection between morality and happiness that hedonistic or eudaemonistic theories of morals ultimately spring, just as the moral theory of Kant is based on the fact, previously explained, that moral precepts have for some people the force of inexorable commands. As each of these theories ignores the fact which lies at the root of the other, both may be criticized as being one-sided; but this is not the main objection to either of them. Their essential defect is that they treat propositions which refer to the causes and attributes of our ethical feelings as if they were definitions of ethical concepts. And thus they fail to recognise that ethical concepts are pseudo-concepts and consequently indefinable.

As we have already said, our conclusions about the nature of ethics apply to aesthetics also. Aesthetic terms are used in exactly the same way as ethical terms. Such aesthetic words as "beautiful" and "hideous" are employed, as ethical words are employed, not to make statements of fact, but simply to express certain feelings and evoke a certain response. It follows, as in ethics, that there is no sense in attributing objective validity to aesthetic judgements, and no possibility of arguing about questions of value in aesthics, but only about questions of fact. A scientific treatment of aesthetics would show us what in general were the causes of aesthetic feeling, why various societies produced and admired the works of art they did, why taste varies as it does within a given society, and so forth. And these are ordinary psychological or sociological questions. They have, of course, little or nothing to do with aesthetic criticism as we understand it. But that is because the purpose of aesthetic criticism is not so much to give knowledge as to communicate emotion. The critic, by calling attention to certain features of the work under review, and expressing his own feelings about them, endeavours to make us share his attitude towards the work as a whole. The only relevant propositions that he formulates are propositions describing the nature of the work. And these are plain records of fact. We conclude, therefore, that there is nothing in aesthetics, any more than there is in ethics, to justify the view that it embodies a unique type of knowledge.

It should now be clear that the only information which we can legitimately derive from the study of our aesthetic and moral experiences is information about our own mental and physical make-up. We take note of these experiences as providing data for our psychological and sociological generalisations. And this is the only way in which they serve to increase our knowledge. It follows that any attempt to make our use of ethical and aesthetic concepts the basis of a metaphysical theory concerning the existence of a world of values, as distinct from the world of facts, involves a false analysis of these concepts. Our own analysis has shown that the phenomena of moral experience cannot fairly be used to support any rationalist or metaphysical doctrine whatsoever. In particular, they cannot, as Kant hoped, be used to establish the existence of a transcendent god.

This mention of God brings us to the question of the possibility of religious knowledge. We shall see that this possibility has already been ruled out by our treatment of metaphysics. But, as this is a point of considerable interest, we may be permitted to discuss it at some length.

It is now generally admitted, at any rate by philosophers, that the existence of a being having the attributes which define the god of any non-animistic religion cannot be demonstratively proved. To see that this is so, we have only to ask ourselves what are the premises from which the existence of such a god could be deduced. If the conclusion that a god exists is to be demonstratively certain, then these premises must be certain; for, as the conclusion of a deductive argument is already contained in the premises, any uncertainty there may be about the truth of the premises is necessarily shared by it. But we know that no empirical proposition can ever be anything more than probable. It is only *a priori* propositions that are logically certain. But we cannot deduce the existence of a god from an *a priori* proposition. For we know that the reason why *a priori* propositions are certain is that they are tautologies. And from a set of tautologies nothing but a further tautology can be validly deduced. It follows that there is no possibility of demonstrating the existence of a god.

What is not so generally recognised is that there can be no way of proving that the existence of a god, such as the God of Christianity, is even probable. Yet this also is easily shown. For if the existence of such a god were probable, then the proposition that he existed would be an empirical hypothesis. And in that case it would be possible to deduce from it, and other empirical hypotheses, certain experiential propositions which were not deducible from those other hypotheses alone. But in fact this is not possible. It is sometimes claimed, indeed, that the existence of a certain sort of regularity in nature constitutes sufficient evidence for the existence of a god. But if the sentence "God exists" entails no more than that certain types of phenomena occur in certain sequences, then to assert the existence of a god will be simply equivalent to asserting that there is the requisite regularity in nature; and no religious man would admit that this was all he intended to assert in asserting the existence of a god. He would say that in talking about God, he was talking about a transcendent being who might be known through certain empirical manifestations, but certainly could not be defined in terms of those manifestations. But in that case the term "god" is a metaphysical term. And if "god" is a metaphysical term, then it cannot be even probable that a god exists. For to say that "God exists" is to make a metaphysical utterance which cannot be either true or false. And by the same criterion, no sentence which purports to describe the nature of a transcendent god can possess any literal significance.

It is important not to confuse this view of religious assertions with the view that is adopted by atheists, or agnostics.[2] For it is characteristic of an agnostic to hold that the existence of a god is a possibility in which there is no good reason either to believe or disbelieve; and it is characteristic of an atheist to hold that it is at least probable that no god exists. And our view that all utterances about the nature of God are nonsensical, so far from being identical with, or even lending any support to, either of these familiar contentions, is actually incompatible with them. For if the assertion that there is a god is nonsensical, then the atheist's assertion that there is no god is equally nonsensical, since it is only a significant proposition that can be significantly contradicted. As for the agnostic, although he refrains from saying either that there is or that there is not a god, he does not deny that the question whether a transcendent god exists is a genuine question. He does not deny that the two sentences "There is a transcendent god" and "There is no transcendent god" express propositions one of which is actually true and the other false. All he says is that we have no means of telling which of them is true, and therefore ought not to commit ourselves to either. But we have seen that the sentences in question do not express propositions at all. And this means that agnosticism also is ruled out.

Thus we offer the theist the same comfort as we gave to the moralist. His assertions cannot possibly be valid, but they cannot be invalid either. As he says nothing at all about the world, he cannot justly be accused of saying anything false, or anything for which he has insufficient grounds. It is only when the theist claims that in asserting the existence of a transcendent god he is expressing a genuine proposition that we are entitled to disagree with him.

It is to be remarked that in cases where deities are identified with natural objects, assertions concerning them may be allowed to be significant. If, for example, a man tells me that the occurrence of thunder is alone both necessary and sufficient to establish the truth of the proposition that Jehovah is angry, I may conclude that, in his usage of words, the sentence "Jehovah is angry" is equivalent to "It is thundering." But in sophisticated religions, though they may be to some extent based on men's awe of natural process which they cannot sufficiently understand, the "person" who is supposed to control the empirical world is not himself located in it; he is held to be superior to the empirical world, and so outside it; and he is endowed with super-empirical attributes. But the notion of a person whose essential attributes are nonempirical is not an intelligible notion at all. We may have a word which is used as if it named this "person," but, unless the sentences in which it occurs express propositions which are empirically verifiable, it cannot be said to symbolize anything. And this is the case with regard to the word "god," in the usage in which it is intended to refer to a transcendent

2. This point was suggested to me by Professor H. H. Price.

object. The mere existence of the noun is enough to foster the illusion that there is a real, or at any rate a possible entity corresponding to it. It is only when we enquire what God's attributes are that we discover that "God," in this usage, is not a genuine name.

It is common to find belief in a transcendent god conjoined with belief in an after-life. But, in the form which it usually takes, the content of this belief is not a genuine hypothesis. To say that men do not ever die, or that the state of death is merely a state of prolonged insensibility, is indeed to express a significant proposition, though all the available evidence goes to show that it is false. But to say that there is something imperceptible inside a man, which is his soul or his real self, and that it goes on living after he is dead, is to make a metaphysical assertion which has no more factual content than the assertion that there is a transcendent god.

It is worth mentioning that, according to the account which we have given of religious assertions, there is no logical ground for antagonism between religion and natural science. As far as the question of truth or falsehood is concerned, there is no opposition between the natural scientist and the theist who believes in a transcendent god. For since the religious utterances of the theist are not genuine propositions at all, they cannot stand in any logical relation to the propositions of science. Such antagonism as there is between religion and science appears to consist in the fact that science takes away one of the motives which make men religious. For it is acknowledged that one of the ultimate sources of religious feeling lies in the inability of men to determine their own destiny; and science tends to destroy the feeling of awe with which men regard an alien world, by making them believe that they can understand and anticipate the course of natural phenomena, and even to some extent control it. The fact that it has recently become fashionable for physicists themselves to be sympathetic towards religion is a point in favour of this hypothesis. For this sympathy towards religion marks the physicists' own lack of confidence in the validity of their hypotheses, which is a reaction on their part from the anti-religious dogmatism of nineteenth-century scientists, and a natural outcome of the crisis through which physics has just passed.

It is not within the scope of this enquiry to enter more deeply into the causes of religious feeling, or to discuss the probability of the continuance of religious belief. We are concerned only to answer those questions which arise out of our discussion of the possibility of religious knowledge. The point which we wish to establish is that there cannot be any transcendent truths of religion. For the sentences which the theist uses to express such "truths" are not literally significant.

An interesting feature of this conclusion is that it accords with what many theists are accustomed to say themselves. For we are often told that the nature of God is a mystery which transcends the human understanding. But to say that something transcends the human understanding is to say that it is unintelligible.

And what is unintelligible cannot significantly be described. Again, we are told that God is not an object of reason but an object of faith. This may be nothing more than an admission that the existence of God must be taken on trust, since it cannot be proved. But it may also be an assertion that God is the object of a purely mystical intuition, and cannot therefore be defined in terms which are intelligible to the reason. And I think there are many theists who would assert this. But if one allows that it is impossible to define God in intelligible terms, then one is allowing that it is impossible for a sentence both to be significant and to be about God. If a mystic admits that the object of his vision is something which cannot be described, then he must also admit that he is bound to talk nonsense when he describes it.

For his part, the mystic may protest that his intuition does reveal truths to him, even though he cannot explain to others what these truths are; and that we who do not possess this faculty of intuition can have no ground for denying that it is a cognitive faculty. For we can hardly maintain *a priori* that there are no ways of discovering true propositions except those which we ourselves employ. The answer is that we set no limit to the number of ways in which one may come to formulate a true proposition. We do not in any way deny that a synthetic truth may be discovered by purely intuitive methods as well as by the rational method of induction. But we do say that every synthetic proposition, however it may have been arrived at, must be subject to the test of actual experience. We do not deny *a priori* that the mystic is able to discover truths by his own special methods. We wait to hear what are the propositions which embody his discoveries, in order to see whether they are verified or confuted by our empirical observations. But the mystic, so far from producing propositions which are empirically verified, is unable to produce any intelligible propositions at all. And therefore we say that his intuition has not revealed to him any facts. It is no use his saying that he has apprehended facts but is unable to express them. For we know that if he really had acquired any information, he would be able to express it. He would be able to indicate in some way or other how the genuineness of his discovery might be empirically determined. The fact that he cannot reveal what he "knows," or even himself devise an empirical test to validate his "knowledge," shows that his state of mystical intuition is not a genuinely cognitive state. So that in describing his vision the mystic does not give us any information about the external world; he merely gives us indirect information about the condition of his own mind.

These considerations dispose of the argument from religious experience, which many philosophers still regard as a valid argument in favour of the existence of a god. They say that it is logically possible for men to be immediately acquainted with God, as they are immediately acquainted with a sense-content, and that there is no reason why one should be prepared to believe a man when he says that he is seeing a yellow patch, and refuse to believe him when he says that he is seeing God. The answer to this is that if the man who asserts that he is seeing God is merely asserting that he is experiencing a peculiar kind of sense-content, then we

[39]

do not for a moment deny that his assertion may be true. But, ordinarily, the man who says that he is seeing God is saying not merely that he is experiencing a religious emotion, but also that there exists a transcendent being who is the object of this emotion; just as the man who says that he sees a yellow patch is ordinarily saying not merely that his visual sense-field contains a yellow sense-content, but also that there exists a yellow object to which the sense-content belongs. And it is not irrational to be prepared to believe a man when he asserts the existence of a yellow object, and to refuse to believe him when he asserts the existence of a transcendent god. For whereas the sentence "There exists here a yellow-coloured material thing" expresses a genuine synthetic proposition which could be empirically verified, the sentence "There exists a transcendent god" has, as we have seen, no literal significance.

We conclude, therefore, that the argument from religious experience is altogether fallacious. The fact that people have religious experiences is interesting from the psychological point of view, but it does not in any way imply that there is such a thing as religious knowledge, any more than our having moral experiences implies that there is such a thing as moral knowledge. The theist, like the moralist, may believe that his experiences are cognitive experiences, but, unless he can formulate his "knowledge" in propositions that are empirically verifiable, we may be sure that he is deceiving himself. It follows that those philosophers who fill their books with assertions that they intuitively "know" this or that moral or religious "truth" are merely providing material for the psycho-analyst. For no act of intuition can be said to reveal a truth about any matter of fact unless it issues in verifiable propositions. And all such propositions are to be incorporated in the system of empirical propositions which constitutes science.

2

Ethical Consistency

Bernard Williams

I shall not attempt any discussion of ethical consistency in general. I shall consider one question that is near the centre of that topic: the nature of moral conflict. I shall bring out some characteristics of moral conflict that have bearing, as I think, on logical or philosophical questions about the structure of moral thought and language. I shall centre my remarks about moral conflict on certain comparisons between this sort of conflict, conflicts of beliefs, and conflicts of desires; I shall start, in fact, by considering the latter two sorts of conflict, that of beliefs very briefly, that of desires at rather greater length, since it is both more pertinent and more complicated.

Some of what I have to say may seem too psychological. In one respect, I make no apology for this; in another, I do. I do not, in as much as I think that a neglect of moral psychology and in particular of the role of emotion in morality has distorted and made unrealistic a good deal of recent discussion; having disposed of emotivism as a theory of the moral judgment, philosophers have perhaps tended to put the emotions on one side as at most contingent, and therefore philosophically uninteresting, concomitants to other things which are regarded as alone essential This must surely be wrong: to me, at least, the question of what emotions a man feels in various circumstances seems to have a good deal to do, for instance, with whether he is an admirable human being or not. I do apologise, however, for employing in the following discussion considerations about emotion (in particular, *regret*) in a way which is certainly less clear than I should like.

1. It is possible for a man to hold inconsistent beliefs, in the strong sense that the statements which would adequately express his beliefs involve a logical contradiction. This possibility, however, I shall not be concerned with, my interest being rather in the different case of a man who holds two beliefs which are

This essay was originally published in *Proceedings of the Aristotelian Society*, supplementary volume 39 (1965), 103–24.

not inconsistent in this sense, but which for some empirical reason cannot both be true. Such beliefs I shall call 'conflicting'. Thus a man might believe that a certain person was a Minister who took office in October 1964 and also that the person was a member of the Conservative Party. This case will be different from that of inconsistent beliefs, of course, only if the man is ignorant of the further information which reveals the two beliefs as conflicting, viz. that no such Minister is a Conservative. If he is then given this information, and believes it, then either he becomes conscious of the conflict between his original beliefs[1], or, if he retains all three beliefs (for instance, because he has not 'put them together'), then he is in the situation of having actually inconsistent beliefs. This shows a necessary condition of beliefs conflicting: that if a pair of beliefs conflict, then (a) they are consistent, and (b) there is a true factual belief which, if added to the original pair, will produce a set that is inconsistent.

2. What is normally called conflict of *desires* has, in many central cases, a feature analogous to what I have been calling conflict of beliefs: that the clash between the desires arises from some contingent matter of fact. This is a matter of fact that makes it impossible for both the desires to be satisfied; but we can consistently imagine a state of affairs in which they could both be satisfied. The contingent root of the conflict may, indeed, be disguised by a use of language that suggests logical impossibility of the desires being jointly satisfied; thus a man who was thirsty and lazy, who was seated comfortably, and whose drinks were elsewhere, might perhaps represent his difficulty to himself as his both wanting to remain seated and wanting to get up. But to put it this way is for him to hide the roots of his difficulty under the difficulty itself; the second element in the conflict has been so described as to reveal the obstacle to the first, and not its own real object. The sudden appearance of a friend or servant, or the discovery of drinks within arm's reach, would make all plain.

While many cases of conflict of desires are of this contingent character, it would be artificial or worse to try to force all cases into this mould, and to demand for every situation of conflict an answer to the question ''what conceivable change in the contingent facts of the world would make it possible for both desires to be satisfied?'' Some cases involving difficulties with space and time, for instance, are likely to prove recalcitrant: can one isolate the relevant contingency in the situation of an Australian torn between spending Christmas in Christmassy surroundings in Austria, and spending it back home in the familiar Christmas heat of his birthplace?

A more fundamental difficulty arises with conflicts of desire and aversion towards one and the same object. Such conflicts can be represented as conflicts of two desires: in the most general case, the desire to have and the desire not to have

---

1. I shall in the rest of this paper generally use the phrase 'conflict of beliefs' for the situation in which a man has become conscious that his beliefs conflict.

the object, where 'have' is a variable expression which gets a determinate content from the context and from the nature of the object in question[2]. There are indeed other cases in which an aversion to *x* does not merely take the form of a desire *not to have x* (to avoid it, reject it, to be elsewhere, etc.), but rather the form of a desire that *x should not exist*—in particular, a desire to destroy it. These latter cases are certainly different from the former (aversion here involves advancing rather than retreating), but I shall leave these, and concentrate on the former type. Conflicts of desire and aversion in this sense differ from the conflicts mentioned earlier, in that the most direct characterization of the desires—'I want to have x' and 'I want not to have x'—do not admit an imaginable contingent change which would allow both the desires to be satisfied, the descriptions of the situations that would satisfy the two desires being logically incompatible. However, there is in many cases something else that can be imagined which is just as good: the removal from the object of the disadvantageous features which are the ground of the aversion or (as I shall call aversions which are merely desires *not to have*) negative desire. This imaginable change would eliminate the conflict, not indeed by satisfying, but by eliminating, the negative desire.

This might be thought to be cheating, since any conflict of desires can be imagined away by imagining away one of the desires. There is a distinction, however, in that the situation imagined without the negative desire involves no loss of utility: no greater utility can be attached to a situation in which a purely negative desire is satisfied, than to one in which the grounds of it were never present at all. This does not apply to desires in general (and probably not to the more active, destructive, type of aversion distinguished before). Admittedly, there has been a vexed problem in this region from antiquity on, but (to take the extreme case) it does seem implausible to claim that there is no difference of utility to be found between the lives of two men, one of whom has no desires at all, the other many desires, all of which are satisfied.

Thus it seems that for many cases of conflict of desire and aversion towards one object, the basis of the conflict is still, though in a slightly different way, contingent, the contingency consisting in the co-existence of the desirable and the undesirable features of the object. Not all cases, however, will yield to this treatment, since there may be various difficulties in representing the desirable and undesirable features as only contingently co-existing. The limiting case in this direction is that in which the two sets of features are identical (the case of ambivalence)—though this will almost certainly involve the other, destructive, form of aversion.

This schematic discussion of conflicts between desires is meant to apply only to non-moral desires; that is to say, to cases where the answer to the question "why do you want x?" does not involve expressing any moral attitude. If this limitation

2. For a discussion of a similar notion, see A. Kenny, *Action, Emotion and Will,* chap. 5.

is removed, and moral desires are considered, a much larger class of non-contingently based conflicts comes into view, since it is evidently the case that a moral desire and a non-moral desire which are in conflict may be directed towards exactly the same features of the situation.[3] Leaving moral desires out of it, however, I think we find that a very large range of conflicts of desires have what I have called a contingent basis. Our desires that conflict are standardly like beliefs that conflict, not like beliefs that are inconsistent; as with conflicting beliefs it is the world, not logic, that makes it impossible for them both to be true, so with most conflicting desires, it is the world, not logic, that makes it impossible for them both to be satisfied.

3. There are a number of interesting contrasts between situations of conflict with beliefs and with desires; I shall consider two.

(a) If I discover that two of my beliefs conflict, at least one of them, by that very fact, will tend to be weakened; but the discovery that two desires conflict has no tendency, in itself, to weaken either of them. This is for the following reason: while satisfaction is related to desire to some extent as truth is related to belief, the discovery that two desires cannot both be satisfied is not related to those desires as the discovery that two beliefs cannot both be true is related to those beliefs. To believe that p is to believe that p is true, so the discovery that two of my beliefs cannot both be true is itself a step on the way to my not holding at least one of them; whereas the desire that I should have such-&-such, and the belief that I will have it, are obviously not so related.

(b) Suppose the conflict ends in a decision, and, in the case of desire, action; in the simplest case, I decide that one of the conflicting beliefs is true and not the other, or I satisfy one of the desires and not the other. The rejected belief cannot substantially survive this point, because to decide that a belief is untrue *is* to abandon, i.e. no longer to have, that belief. (Of course, there are qualifications to be made here: it is possible to say "I know that it is untrue, but I can't help still believing it". But it is essential to the concept of belief that such cases are secondary, even peculiar.) A rejected desire, however, can, if not survive the point of decision, at least reappear on the other side of it on one or another guise. It may reappear, for instance, as a general desire for something of the same sort as the object rejected in the decision; or as a desire for another particular object of the same sort; or—and this is the case that will concern us most—if there are no substitutes, the opportunity for satisfying that desire having irrevocably gone, it may reappear in the form of a *regret* for what was missed.

It may be said that the rejection of a belief may also involve regret. This is

---

3. Plato, incidentally, seems to have thought that all conflicts that did not involve a moral or similar motivation had a contingent basis. The argument of *Republic* IV which issues in the doctrine of the divisions of the soul bases the distinction between the rational and epithymetic parts on conflicts of desire and aversion directed towards the same object in the same respects. But not all conflicts establish different parts of the soul: the epithymetic part can be in conflict with itself. These latter conflicts, therefore, cannot be of desires directed towards the same object in the same respects; that is to say, purely epithymetic conflicts have a contingent basis.

indeed true, and in more than one way: if I have to abandon a belief, I may regret this either because it was a belief of mine (as when a scientist or an historian loses a pet theory), or—quite differently—because it would have been more agreeable if the world had been as, when I had the belief, I thought it was (as when a father is finally forced to abandon the belief that his son survived the sinking of the ship). Thus there are various regrets possible for the loss of beliefs. But this is not enough to reinstate a parallelism between beliefs and desires in this respect. For the regret that can attach to an abandoned belief is never sufficiently explained just by the fact that the man did have the belief; to explain this sort of regret, one has to introduce something else—and this is, precisely, a desire, a desire for the belief to be true. That a man regrets the falsification of his belief that p shows not just that he believed that p, but that he wanted to believe that p: where ''wanting to believe that p'' can have different sorts of application, corresponding to the sorts of regret already distinguished. That a man regrets not having been able to satisfy a desire, is sufficiently explained by the fact that he had that desire.

4. I now turn to moral conflict. I shall discuss this in terms of 'ought', not because 'ought' necessarily figures in the expression of every moral conflict, which is certainly not true, but because it presents the most puzzling problems. By 'moral conflict' I mean only cases in which there is a conflict between two moral judgments that a man is disposed to make relevant to deciding what to do; that is to say, I shall be considering what has traditionally, though misleadingly, been called 'conflict of obligations', and not, for instance, conflicts between a moral judgment and a non-moral desire, though these, too, could naturally enough be called 'moral conflicts'. I shall further omit any discussion of the possibility (if it exists) that a man should hold moral principles or general moral views which are intrinsically inconsistent with one another, in the sense that there could be no conceivable world in which anyone could act in accordance with both of them; as might be the case, for instance, with a man who thought both that he ought not to go in for any blood-sport (as such), and that he ought to go in for foxhunting (as such). I doubt whether there are any interesting questions that are peculiar to this possibility. I shall confine myself, then, to cases in which the moral conflict has a contingent basis, to use a phrase that has already occurred in the discussion of conflicts of desires. Some real analogy, moreover, with those situations emerges if one considers two basic forms that the moral conflict can take. One is that in which it seems that I ought to do each of two things, but I cannot do both. The other is that in which something which (it seems) I ought to do in respect of certain of its features also has other features in respect of which (it seems) I ought not to do it. This latter bears an analogy to the case of desire and aversion directed towards the same object. These descriptions are of course abstract and rather artificial; it may be awkward to express in many cases the grounds of the 'ought' or 'ought not' in terms of features of the thing I ought or ought not to do, as suggested in the general description. I only hope that the simplification achieved by this compensates for the distortions.

The two situations, then, come to this: in the first, it seems that I ought to do a and that I ought to do b, but I cannot do both a and b; in the second, it seems that I ought to do c and that I ought not to do c. To many ethical theorists it has seemed that actually to accept these seeming conclusions would involve some sort of logical inconsistency. For Ross, it was of course such situations that called for the concept of *prima facie* obligations: two of these are present in each of these situations, of which at most one in each case can constitute an actual obligation. On Mr. Hare's views, such situations call (in some logical sense) for a revision or qualification of at least one of the moral principles that give rise, in their application, to the conflicting 'ought's. It is the view, common to these and to other theorists, that there is a logical inconsistency of some sort involved here, that is the ultimate topic of this paper.

5. I want to postpone, however, the more formal sorts of consideration for a while, and try to bring out one or two features of what these situations are, or can be, like. The way I shall do this is to extend further the comparison I sketched earlier, between conflicts of beliefs and conflicts of desires. If we think of it in these terms, I think it emerges that there are certain important respects in which these moral conflicts are more like conflicts of desires than they are like conflicts of beliefs.

(a) The discovery that my factual beliefs conflict *eo ipso* tends to weaken one or more of the beliefs; not so, with desires; not so, I think, with one's conflicting convictions about what one ought to do. This comes out in the fact that conflicts of 'ought's, like conflicts of desires, can readily have the character of a struggle, whereas conflicts of beliefs scarcely can, unless the man not only believes these things, but wants to believe them. It is of course true that there are situations in which, either because of some practical concern connected with the beliefs, or from an intellectual curiosity, one may get deeply involved with a conflict of beliefs, and something rather like a struggle may result: possibly including the feature, not uncommon in the moral cases, that the more one concentrates on the dilemma, the more pressing the claims of each side become. But there is still a difference, which can be put like this: that in the belief case my concern to get things straight is a concern both to find the right belief (whichever it may be) and to be disembarrassed of the false belief (whichever it may be), whereas in the moral case my concern is not in the same way to find the right item and be rid of the other. I may wish that the facts had been otherwise, or that I had never got into the situation; I may even, in a certain frame of mind, wish that I did not have the moral views I have. But granted that it is all as it is, I do not think in terms of banishing error. I think, if constructively at all, in terms of acting for the best, and this is a frame of mind that *acknowledges* the presence of both the two 'ought's.

(b) If I eventually choose for one side of the conflict rather than the other, this is a possible ground of regret—as with desires, although the regret, naturally, is a different sort of regret. As with desires, if the occasion is irreparably past, there may be room for nothing but regret. But it is also possible (again like desires) that the moral impulse that had to be abandoned in the choice may find a new object, and I may try to 'make up' to the people involved for the claim that was neglected. These states of mind do not depend, it seems to me, on whether I am convinced that in the choice I made I acted for the best; I can be convinced of this, yet have these regrets, ineffectual or possibly effective, for what I did not do.

It may be said that if I am convinced that I acted for the best; if, further, the question is not the different one of self-reproach for having got into the conflict-situation in the first place; then it is merely irrational to have any regrets. The weight of this comment depends on what it is supposed to imply. Taken most naturally, it would seem at least to imply that these reactions are a bad thing, which a fully admirable moral agent (taken, presumably, to be rational) would not display. In this sense, the comment seems to me to be just false; such reactions do not appear to me to be necessarily a bad thing, nor an agent who displays them *pro tanto* less admirable than one who does not. But I do not have to rest much on my thinking that this is so; only on the claim that it is not inconsistent with the nature of morality to think that this is so. This modest claim seems to me undeniable; it is possible to think, for instance, that the notion of an admirable moral agent cannot be all that remote from that of a decent human being, and decent human beings are disposed in some situations of conflict to have the sort of reactions I am talking about.

Some light, though necessarily a very angled one, is shed on this point by the most extreme cases of moral conflict, tragic cases. One peculiarity of these is that the notion of 'acting for the best' may very well lose its content: Agamemnon at Aulis may have said ''May it be well''[4], but he is neither convinced nor convincing. The agonies that a man will experience after acting in full consciousness of such a situation are not to be traced to a persistent doubt that he may not have chosen the better thing; but, for instance, to a clear conviction that he has not done the better thing because there was no better thing to be done. It may, on the other hand, even be the case that by some not utterly irrational criteria of 'the better thing', he is convinced that he did the better thing: rational men no doubt pointed out to Agamemnon his responsibilities as a commander, the many people involved, the considerations of honour, and so forth. If he accepted all this, and acted accordingly: it would seem a glib moralist who said, as some sort of criticism, that he must be irrational to lie awake at night, having killed his daughter. And he lies awake, not because of a doubt, but because of a certainty. Some may say that the mythology of Agamemnon and his choice is nothing to us, because we do not move in a world in which irrational gods order men to kill their own children. But there is no need of irrational gods, to give rise to tragic situations.

Perhaps, however, it might be conceded that men may have regrets in these situations; it might even be conceded that a fully admirable moral agent would, on

4. Aeschylus, *Agamemnon* 217.

occasion, have such regrets; but nevertheless (it may be said) this is not to be connected directly with the structure of the moral conflict. The man may have regrets because he has had to do something distressing or appalling or which in some way goes against the grain, but this is not the same as having regrets because he thinks that he has done something that he ought not to have done, or not done something that he ought to have done; but it is only the latter that can be relevant to the interpretation of the moral conflict. This point might be put, in terms which I hope will be recognizable, by saying that regrets may be experienced in terms of purely *natural* motivations, and these are not to be confused, whether by the theorist or by a rational moral agent, with *moral* motivations, i.e. motivations that spring from thinking that a certain course of action is one that one ought to take.

There are three things I should like to say about this point. First, if it does concede that a fully admirable moral agent might be expected to experience such regrets on occasion, then it concedes that the notion of such an agent involves his having certain natural motivations as well as moral ones. This concession is surely correct, but it is unclear that it is allowed for in many ethical theories. Apart from this, however, there are two other points that go further. The sharp distinction that this argument demands between these natural and moral motivations is unrealistic. Are we really to think that if a man (a) thinks that he ought not to cause needless suffering and (b) is distressed by the fact of prospect of his causing needless suffering, then (a) and (b) are just two separate facts about him? Surely (b) can be one expression of (a), and (a) one root of (b)? And there are other possible connections between (a) and (b) besides these. If such connections are admitted, then it may well appear absurdly unrealistic to try to prise apart a man's feeling regrets about what he has done and his thinking that what he has done is something that he ought not to have done, or constituted a failure to do what he ought to have done. This is not, of course, to say that it is impossible for moral thoughts of this type, and emotional reactions or motivations of this type, to occur without each other; this is clearly possible. But it does not follow from this that if a man does both have moral thoughts about a course of action and certain feelings of these types related to it, then these items have to be clearly and distinctly separable one from another. If a man in general thinks that he ought not to do a certain thing, and is distressed by the thought of doing that thing; then if he does it, and is distressed at what he has done, this distress will probably have the shape of his thinking that in doing that thing, he has done something that he ought not to have done.

The second point of criticism here is that even if the sharp distinction between natural and moral motivations were granted, it would not, in the matter of regrets, cover all the cases. It will have even the appearance of explaining the cases only where the man can be thought to have a ground of regret or distress independently of his moral opinions about the situation; thus if he has caused pain, in the course of acting (as he sincerely supposes) for the best, it might be said that any regret or distress he feels about having caused the pain is independent of his views of whether in doing this, he did something that he ought not to have done: he is just naturally distressed by the thought of having caused pain. I have already said that I find this account unrealistic, even for such cases. But there are other cases in which it could not possibly be sustained. A man may, for instance, feel regret because he has broken a promise in the course of acting (as he sincerely supposes) for the best; and his regret at having broken the promise must surely arise *via* a moral thought. Here we seem just to get back to the claim that such regret in such circumstances would be irrational, and to the previous answer that if this claim is intended pejoratively, it will not stand up. A tendency to feel regrets, particularly creative regrets, at having broken a promise even in the course of acting for the best might well be considered a reassuring sign that an agent took his promises seriously. At this point, the object might say that he still thinks the regrets irrational, but that he does not intend 'irrational' pejoratively: we must rather admit that an admirable moral agent is one who on occasion is irrational. This, of course, is a new position: it may well be correct.

6. It seems to me a fundamental criticism of many ethical theories that their accounts of moral conflict and its resolution do not do justice to the facts of regret and related considerations: basically because they eliminate from the scene the 'ought' that is not acted upon. A structure appropriate to conflicts of belief is projected on to the moral case; one by which the conflict is basically adventitious, and a resolution of it disembarrasses one of a mistaken view that for a while confused the situation. Such an approach must be inherent in purely cognitive accounts of the matter; since it is just a question of which of the conflicting 'ought' statements is true, and they cannot both be true, to decide correctly for one of them must be to be rid of error with respect to the other, an occasion, if for any feelings, then for such feelings as relief (at escaping mistake), self-congratulation (for having got the right answer), or possibly self-criticism (for having so nearly been misled). Ross—whom unfairly I shall mention without discussing in detail— makes a valiant attempt to get nearer to the facts than this, with his doctrine that the *prima facie* obligations are not just *seeming* obligations, but more in the nature of a claim, which can generate residual obligations if not fulfilled.[5] But it remains obscure how all this is supposed to be so within the general structure of his theory; a claim, on these views, must surely be a claim for consideration as the only thing that matters, a duty, and if a course of action has failed to make good this claim in a situation of conflict, how can it maintain in that situation some residual influence on my moral thought?

A related inadequacy on this issue emerges also, I think, in certain prescriptivist theories. Mr. Hare, for instance, holds that when I encounter a situation of

---

5. Cf. *Foundations of Ethics*, p. 84 seq. The passage is full of signs of unease; he uses, for instance, the unhappy expression "the most right of the acts open to us", a strong indication that he is trying to have it both ways at once. Most of the difficulties, too, are wrapped up in the multiply ambiguous phrase "laws stating the tendencies of actions to be obligatory in virtue of this characteristic or of that" (p. 86).

conflict, what I have to do is modify one or both of the moral principles that I hold, which, in conjunction with the facts of the case, generated the conflict. The view has at least the merit of not representing the conflict as entirely adventitious, a mere misfortune that befalls my moral faculties. But the picture that it offers still seems inadequate to one's view of the situation *ex post facto*. It explains the origin of the conflict as my having come to the situation insufficiently prepared, as it were, because I had too simple a set of moral principles; and it pictures me as emerging from the situation better prepared, since I have now modified them—I can face a recurrence of the same situation without qualms, since next time it will not present me with a conflict. This is inadequate on two counts. First, the only focus that it provides for retrospective regret is that I arrived unprepared, and not that I did not do the thing rejected in the eventual choice. Second, there must surely be something wrong with the consequence that if I do not go back on the choice I make on this occasion, no similar situation later can possibly present me with a conflict. This may be a not unsuitable description of *some* cases, since one thing I may learn from such experiences is that some moral principle or view that I held was too naive or *simpliste*. But even among lessons, this is not the only one that may be learned: I may rather learn that I ought not to get into situations of this kind—and this lesson seems to imply very much the opposite of the previous one, since my reason for avoiding such situations in the future is that I have learned that in them both 'ought's *do* apply. In extreme cases, again, it may be that there is no lesson to be learned at all, at least of this practical kind.

7. So far I have been largely looking at moral conflict in itself; but this last point has brought us to the question of avoiding moral conflict, and this is something that I should like to discuss a little further. It involves, once more, but in a different aspect, the relations between conflict and rationality. Here the comparison with beliefs and desires is once more relevant. In the case of beliefs, we have already seen how it follows from the nature of beliefs that a conflict presents a problem, since conflicting beliefs cannot both be true, and the aim of beliefs is to be true. A rational man in this respect is one who (no doubt among other things) so conducts himself that this aim is likely to be realised. In the case of desires, again, there is something in the nature of desires that explains why a conflict essentially presents a problem: desires, obviously enough, aim at satisfaction, and conflicting desires cannot both be satisfied. Corresponding to this there will be a notion of practical rationality, by which a man will be rational who (no doubt among other things) takes thought to prevent the frustration of his desires. There are, however, two sides to such a policy: there is a question, not only of how he satisfies the desires he has, but of what desires he has. There is such a thing as abandoning or discouraging a desire which in conjunction with others leads to frustration, and this a rational man will sometimes do. This aspect of practical rationality can be exaggerated, as in certain moralities (some well known in antiquity) which avoid frustration of desire by reducing desire to a minimum: this can lead to the result that, in pursuit of a coherent life, a man misses out on the more elementary

requirement of having a life at all. That this is the type of criticism appropriate to this activity is important: it illustrates the sense in which a man's policy for organizing his desires is *pro tanto* up to him, even though some ways a man may take of doing this constitute a disservice to himself, or may be seen as, in some rather deeper way, unadmirable.

There are partial parallels to these points in the sphere of belief. I said just now that a rational man in this sphere was (at least) one who pursued as effectively as possible truth in his beliefs. This condition, in the limit, could be satisfied by a man whose sole aim was to avoid falsity in his beliefs, and this aim he might pursue by avoiding, so far as possible, belief: by cultivating scepticism, or ignorance (in the sense of never having heard of various issues), and of the second of these, at least, one appropriate criticism might be similar to one in the case of desires, a suggestion of self-impoverishment. There are many other considerations relevant here, of course; but a central point for our present purpose does stand, that from the fact that given truths or a given subject-matter exist, it does not follow that a given man ought to have beliefs about them: though it does follow that if he is to have beliefs about them, there are some beliefs rather than others that he ought to have.

In relation to these points, I think that morality emerges as different from both belief and desire. It is not an option in the moral case that possible conflict should be avoided by way of scepticism, ignorance, or the pursuit of *ataraxia*—in general, by indifference. The notion of a moral claim is of something that I may not ignore: hence it is not up to me to give myself a life free from conflict by withdrawing my interest from such claims.

It is important here to distinguish two different questions, one moral and one logical. On the one hand, there is the question whether extensive moral indifference is morally deplorable, and this is clearly a moral question, and indeed one on which two views are possible: *pas trôp de zèle* could be a moral maxim. That attitude, however, does not involve saying that there are moral claims, but it is often sensible to ignore them; it rather says that there are fewer moral claims than people tend to suppose. Disagreement with this attitude will be moral disagreement, and will involve, among other things, affirming some of the moral claims which the attitude denies. The logical question, on the other hand, is whether the relation of moral indifference and moral conflict is the same as that of desire-indifference and desire-conflict, or, again, belief-indifference and belief-conflict. The answer is clearly 'no'. After experience of these latter sorts of conflict, a man may try to cultivate the appropriate form of indifference while denying nothing about the nature of those conflicts as, at the time, he took them to be. He knows them to have been conflicts in believing the truth or pursuing what he wanted, and, knowing this, he tries to cut down his commitment to believing or desiring things. This may be sad or even dotty, but it is not actually inconsistent. A man who retreats from moral conflict to moral indifference, however, cannot at the same time admit that those conflicts were what, at the time, he took them to be, viz. conflicts of moral claims, since to admit that there exist moral claims in situations of that sort is incompatible with moral indifference towards those situations.

The avoidance of moral conflict, then, emerges in two ways as something for which one is not merely free to devise a policy. A moral observer cannot regard another agent as free to restructure his moral outlook so as to withdraw moral involvement from the situations that produce conflict; and the agent himself cannot try such a policy, either, so long as he regards the conflicts he has experienced as conflicts with a genuine moral basis. Putting this together with other points that I have tried to make earlier in this paper, I reach the conclusion that a moral conflict shares with a conflict of desires, but not with a conflict of beliefs, the feature that to end it in decision is not necessarily to eliminate one of the conflicting items: the item that was not acted upon may, for instance, persist as regret, which may (though it does not always) receive some constructive expression. Moral conflicts do not share with conflicts of desire (nor yet with conflicts of belief) the feature that there is a general freedom to adopt a policy to try to eliminate their occurrence. It may well be, then, that moral conflicts are in two different senses ineliminable. In a particular case, it may be that neither of the 'ought's is eliminable; and the tendency of such conflicts to occur may itself be ineliminable, since, first, the agent cannot feel himself free to restructure his moral thought in a policy to eliminate them; and, second, while there are *some* cases in which the situation was his own fault, and the correct conclusion for him to draw was that he ought not to get into situations of that type, it can scarcely be believed that all genuine conflict situations are of that type.

Moral conflicts are neither systematically avoidable, nor all soluble without remainder.

8. If we accept these conclusions, what consequences follow for the logic of moral thought? How, in particular, is moral conflict related to logical inconsistency? What I have to say is less satisfactory than I should like; but I hope that it may help a little.

We are concerned with conflicts that have a contingent basis, with conflict *via* the facts. We distinguished earlier two types of case: that in which it seems that I ought to do a and that I ought to do b, but I cannot do both; and that in which it seems that I ought to do c in respect of some considerations, and ought not to do c in respect of others. To elicit something that looks like logical inconsistency here obviously requires in the first sort of case extra premisses, while extra premisses are at least not obviously required in the second case. In the second case, the two conclusions 'I ought to do c' and 'I ought not to do c' already wear the form of logical inconsistency. In the first case, the pair 'I ought to do a' and 'I ought to do b' do not wear it at all. This is not surprising, since the conflict arises not from these two alone, but from these together with the statement that I cannot do both a and b. How do these three together acquire the form of logical inconsistency? The most natural account is that which invokes two further premisses or rules: that 'ought' implies 'can', and that 'I ought to do a' and 'I ought to do b' together imply 'I ought to do a and b' (which I shall call the *agglomeration principle*). Using these, the conflict can be represented in the following form:

(i) I ought to do a
(ii) I ought to do b
(iii) I cannot do a and b.

From (i) and (ii), by agglomeration

(iv) I ought to do a and b;

from (iii) by *'ought' implies 'can'* used contrapositively,

(v) It is not the case that I ought to do a and b.

This produces a contradiction; and since one limb of it, (v), has been proved by a valid inference from an undisputed premiss, we accept this limb, and then use the agglomeration principle contrapositively to unseat one or other of (i) and (ii).

This formulation does not, of course, produce an inconsistency of the 'ought'– 'ought not' type, but of the 'ought'–'not ought' type, i.e. a genuine *contradiction*. It might be suggested, however, that there is a way in which we could, and perhaps should, reduce cases of this first type to the 'ought'–'ought not' kind, i.e. to the pattern of the second type of case. We might say that 'I ought to do b', together with the empirical statement that doing a excludes doing b, jointly yield the conclusion that I ought to do something which, if I do a, I shall not do; hence that I ought to refrain from doing a; hence that I ought not to do a. This, with the original statement that I ought to do a, produces the 'ought'–'ought not' form of inconsistency. A similar inference can also be used, of course, to establish that I ought not to do b, a conclusion which can be similarly joined to the original statement that I ought to do b. To explore this suggestion thoroughly would involve an extensive journey on the troubled waters of deontic logic; but I think that there are two considerations that suggest that it is not to be preferred to the formulation that I advanced earlier. The first is that the principle on which it rests looks less than compelling in purely logical terms: it involves the substitution of extensional equivalences in a modal context, and while this might possibly fare better with 'ought' than it does elsewhere, it would be rash to embrace it straight off. Second, it suffers from much the same defect as was noticed much earlier with a parallel situation with conflicts of desires: it conceals the real roots of the conflict. The formulation with *'ought' implies 'can'* does not do this, and offers a more realistic picture of how the situation is.

Indeed, so far from trying to assimilate the first type of case to the second, I am now going to suggest that it will be better to assimilate the second to the first, as now interpreted. For while 'I ought to do c' and 'I ought not to do c' do indeed wear the form of logical inconsistency, the blank occurrence of this form itself depends to some extent on our having left out the real roots of the conflict—the consideration or aspects that lead to the conflicting judgments. Because of this, it

[53]

conceals the element that it is in common between the two types of case: that in both, the conflict arises from a contingent impossibility. To take Agamemnon's case as example, the basic 'ought's that apply to the situation are presumably that he ought to discharge his responsibilities as a commander, further the expedition, and so forth; and that he ought not to kill his daughter. Between these two there is no inherent inconsistency. The conflict comes, once more, in the step to action: that as things are, there is no way of doing the first without doing the second. This should encourage us, I think, to recast it all in a more artificial, but perhaps more illuminating way, and say that here again there is a double 'ought': the first, to further the expedition, the second, to refrain from the killing; and that as things are he cannot discharge both.

Seen in this way, it seems that the main weight of the problem descends on to '*ought*' *implies* '*can*' and its application to these cases; and from now on I shall consider both types together in this light. Now much could be said about '*ought*' *implies* '*can*', which is not a totally luminous principle, but I shall forgo any general discussion of it. I shall accept, in fact, one of its main applications to this problem, namely that from the fact that I cannot do both a and b it follows contrapositively that it is not the case that I ought to do both a and b. This is surely sound, but it does not dispose of the logical problems: for no agent, conscious of the situation of conflict, in fact thinks that he ought to do *both* of the things. What he thinks is that he ought to do *each* of them; and this is properly paralleled at the level of 'can' by the fact that while he cannot do both of the things, it is true of each of the things, taken separately, that he can do it.

If we want to emphasise the distinction between 'each' and 'both' here, we shall have to look again at the principle of agglomeration, since it is this that leads us from 'each' to 'both'. Now there are certainly many characterizations of actions in the general field of evaluation for which agglomeration does not hold, and for which what holds of each action taken separately does not hold for both taken together: thus it may be *desirable,* or *advisable,* or *sensible,* or *prudent,* to do a, and again desirable or advisable etc. to do b, but not desirable etc. to do both a and b. The same holds, obviously enough, for what a man wants; thus marrying Susan and marrying Joan may be things each of which Tom wants to do, but he certainly does not want to do both. Now the mere existence of such cases is obviously not enough to persuade anyone to give up agglomeration for 'ought', since he might reasonably argue that 'ought' is different in this respect; though it is worth noting that anyone who is disposed to say that the sorts of characterizations of actions that I just mentioned are evaluative *because they entail 'ought'-statements* will be under some pressure to reconsider the agglomerative properties of 'ought'. I do not want to claim, however, that I have some knock-down disproof of the agglomeration principle; I want to claim only that it is not a self-evident datum of the logic of 'ought', and that if a more realistic picture of moral thought emerges from abandoning it, we should have no qualms in abandoning it. We can in fact see the problem the other way round: the very fact that there can be two things, each of which I ought to do and each of which I can do, but of which I cannot do both, shows the weakness of the agglomeration principle.

Let us then try suspending the agglomeration principle, and see what results follow for the logical reconstruction of moral conflict. It is not immediately clear how *'ought' implies 'can'* will now bear on the issue. On the one hand, we have the statement that I cannot do both a and b, which indeed disproves that I ought to do both a and b, but this is uninteresting: the statement it disproves is one that I am not disposed to make in its own right, and which does not follow (on the present assumptions) from those that I am disposed to make. On the other hand, we have the two 'ought' statements and their associated 'can' statements, each of which, taken separately, I can assert. But this is not enough for the conflict, which precisely depends on the fact that I cannot go on taking the two sets separately. What we need here, to test the effect of *'ought' implies 'can'*, is a way of applying to each side the fact that I cannot satisfy both sides. Language provides such a way very readily, in a form which is in fact the most natural to use in such deliberations:

(i) If I do b, I will not be able to do a;
(ii) If I do a, I will not be able to do b.

Now (i) and (ii) appear to be genuine conditional statements; with suitable adjustment of tenses, they admit both of contraposition and of use in *modus ponens*. They are thus not like the curious non-conditional cases discussed by Austin.[6]

Consider now two apparently valid applications of *'ought' implies 'can'*:

(iii) If I will not be able to do a, it will not be the case that I ought to do a;
(iv) If I will not be able to do b, it will not be the case that I ought to do b.

Join (iii) and (iv) to (i) and (ii) respectively, and one reaches by transitivity:

(v) If I do b it will not be the case that I ought to do a;
(vi) If I do a, it will not be the case that I ought to do b.

At first glance, (v) and (vi) appear to offer a very surprising and reassuring result: that whichever of a and b I do, I shall get off the moral hook with respect to the other. This must surely be too good to be true; and suspicion that this is so must turn to certainty when one considers that the previous argument would apply just as well if the conflict between a and b were not a conflict between two 'ought's at all, but, say, a conflict between an 'ought' and some gross inclination; the argument depends solely on the fact that a and b are empirically incompatible.

6. *Ifs and Cans*, reprinted in his *Philosophical Papers*.

This shows that the reassuring interpretation of (v) and (vi) must be wrong. There is a correct interpretation, which reveals (v) and (vi) as saying something true but less interesting: (taking (v) as example), that if I do b, it will then not be correct to say that I ought (then) to do a. And this is correct, since a will *then* not be a course of action open to me. It does not follow from this that I cannot correctly say then that *I ought to have done a;* nor yet that I was wrong in thinking earlier that a was something I ought to do. It seems, then, that if we waive the agglomeration principle, and just consider a natural way of applying to each course of action the consideration that I cannot do both it and the other one, we do not get an application of *'ought' implies 'can'* that necessarily cancels out one or other of the original 'ought's regarded retrospectively. And this seems to me what we should want.

As I have tried to argue throughout, it is surely falsifying of moral thought to represent its logic as demanding that in a conflict situation one of the conflicting 'ought's must be totally rejected. One must, certainly, be rejected in the sense that not both can be acted upon; and this gives a (fairly weak) sense to saying that they are incompatible. But this does not mean they do not both (actually) *apply* to the situation; or that I was in some way mistaken in thinking that these were both things that I ought to do. I may continue to think this retrospectively, and hence have regrets; and I may even do this when I have found some moral reason for acting on one in preference to the other. For while there are some cases in which finding a moral reason for preference *does* cancel one of the 'ought's, this is not always so. I may use some emergency provision, of a utilitarian kind for example, which deals with the conflict of choice, and gives me a way of ''acting for the best''; but this is not the same as to revise or reconsider the reasons for the original 'ought's, nor does it provide me with the thought ''If that had occurred to me in the first place, there need have been no conflict''. It seems to me impossible, then, to rest content with a logical picture which makes it a necessary consequence of conflict that one 'ought' must be totally rejected in the sense that one becomes convinced that it did not actually apply. The condition of moving away from such a picture appears to be, at least within the limits of argument imposed by my rather crude use of *'ought' implies 'can'*, the rejection of the agglomeration principle.

I have left until last what may seem to some the most obvious objection to my general line of argument. I have to act in the conflict; I can choose one course rather than the other; I can think about which to choose. In thinking about this, or asking another advice on it, the question I may characteristically ask is ''what ought I to do?'' The answer to this question, from myself or another, cannot be ''both'', but must rather be (for instance) ''I (or you) ought to do a''. This (it will be said) just shows that to choose in a moral conflict, or at least to choose as a result of such deliberation, is to give up one of the 'ought's completely, to arrive at the conclusion that it does not apply; and that it cannot be, as I have been arguing that it may be, to decide not to act on it, while agreeing that it applies.

[56]

This objection rests squarely on identifying the 'ought' that occurs in ments of moral principle, and in the sorts of moral judgments about partic situations that we have been considering, with the 'ought' that occurs in th deliberative question "what ought I to do?" and in answers to this question, given by myself or another. I think it can be shown that this identification is a mistake, and on grounds independent of the immediate issue. For suppose I am in a situation in which I think that I ought (morally) to do a, and would merely very much like to do b, and cannot do both. Here, too, I can presumably ask the deliberative question "what ought I to do?" and get an answer to it. If this question meant "Of which course of action is it the case that I ought (morally) to do it?", the answer is so patent that the question could not be worth asking: indeed, it would not be a deliberative question at all. But the deliberative question can be worth asking, and I can, moreover, intelligibly arrive at a decision, or receive advice, in answer to it which is offensive to morality. To identify the two 'ought's in this sort of case commits one to the necessary supremacy of the moral; it is not surprising if theories that tend to assimilate the two end up with the Socratic paradox. Indeed, one is led on this thesis not only to the supremacy, but to the ubiquity, of the moral; since the deliberative question can be asked and answered, presumably, in a situation where neither course of action involves originally a moral 'ought'.

An answer to the deliberative question, by myself or another, can of course be supported by moral reasons, as by other sorts; but its role as a deliberative 'ought' remains the same, and this role is not tied to morality. This remains so even in the case in which both the candidates for action that I am considering involve moral 'ought's. This, if not already clear, is revealed by the following possibility. I think that I ought to do a and that I ought to do b, and I ask of two friends "what ought I to do?". One says "You ought to do a", and gives such-&-such moral reasons. The other says "You ought to do neither: you ought to go to the pictures and give morality a rest". The sense of 'ought' in these two answers is the same: they are both answers to the unambiguous question that I asked. All this makes clear, I think, that if I am confronted with two conflicting 'ought's, and the answer to the deliberative question by myself or another *coincides* with one of the original 'ought's, it does not represent a mere *iteration* of it. The decision or advice is decision or advice to act on that one; not a re-assertion of that one with an implicit denial of the other. This distinction may also clear up what may seem troubling on my approach, that a man who has had a moral conflict, has acted (as he supposes) for the best, yet has the sorts of regrets that I have discussed about the rejected course of action, would not most naturally express himself with respect to that course of action by saying "I ought to have done the other". This is because the standard function of such an expression in this sort of situation would be to suggest a deliberative mistake, and to imply that if he had the decision over again he would make it differently. That he cannot most naturally say this in the

[57]

t mean that he cannot think of the rejected action as
different sense, he ought to have done; that is to say, as
as not wrong at the time in thinking that he ought to do it.
ot even true that *the* deliberative question is "what ought
for instance, "what am I to do?"; and that question, and
s "do a", or "if I were you, I should . . . "—do not
gh decision or advice to act on one of the 'ought's in a
necessarily involves deciding that the other one had no application.

3

Supervenience Revisited

Simon Blackburn

I

A decade ago, in an article entitled 'Moral Realism' I presented an argument intended to show that two properties, which I called supervenience and lack of entailment, provided together an unpleasant mystery for moral realism (Blackburn, 1971). This argument was originally suggested to me in a discussion with Casimir Lewy, which in turn was directed at the paper of G. E. Moore, entitled 'The Conception of Intrinsic Value' (1922). The intervening decade has provided a number of reasons for revisiting my argument. First of all, it was couched in an idiom which subsequent work on modal logic—particularly the distinctions of various kinds of necessity and the general use of possible worlds as models—has made a little quaint. It would be desirable to see if the new notions allow the argument to stand. Secondly, we have seen a great deal of interest in supervenience, as a notion of importance beyond moral philosophy. Thus in conversation and correspondence I have heard it suggested that my argument must be flawed, because exactly the same combination of properties that I found mysterious occurs all over the place: for example, in the philosophy of mind, in the relationship between natural kind terms and others, in the relation between colours and primary properties, and so on. Since anti-realism in these other areas is not attractive, this casts doubt upon my diagnosis of the moral case. Finally, moral realism is again an attractive option to some philosophers, so that although when I wrote I might have seemed to be shadow boxing, the argument is just now becoming relevant again. In any case, enough puzzles seem to me to surround a proper analysis of supervenience to warrant a fresh look at it.

This essay was originally published in *Exercises in Analysis: Essays by Students of Casimir Levy*, ed. Ian Hacking, copyright © Cambridge University Press, 1985, reprinted by permission of Cambridge University Press.

Suppose we have an area of judgments, such as those involving moral commitments, or attributions of mental states. I shall call these F judgments, and I shall also talk of F truths and F facts: this is not intended to imply any view at all about whether the commitments we express in the vocabulary are beyond question genuine judgments, nor that there is a real domain of truths or facts in the area. Indeed part of the purpose of my argument was to find a way of querying just these ideas. At this stage, all this terminology is entirely neutral. Now suppose that we hold that the truths expressible in this way supervene upon the truths expressed in an underlying G vocabulary. For example, moral judgments supervene upon natural judgments, or mental descriptions of people upon physical ones (either of the people themselves or of some larger reality which includes them). This supervenience claim means that in *some* sense of 'necessary' it is necessarily true that if an F truth changes, then some G truth changes, or necessarily, if two situations are identical in point of G facts, then they are identical in terms of F facts as well. To analyse this more closely, I shall make free use of the possible worlds idiom. But it must be emphasised that this is merely a heuristic device, and implies no theory about the status of the possible worlds. Let us symbolise the kind of necessity in question by 'N' and possibility by 'P': for the present it does not matter whether these are thought of as logical, metaphysical, physical or other kinds of modalities. We are now to suppose that some truth about a thing or event or state, that it is F, supervenes upon some definite total set of G truths, which we can sum up by saying that it is G^*. Of course, G^* can contain all kinds of relational truths about the subject, truths about other things, and so on. In fact, one of the difficulties of thinking about all this properly is that it rapidly becomes unclear just what can be allowed in our conception of a totality of G states. But intuitively it is whatever it is by way of natural or physical states that bring it about that the subject is F. I shall express this by talking of the set of G states which 'underlies' an F state. Belief in supervenience is then at least the belief that whenever a thing is in some F state, this is because it is in some underlying G state, or is in virtue of its being in some underlying G state. This is the minimal sense of the doctrine. But I am interested in something stronger, which ties the particular truth that a thing is F to the fact that it is in some particular G state. We can present the general form of this doctrine as characterising, the relation 'U' that holds when one 'underlies' the other:

(S) $\quad N\,((\exists x)\,(Fx \,\&\, G^*x \,\&\, (G^*xUFx)) \supset (y)(G^*y \supset Fy))$

The formula says that as a matter of necessity, if something x is F, and G^* underlies this, then anything else in the physical or natural or whatever state G^* is F as well. There is no claim that G^* provides the only way in which things can become F: intuitively something might be, say, evil in a number of different ways, and something in one given physical state might posess some mental property which it could equally have possessed by being in any of a family of related physical states. The supervenience claim (S) is thus in no opposition to doctrines which now go under the heading of 'variable realisation'. To get the claim which these doctrines deny, we would need to convert the final conditional: . . . (y) $(Fy \supset G^*y)$). But the resulting doctrine is not in which we shall be interested.

I now want to contrast (S) with a much stronger necessity:

(N) $N(x)(G^*x \supset Fx)$

Of course (N) does not follow from (S). Formally they are merely related like this: (S) necessitates an overall conditional, and (N) necessitates the consequence of that conditional. So it would appear there is no more reason to infer (N) from (S) than there would be to infer Nq from N(p $\supset$ q). Hence also there is no inconsistency in a position which affirms (S), but also affirms:

(P) $P(\exists x)(G^*x \ \& \ {\sim}Fx)$

At least, this is the immediate appearance. In my original paper it was the nature of theories which hold both (S) and (P) (which I shall call the (S)/(P) combination) which occupied me. Such a theory would think it possible (in some sense commensurate with that of the original claim) that any given G state which happens to underlie a certain F state, nevertheless, might not have done so. In other words, even if some G set-up in our world is the very state upon which some F state supervenes, nevertheless, it might not have been *that* F state which supervened upon it. There was the possibility (again, in whatever modal dimension we are working) that the actually arising or supervening F state might not have been the one which supervened upon that particular G set-up. My instinct was that this combination provided a mystery for a realist about judgments made with the F vocabulary, and that the mystery would best be solved by embracing an anti-realist (or as I now prefer to call it, a projectivist) theory about the F judgments.

To pursue this further, we might question whether there could be any motivation for holding the (S)/(P) combination. Consider the following possible doctrine about 'underlying', and about the notion of the complete specification of an underlying state, G^*:

(?) $N((\exists x)(Fx \ \& \ G^*x \ \& \ (G^*x \ U \ Fx)) \supset N(y)(G^*y \supset Fy))$

The rationale for (?) would be this. Suppose there were a thing which was G^* and F, so that we were inclined to say that its being in the G state underlies its F-ness. But suppose there were also a thing which is G^* and $\sim F$. Then would we not want to deny that it was $x's$ being G^* which underlies its being F? Wouldn't it be its being G^* *and* its being different from this other thing in some further respect—

[61]

one which explains why the other thing fails to be *F?* We can call that a *releasing* property, *R,* and then *F* will supervene only on *G** and ~*R*. More accurately, *G** would denote a set of properties which do not really deserve the star. We would be wrong to locate in them a *complete* underlying basis for *F.*

This raises quite complicated questions about the form of these various doctrines. Let me put aside one problem right at the beginning. Since (S) is a conditional and contains an existential clause as part of the antecedent, it will be vacuously true if nothing is *G** and *F;* the necessitation will likewise be vacuously true if nothing could be *G** and *F.* So if (S) captured all that was meant by supervenience, we could say, for instance, that being virtuous supervened upon being homogeneously made of granite. Necessarily, if one thing homogeneously made of granite were virtuous, and this underlay the virtuousness, then anything so made would be. But this is just because it is impossible that anything of this constitution should be virtuous. I am going to sidestep this problem simply by confining the scope of *F* and *G* henceforwards, to cases where it is possible that something with a set of *G* properties, denoted by *G**, should be *F.* In fact, we are soon to deal with different strengths of necessity and possibility, and I shall suppose that this thesis is always strong enough to stop the conditional being satisfied in this vacuous way. The next problem of logical form is quite how we construe the denotation of a set of properties made by the term '*G**'. Firstly, we do not want the supervenience thesis to be made vacuously true through it being impossible that any two distinct things should be *G**—it then following that if one *G** thing is *F*, they all are. And the threat here is quite real. If, for instance, *G** were held to include all the physical properties and relations of a thing—if it were that and nothing less which some property *F* supervened upon—then assuming the identity of indiscernibles, we would have (S) satisfied vacuously again. To get around this I am going to assume a *limitation* thesis. This will say that whenever a property *F* supervenes upon some basis, there is necessarily a boundary to the kind of *G* properties which it can depend upon. For example, the mental may supervene upon the physical, in which case the thesis asserts that necessarily there are physical properties of a thing which are not relevant to its mental ones. A plausible example might be its relations to things with which it is in no kind of causal connection (such as future things). Again, the moral supervenes upon the natural, and the thesis will tell us that there are some natural properties which necessarily have no relevance to moral ones—pure spatial position, perhaps, or date of beginning in time. Given the limitation thesis, (S) will not be trivialised by the identity of indiscernibles. The last problem of form which arises is whether '*G**' is thought of as a name for some particular set of properties (which form a complete basis for *F*), or whether it is built into the sense of '*G**' that any set of properties it denotes is complete. The difference is easily seen if we consider a very strong kind of necessity—say, conceptual (logical or analytic) necessity. It is unlikely to be thought analytic that being made of H_2O underlies being water. One reason is that it is not analytic that being made of H_2O exhausts the kind of physical basis which may affect the kind to which a substance belongs. That is a substantive scientific truth, not one guaranteed in any more *a priori* way. I am going to build it into the sense of '$G*$' that at least in one possible world, the set of properties it denotes is sufficient to underlie F. I do not want it to follow that this is true in all worlds, although that is a very delicate matter. Fortunately, so far as I can see, it does not occupy the centre of the stage I am about to set.

If we accept (?) as a condition on what it is for a set of properties to underlie another, and hence on what it is for a property to supervene upon such a set, the relationship between supervenience and (E) changes. Suppose that there is something whose $G*$-ness underlies its F-ness:

(E) $(\exists x)(Fx \ \& \ G*x \ \& \ (G*x \ U \ Fx))$

then we can now derive (N). In other words, (?) and (E) together entail (N). And as I have already said, (?) is an attractive doctrine. But it does mean that supervenience becomes in effect nothing but a roundabout way of committing ourselves to (N); the *prima facie* simpler doctrine that some set of underlying truths necessitates the F truth. This is in fact the way that supervenience is taken by Kim (1978): it enables Kim to suppose that where we have supervenience, we also have reductionism. Another way of getting at the attractions of (?) would be to cease from mentioning the requirement that there is something which is $G*$ and F altogether. After all, surely some moral property might supervene upon a particular configuration of natural properties, regardless of whether there actually is anything with that set. Or, some mental property might supervene upon a particular physical make-up which nobody actually has. If we took this course, we would replace (S) by a doctrine:

$(G*x \ U \ Fx) \supset (y)(G*y \supset Fy)$

and then the doctrine which would give us (N) immediately would be:

$(G*x \ U \ Fx) \supset N \ (G*x \ U \ Fx)$

Yet supervenience claims are popular at least partly because they offer some of the metaphysical relief of reductions, without incurring the costs; I want therefore to preserve any gap that there may be for as long as possible. This is particularly important in the moral case, where supervenience is one thing, but reductionism is markedly less attractive. So I am going to stick with the original formulation, subject to the caveats already entered, and while we should remain well aware of (?), I do not want to presuppose a verdict on it.

If we put (?) into abeyance we should be left with a possible form of doctrine

[63]

which accepts both (S) and (P): the (S)/(P) combination. It is this which I originally claim to make a mystery for realism. If there is to be a mystery, it is not a formal one, and I actually think that with suitable interpretations, there are relations between F and G vocabularies which are properly characterised by (S) and (P). It is just that when this combination is to be affirmed, I believe it needs explanation. In the moral case I think that it is best explained by a projective theory of the F predicates. But in other cases, with different interpretations of the modalities, other explanations are also possible. I shall argue this later.

Here, then is a way of modelling (S) and (P) together. In any possible world, once there is a thing which is F, and whose F-ness is underlain by G^*, then anything else which is G^* is F as well. However, there are possible worlds in which things are G^* but not F. Call the former worlds G^*/F worlds, and the latter, G^*/O worlds. The one thing we do not have is any *mixed* world, where some things are G^* and F, and some are G^* but not F. We can call mixed worlds G^*/FvO worlds. These are ruled out by the supervenience claim (S): they are precisely the kind of possible world which would falsify that claim. My form of problem, or mystery, now begins to appear. Why should the possible worlds partition into only the two kinds, and not into the three kinds? It seems on the face of it to offend against a principle of plenitude with respect to possibilities, namely that we should allow any which we are not constrained to disallow. Imagine it spatially. Here is a possible world w_1 which is G^*/F. Here is another, w_2 which we can make as much like w_1 as possible, except that it is G^*/O. But there is no possible world anywhere which is just like one of these, except including just one element with its G^* and F properties conforming to the pattern found in the other. Why not? Or, to make the matter yet more graphic, imagine a time element. Suppose our possible worlds are thought of as having temporal duration. A mixed world would be brought about if w_1 starts off as a G^*/F world at some given time, but then at a later time becomes a G^*/O world. For then, overall, it would be mixed and the supervenience claim would be falsified by its existence. This kind of world then cannot happen, although there can be worlds which are like it in respect of the first part of its history, and equally worlds which are like it in respect of the second part of its history.

This is the ban on mixed worlds: it is a ban on inter-world travel by things which are, individually, at home. The problem which I posed is that of finding out the authority behind this ban. Why the embargo on travel? The difficulty is that once we have imagined a G^*/F world, and a G^*/O world, it is as if we have done enough to imagine a G^*/FvO world, and have implicitly denied ourselves a right to forbid its existence. At least, if we are to forbid its existence, we need some explanation of why we can do so. The positive part of my contention was that in the moral case, projectivists can do this better than realists. In the next section I rehearse briefly why this still seems to me to be so, if we make some important distinctions, and then I turn to consider related examples. And in time we have to return to the difficult claim (?), to assess its role in this part of metaphysics.

[64]

II

Necessities can range from something very strict, approximating to 'analytically true' through a metaphysical and physical necessity, to someting approximating to 'usually true'. Then anyone who sympathises a little with the puzzle in a (S)/(P) combination can quickly see that it will remain not only if there is one fixed sense of necessity and possibility involved, but also in a wider class of cases. For whenever the supervenience claim (S) involves a strong sense of necessity, then it will automatically entail any version with a weaker sense of necessity. Hence, we will get the same structure at the lower level, when the possibility is affirmed in that corresponding, weaker, sense. Thus if (S) took the form of claiming that it is metaphysically necessary that . . . , and (P) took the form of claiming that it is physically possible . . . , and if we also suppose (as we surely should) that metaphysical necessity entails physical necessity, then we would have the (S)/(P) combination at the level of physical necessity. On the other hand we will not get the structure if the relation is reversed. If the possibility claim (P) is made, say, in the sense of 'possible as far as conceptual constraints go', this does not entail, say, 'metaphysically possible'. And then if (S) just reads 'metaphysically necessary' there is no mystery: we would just have it that it is metaphysically necessary that F supervenes upon G^*, but not an analytic or conceptual truth; equally, it would not be an analytic or conceptual truth that any given G^* produces F, and there is no puzzle there. For the puzzle to begin to arise, we need to bring the modalities into line.

I mention this because it affects the moral case quite closely. Suppose we allow ourselves a notion of 'analytically necessary' applying to propositions which, in the traditional phrase, can be seen to be true by conceptual means alone. Denying one of these would be exhibiting a conceptual confusion: a failure to grasp the nature of the relevant vocabulary, or to follow out immediate implications of that grasp. In a slightly more modern idiom, denying one of these would be 'constitutive' of lack of competence with the vocabulary. We may contrast this with metaphysical necessity: a proposition will be this if it is true in all the possible worlds which, as a matter of metaphysics, could exist. Of course, we may be sceptical about this division, but I want to respect it at least for the sake of argument. For the (S)/(P) combination in moral philosophy provides a nice example of a *prima facie* case of the difference, and one which profoundly affects my original argument. This arises because someone who holds that a particular natural state of affairs, G^*, underlies a moral judgment, is very likely to hold that this is true as a matter of metaphysical necessity. For example, if I hold that the fact that someone enjoys the misery of others underlies the judgment that he is evil, I should also hold that in any possible world, the fact that someone is like this is enough to make him evil. Using 'MN' for metaphysical necessity, I would have both:

(S_m) $MN((\exists x)(Fx \ \& \ G^*x \ \& \ (G^*x \ U \ Fx)) \supset (y)(G^*y \supset Fy))$

and

(N_m) $MN(x)(G^*x \supset Fx)$

and I would evade the original argument by disallowing the metaphysical possibility of a world in which people like that were not evil. This, it might be said, is part of what is involved in having a genuine standard, a belief that some natural state of affairs is sufficient to warrant the moral judgment. For, otherwise, if in some metaphysically possible worlds people like that were evil and in others not, surely this would be a sign that we hadn't yet located the natural basis for the judgment properly. For instance, if I did allow a possible world in which some people like that were not evil, it might be because (for instance) they believe that misery is so good for the soul that it is a cause of congratulation and rejoicing to find someone miserable. But then this fact becomes what I earlier called a releasing fact and the real underlying state of affairs is now not just that someone enjoys the misery of others, but that he does so not believing that misery is good for the soul.

Because of this the original puzzle does not arise at the level of metaphysical necessity. But now suppose we try analytic necessity. It seems to be a conceptual matter that moral claims supervene upon natural ones. Anyone failing to realise this, or to obey the constraint would indeed lack something constitutive of competence in the moral practice. And there is good reason for this: it would betray the whole purpose for which we moralise, which is to choose, commend, rank, approve, forbid, things on the basis of their natural properties. So we might have:

(S_a) $AN((\exists x)(Fx \ \& \ G^*x \ \& \ (G^*x \ U \ Fx)) \supset (y) \ (G^*y \supset Fy))$

But we would be most unwise to have

(N_a) $AN(x)(G^*x \supset Fx)$

For it is not plausible to maintain that the adoption of some particular standard is 'constitutive of competence' as a moralist. People can moralise in obedience to the conceptual constraints that govern all moralising, although they adopt different standards, and come to different verdicts in the light of a complete set of natural facts. Of course, this can be denied but for the sake of this paper, I shall rely on the common view that it is mistaken. So since we deny (N_a) we have:

(P_a) $AP(\exists x)(G^*x \ \& \ Fx)$

[66]

We then arrive at a $(S_a)/(P_a)$ combination, and my mystery emerges: why the ban on mixed worlds at this level? These would be worlds possible as far as conceptual constraints go, or 'analytically possible' worlds. They conform to conceptual constraints, although there might be metaphysical or physical bars against their actual existence.

Of course, in a sense I have already proposed an answer to this question. By saying enough of what moralising is to make (S_a) plausible, and enough to make (P_a) plausible, I hope to enable us to learn to relax with their combination. It is just that this relaxation befits the anti-realist better. Because the explanation of the combination depended crucially upon the role of moralising being to guide desires and choices amongst the natural features of the world. If, as a realist ought to say, its role is to describe further, moral aspects of morality, there is no explanation at all of why it is constitutive of competence as a moralist to obey the constraint (S_a).

Can this argument be avoided by maintaining (?)? No, because there is no prospect of accepting (?) in a relevantly strong sense. For (Z) to help, we would need it to be read so that, necessarily (in some sense) if something is F and G^*, and the G^*-ness underlies its being F, then it is analytically necessary that anything G^* is F. And this we will not have in the moral case, for we want to say that there are things with natural properties underlying moral ones, but we also deny analyticities of the form (N_a). (?) would not help if the necessity of the consequent were interpreted in any weaker sense. For example, we might want to accept (?) in the form:

$$(?_{MN}) \quad MN((\exists x) \, (Fx \, \& \, G^*x \, \& \, (G^*x \, U \, Fx)) \supset MN(y)(G^*y \supset Fy))$$

and then there will be metaphysical necessities of the form of the consequent, that is, of the form (N_m), but they will not help to resolve the original mystery, since that is now proceeding at the level of analytical necessity. It is the possibility, so far as conceptual constraints go, of mixed worlds, which is to be avoided.

III

The argument above works because we are careful to distinguish the status of the supervenience claim, and in this case its extremely strong status, from that of the related possibility claim. I have done that by indexing the modal operators involved: we have four different forms of modal claim: (S), (N), (P) and (?), and each of them can involve analytic or conceptual necessity $(_a)$, metaphysical necessity $(_m)$, and we come now to physical necessity $(_p)$. For now I want to turn to consider non-moral cases of the same kind of shape. These examples are all going to start life as examples of the joint (S)/(P) combination. They may not finish life like that: it may become obvious, if it is not so already, in the light either of the plausibility of (?) or of the difficulty over banning mixed worlds, that either supervenience is to be abandoned, or (N) accepted. But here are some test cases:

1st example

Suppose that in w_1 a physical set-up G^* underlies some particular mental state F. Suppose G^* is possession of some pattern of neurones or molecules in the head, and F is having a headache. Nowhere in w_1 is there anything unlike x, in being G^* but not F. Next door, in w_2 however, there are things which are G^* but not F. Now we are told that w_1 is acceptable, and that w_2 is acceptable. But nowhere is there a world w_3 which is like w_1 but which changes to become like w_2, or which contains some particular individuals who are like those of w_2.

2d example

Suppose that in w_1 a particular molecular constitution G^* underlies membership of a natural kind, F. G^* consists of a complete physical or chemical breakdown of the constitution of a substance (e.g. being composed of molecules of H_2O) and F is being water. Nowhere in w_1 could there be a substance with that chemistry, which is not water. In w_2 however this combination is found. Once again, although each of these possible worlds exists, there is no G^*/FvO, or mixed world, in which some substances with this chemical constitution are water, and others are not.

3d example

Suppose that in w_1 a particular set of primary qualities, particularly concerning refractive properties of surfaces, G^*, underlies possession of a colour F. Nowhere in w_1 could there be things with that kind of surface, without the particular colour. However, there are possible worlds where this combination is found: G^*/O worlds. Again, there are no mixed worlds, where some things with the primary, surface properties are F, and others are not.

In each of these cases we have the (S)/(P) combination. And I hope it is obvious that each case is at least *prima facie* puzzling—enough so to raise questions about whether the combination is desirable, or whether we should make severe distinctions within the kinds of necessity and possibility involved, to end up avoiding the combination altogether. How would this be done?

First example

How should we interpret the supervenience of the mental on the physical? Perhaps centrally as a metaphysical doctrine. So we shall accept (S_m). Should we accept (S_a)? We should if we can find arguments, as strong as those in the moral case, for claiming that it is constitutive of competence in the mental language that we recognise the supervenience of the mental on the physical. But I doubt if we can do this. For whether or not we are philosophically wedded to the doctrine, we can surely recognise ordinary competence in users who would not agree. One day Henry has a headache, and the next day he does not. Something mental is different. But suppose he simply denies that anything physical is different (giving voice to Cartesianism). Is this parallel to the error of someone who makes the same move in a moral case? I do not think so: Henry is not so very unusual, and if his error is shown to be one because of the 'very meaning' of mental ascriptions, then whole cultures have been prone to denial of an analytic truth. In other words, it seems to me to be over-ambitious to claim that it follows, or follows analytically, from change in mental state, that there is change in an underlying physical state. It makes views conceptually incoherent when enough people have found them perfectly coherent (consider, for example, changes in God's mind).

Let us stick then with (S_m). It would seem to me plausible, if we accept this, to accept the correlative necessities, (N_m), and $(?_m)$. We would then be forced to deny (P_m), and we just do not get involved with the problem of banning mixed worlds. (N_m) does the work for us, by disallowing the metaphysical possibility of G^* without F. However, there is the famous, or notorious, position of Davidson to consider, which accepts some form of supervenience of the mental on the physical, but also denies the existence of lawlike propositions connecting the two vocabularies (Davidson 1980). Davidson is not very explicit about the strength of necessity and possibility involved in his claims. But it can scarcely be intended to be weaker than joint acceptance of (S_m) and of (P_p). And even if supervenience is taken not as a matter of metaphysical, but just of sheer physical necessity (in our physical world, there is no mental change without physical change, even if there *could* be), it does not matter. For from (S_m) we can deduce (S_p), so we have the $(S)/(P)$ combination, at the level of physical necessity. So according to me the position ought to be odd, and indeed it is. Why is it physically impossible for there to be a world which contains some w_1 characters, with headaches, and some w_2 characters, in the same physical state, but without them? Once we have allowed the physical possibility of the w_2 type, how can we disallow the physical possibility of them mingling with w_1 types?

It does not appear to me that light is cast on this by Davidson's reason for allowing (P_p). This reason is that in some sense the mental and the physical belong to different realms of theory: ascriptions of mental properties answer to different constraints from ascriptions of physical ones, and hence we can never be in a position to insist upon a lawlike correlation of any given physical state with any given mental state (this is not just the variable realisation point: here we are told not to insist upon a physically necessary physical-to-mental correlation; *prima facie* we might be allowed to do that in various cases, even if we could never insist upon lawlike connections the other way around). I do not accept that this is a good

[69]

argument, for there can certainly be interesting laws which connect properties whose ascriptions answer to different constraints: temperature and pressure, or colour and primary properties, for instance. However, I do not want to insist upon that. For there remains the oddity that if Davidson's reasoning is good, it should equally apply to the supervenience claim. How can we be in a position to insist upon anything as strong as (S_p), let alone as strong as (S_m)? The freedom which gives us (P_p) is just as effective here. I may coherently and effectively 'rationalise' one person as being in one mental state, and another as being in another, obeying various canonical principles of interpretation, regardless of whether they are in an identical G^* state: I might just disclaim interest in that. Of course, if (N_p) were true, it would be different, but it is precisely this which the anomalous character of the position denies.

So if the mental reality is in no lawlike connection with the physical, as (P_p) claims, I can see no basis for asserting that nevertheless it supervenes upon it. But, again, the word 'reality' matters here. *One* way of thinking that the mental just has to supervene upon the physical, in at least the sense of (S_m), is by convincing ourselves that the physical reality is at bottom the only one: molecules and neurones are all that there are. And then there might be something about the way mental vocabulary relates to this—relates to the only reality there is—which justifies both (S_p) and (P_p). Perhaps there is some argument that obeying the supervenience constraint is required for conceptual coherence, or at any rate for metaphysical coherence; and perhaps Davidson's argument for (P_p) can be put in a better light than I have allowed. I do not want to deny this possibility. But I do want to point out that once more it is bought at the cost of a highly anti-realist, even idealist, view of the mental. The 'truth' about the mental world is not a matter of how some set of facts actually falls out. It is a matter of how we have to relate this particular vocabulary to the one underlying reality. If we thought like this, then we would begin to assimilate the mental/physical case to the moral/natural case. At any rate, it provides no swift model for arguing that anti-realism is the wrong diagnosis of the (S)/(P) combination in that case.

Why do I say that this is an idealist or anti-realist direction? Because the constraint on our theorising is not explained by any constraint upon the way the facts can fall out. It is constrained by the way we 'must' use the vocabulary, but that 'must' is not itself derived from a theory according to which mental facts and events cannot happen in some given pattern; it is derived from constraints on the way in which we must react to a non-mental, physical world. I regard this as a characteristically idealist pattern: the way the facts have to be is explained ultimately by the way we have to describe them as being. Thus I would say that the explanation of moral supervenience is a paradigmatically anti-realist explanation. By way of contrast, and anticipating example 3, we can notice how there cannot be a strong, analytic, version of the doctrine that colours supervene upon primary properties, precisely because it is so obvious that the only conceptual constraint upon using the colour vocabulary is that you react to perceived colour in the right way. Somebody who thinks that a thing has changed colour, but who is perfectly indifferent to any question of whether it has changed in respect of any primary property (or even who believes that it positively has not done so) is quite within his rights. His eyesight may be defective, but his grasp of the vocabulary need not be so.

Second example

In order to avoid unnecessary complexity, I should enter a caveat here. I am going to take being composed of H_2O as a suitable example of G^*; an example that is of the kind of complete physical or chemical basis which results in stuff being of a certain kind, such as water. I am going to take it that this is known to be the case. So I shall not be interested in the kind of gap, which can in principle open up, in which people might allow that something is H_2O and is water, allow that wateriness supervenes upon the chemical or the physical, but deny that some other specimen of H_2O is water. This is a possible position, because it is possible to disbelieve that the facts registered by something's being H_2O exhaust the physical or chemical facts which may be relevant to its kind. I am going to cut this corner by writing as though it is beyond question that molecular constitution is the right candidate for a complete underlying property—a G^* property. I don't think that this affects the argument, although it is a complex area and one in which it is easy to mistake one's bearings.

Once more, it is natural to take the various claims involving the relationship of H_2O and water in a metaphysical sense. It is also natural to me at any rate, to assert (S_m) only if we also assert (N_m), and $(?_m)$. Being water supervenes upon being H_2O only because anything made of H_2O has to be water. And if we had an argument that it does not have to be water, perhaps because we imagine a world in which countervailing circumstance makes substances composed of that molecule quite unlike water at the macro-level (and more unlike it than ice or steam), then we would just change the basis for the supervenience. We would have argued for a releasing property, R, and the true basis upon which being water supervenes would be G^* (being H_2O) and being $\sim R$.

Might someone believe that the (S)/(P) combination arises at some level here? The argument would have to be that in some strong sense it *must* be true that being water supervenes upon physical or chemical constitution; but it need not be true, in this equally strong sense, that H_2O is the particular underlying state. Now I do not think there is any very strong sense in which being water has to be a property underlain by a physical or chemical basis. Of course, *we* are familiar with the idea that any such property must be a matter of chemistry. But there is no good reason for saying that people who fail to realise this are incompetent with the kind term 'water'. They just know less about the true scientific picture of what it is that explains the phenomenologically important, macro-properties of kinds. They are not, in my view, in at all the same boat as persons who fail to respect the supervenience of the moral on the natural. This is because this latter fault breaks up the whole point of moralising. Whereas ignorance of the way in which wateriness is supervenient on the chemical or physical does not at all destroy the point of classifying some stuff as water and other stuff as not. Uneducated people still need to drink and wash. However, it is now commonly held that there is no absolute distinction here: Quine has taught us how fragile any division would be, between conceptual and 'merely' scientific ignorance. So someone might hold that there is an important kind of incompetence, half-conceptual but perhaps half-scientific, which someone would exhibit if he failed to realise the supervenience of being water upon chemistry or physics, and that this is a worse kind of incompetence than any which would be shown by mere failure to realise that it is H_2O which is the relevant molecule. So we might try a notion of 'competently possible worlds' $(_c)$ meaning those which are as a competent person might describe a world as being: then we would have an $(S_c)/(P_c)$ combination. Should this tempt us to an anti-realist theory of 'being water'?

Saying that there are no mixed possible worlds in *this* sense just means that any competent person is going to deny that there are worlds in which some things of a given chemical or physical structure are water, and others are not; but that competent people might allow worlds in which things are H_2O but are not water. The first is a kind of *framework* knowledge, which we might expect everyone to possess; but competence to this degree need not require the *specialist* piece of scientific knowledge, which we might not expect of everyone, and which might even turn out to be false without affecting the framework. We might even suppose that supervenience claims have, characteristically, this framework appearance, and suggest that this is why they do not trail in their wake particular commitments of the (N) form. And now the counterattack against my argument in the case of morals and mind gathers momentum.[1] For if an $(S_c)/(P_c)$ combination works in a harmless case like this, then the shape of that combination cannot in general suggest anti-realism, and something must be wrong in the arguments so far given.

One reaction would be to allow the parallelism, and to grasp the nettle. When I said that we could relax with the $(S_a)/(P_a)$ combination in morals, I tried to explain this by saying that the role of a moral judgment is not to describe further *moral* aspects of reality; it is because the vocabulary must fit the *natural* world in certain ways, that the combination is explicable. I might try the same move here: it is because 'wateriness' is not a further aspect of reality (beyond its containing various stuffs defined in chemical ways) that the combination is permissible at the level of 'competently possible' worlds. But I think this will strike most uncommitted readers as weak: anti-realism has to fight for a place these days even in the philosophy of morals, and is hardly likely to seem the best account of the judgment that I have water in my glass. I think a better reaction is to remember well all that is meant by the notion of a 'competently possible world'. Remember-

---

1. I owe this objection to Michael Smith. I am also greatly indebted to conversations with David Bostock and Elizabeth Fricker.

ing this enables us to say that an $(S_c)/(P_c)$ combination is harmless, and implies no problem of explanation which is best met by anti-realism. This is because the 'ban on mixed possible worlds' which it gives rise to is explicable purely in terms of *beliefs* of ours—in particular, a belief which we suppose competent people to share. We believe, that is, that no two things could be identical physically without also forming the same stuff or kind *and we believe that all competent people will agree*. Whilst we suppose this, but also suppose that competent people may not agree that if a thing is H_2O then it is water (because this requires a higher level of specialised, as opposed to framework knowledge) then we have 'competently possible' worlds of the two kinds, and the ban on mixed worlds. But this has now been explained purely by the structure of beliefs which can coexist with competence. There is indeed no further inference to a metaphysical conclusion about the status of wateriness, because the explanation which, in the other cases, that inference helps to provide, is here provided without it. To put it another way, we could say that in the moral case as well, when we deal with analytically possible worlds, we are dealing with beliefs we have about competence: in this case the belief that the competent person will not flout supervenience. But this belief is only explained by the further, anti-realist, nature of moralising. If moralising were depicting further, moral aspects of reality, there would be no explanation of the conceptual constraint, and hence of our belief about the shape of a competent morality.

It cannot be overemphasised that my original problem is one of *explanation*. So it does not matter if sometimes an $(S)/(P)$ combination is explained in some ways, and sometimes in others. I do not suppose that there is one uniform pattern of explanation, suitable for all examples and for all strengths of modality (particularly if we flirt with hybrids like the present one). The explanation demanded in the moral case is, according to me, best met by recognising that moralising is an activity which cannot proceed successfully without recognition of the supervenience constraint, but this in turn is best explained by projectivism. In the present case the best explanation of why competent people recognise the supervenience of kinds upon physical or chemical structure is that we live in a culture in which science has found this out. I don't for a moment believe that *this* suggests any metaphysical conclusions. If this is right it carries a small bonus. It means that the argument in the moral case does not depend upon drawing a hard and controversial distinction between 'conceptual' and other kinds of incompetence. It merely requires us to realise that there can be good explanations of our beliefs about the things which reveal incompetence. Anti-realism is one of them, in the case of morals, and awareness of the difference between framework scientific beliefs, and specific realisations of them, is another and works in the case of natural kinds.

Third example

The previous case posed the only real challenge which I know to the original mystery. By contrast the case of colours reinforces the peculiarity in the case of morals. For it would be highly implausible to aim for colour/primary property supervenience as an analytic truth, or one constitutive of competence with a colour vocabulary. Intuitively we feel that it is very nice and satisfying that colours do indeed supervene upon primary properties, and that there would be scientific havoc if they did not. But anybody who believes that they do not (mightn't God live in a world where displays reveal different colours to him, although there are no physical properties of surfaces of the things displayed?) can recognise colours and achieve all the point and subtlety of colour classification for all that.

Recent empirical work casts doubt even on the fact that 'everybody knows', that colours of surfaces are caused by the wavelength of reflected light. Other relational properties may matter. So it is wise to be cautious before putting any advanced modal status on supervenience or necessitation claims in this area. Certainly we expect there to be *some* complete primary property story, G^*, upon which colour supervenes as a matter of physical necessity. But then we would also immediately accept the corresponding thesis (N_p), and there is no problem about mixed worlds. Similarly if we bravely elevated the supervenience (S) into a metaphysical thesis, there would be no good reason why (N) should not follow suit. (N) will not rise into the realms of analytic necessity, but then neither will (S). So at no level is there a mystery parallel to the one which arose with morals, and with Davidson's position on the mental and the physical. Of course, an (S)/(P) combination could be manufactured at the level of 'competently possible worlds', as in the last example, but once more it would avail nothing, because it would be explained simply by the shape of the beliefs which we have deemed necessary for competence.

IV

I have now said enough by way of exploring the original argument and its near neighbours. It would be nice to conclude with an estimate of the importance of supervenience claims in metaphysics. Here I confess I am pessimistic. It seems to me that (?) is a plausible doctrine, and in every case in which we are dealing with metaphysical or physical necessity, it seems to me that we could cut through talk of supervenience, and talk directly of propositions of the form (N). This makes it clear, for example, that we may be dealing with 'nomological danglers' or necessities which connect together properties of very different kinds, and it may lessen our metaphysical pride to remember that it is one thing to assert such necessities, but quite another thing to have a theory about why we can do so. Like many philosophers, I believe many supervenience claims in varying strengths; perhaps unlike them I see them as part of the problem—in the philosophy of mind, or of secondary properties, or of morals or kinds—and not part of the solution.

References

Blackburn, S. 1971. 'Moral Realism', in *Morality and Moral Reasoning,* ed. J. Casey, London: Methuen.

Davidson, D. 1980. 'Mental Events' in *Essays on Actions and Events,* Oxford: Clarendon Press.

Kim, J. 1978. 'Supervenience and Nomological Incommensurables', *American Philosophical Quarterly,* 15: 149–56.

Moore, G. E. 1922. 'The Conception of Intrinsic Value' in *Philosophical Studies,* London: Routledge and Kegan Paul.

4

Ethics, Mathematics
and Relativism

Jonathan Lear

1. The belief that reflection on mathematical practice will lend insight into the nature of ethics has exercised a strong hold on the philosophical imagination. Plato argued that training in mathematics was important preparation for our ascent toward perception of the Form of the Good.[1] In mathematics, according to him, we come to perceive that which is universal, immutable and abstract: this is supposed to be relevantly similar to perception of the Good. Both mathematical and moral reality were conceived as being very 'hard', deeply real. Though platonism in both ethics and mathematics has been the target of sustained attack, the belief that there are significant analogies between these areas of discourse persists. The similarity between moral and mathematical discourse is supposed to save ethics from the threat of non-cognitivism. By reflecting on the fact that mathematics is 'soft'—that is, that it does not sustain a platonist interpretation—and yet seems to satisfy any reasonable demand for objectivity, we are supposed to see that there is a way in which ethics can be objective, even though it is only a 'human construction'. The ethical non-cognitivist alleges that values are a human creation: that we make them up and apply them to a morally inert world. The sophisticated cognitivist tries to deflect this charge by exhibiting mathematics as a paradigm of a human creation, impossible and unintelligible outside a shared form of life, that nevertheless lays strong claims to truth and objectivity. So if one can show that a moral outlook is a human creation in *this* sense, one will have sufficiently established its objectivity.

2. *Sophisticated cognitivism*, as I shall call it, is a position that combines two theses, one negative, one positive.[2] The negative thesis is that Wittgensteinian

This essay first appeared in *Mind* 92 (1983), 38–60. I would like to thank Cynthia Farrar, Christopher Kirwan, John McDowell, and Bernard Williams for criticisms of a previous draft.

1. Plato, *Republic* VII.

2. It is not clear that any one person is a sophisticated cognitivist, but the theses of sophisticated cognitivism are argued for, at least separately, in two deep and suggestive papers. For the negative

considerations about following a rule can be used to undermine non-cognitivism, at least as it is traditionally conceived. The positive thesis is that Wittgenstein's philosophy of mathematics, in which mathematics is shown to be a human creation that nevertheless lays claims to truth, objectivity and necessity, provides a model for a satisfying form of ethical realism. In this paper I shall argue, first, that though the argument for the negative thesis is successful against *certain* traditional forms of non-cognitivism, other forms of non-cognitivism survive. Thus non-cognitivism as such is not undermined. Second, it is doubtful whether mathematics can provide a model for the construal of ethical practice. The lesson that emerges for those of us who would like to defend some form of ethical cognitivism is that more work needs to be done. Finally, by a discussion of problems that face both cognitivist and non-cognitivist, I shall try to make clear the challenge that the cognitivist must meet.

3. Humean non-cognitivism stems from two beliefs: First, that moral judgements motivate actions; second, that no strictly cognitive belief about the world could alone motivate action: some non-cognitive desire or volition must also be present.[3] Thus if we encounter others who happen to lack our interests and desires, we will recognize, if we are non-cognitivists, that they do not have reason to act as we do.

According to the sophisticated cognitivist, there are two reasons why people tend to be persuaded by non-cognitivism. First, they tend to accept a philosophy of mind which separates cognitive from appetitive faculties. Since this is equivalent to assuming non-cognitivism outright, it does not constitute an argument for non-cognitivism unless there is an independent argument for this anatomy of the mind. Second, and more important, cognitivism is often conceived in such a way as to make it ripe for sceptical attack. Cognitivism is a position that results from stepping back and reflecting on our everyday moral practices: it aims to provide a justification for them. Sometimes it is conceived as an attempt to step outside the moral universe and, from a totally detached perspective, to tell us how things really are inside.[4]

There is an obvious parallel with the traditional construal of platonism in the philosophy of mathematics. The platonist, so the story goes, tries to step outside mathematical practice to be able to tell us in what it really consists. The platonist answer is that mathematical practice consists in the apprehension of a mathematical reality, existing independently of us, in virtue of which our mathematical statements are true or false. Because of this reality we can, according to the

---

thesis, see John McDowell, 'Non-cognitivism and Rule-Following', in *Wittgenstein: To Follow a Rule*, S. Holtzman and C. Leich (eds.) (London, 1981). For the positive thesis, see David Wiggins, 'Truth, Invention and the Meaning of Life', *Proceedings of the British Academy* 67, 1976; especially pp. 368–372 [reprinted in this volume]. Wiggins and McDowell differ in various ways, for example, on the nature and strength of the is/ought distinction.

3. See McDowell, p. 154
4. See McDowell, *ibid.*

platonist, be said to have *knowledge* of mathematical *truths*. Both the platonist and the cognitivist purport to explain how our mathematical or moral beliefs constitute knowledge of a reality that is there 'any way'.

It is fair to say that Wittgenstein brought about a revolution in philosophical perspective, making platonism—as traditionally presented—impossible. When a person can carry out a mathematical procedure correctly—for example, he can extend the series 2, 4, 6, 8 . . . the platonist says that he has learned how to follow a rule or has apprehended a universal. We manifest our apprehension of a universal, our grasp of a rule, by at most a finite amount of behaviour, yet the grasp of a universal or rule guarantees a potentially infinite amount of behaviour. Wittgenstein's point is, roughly, that it is not because we have grasped a universal or rule that we behave in a certain way, but because we act in certain ways that we say that we have grasped a rule or a universal. No rule or universal will guarantee that someone will not extend the series 2, 4, 6, 8, 17, 28, 1002, =, Δ . . . saying that at each step he is doing exactly what he did before. There is no such thing, Wittgenstein argued, as absolute sameness. A procedure is a case of doing the same thing again if it is a practice within what Wittgenstein called a 'form of life'—a community that shares perceptions of salience, routes of interest, feelings of naturalness—in which it is perceived as the same. If the platonist tries to step outside the form of life in order to tell those within how things really are, then he must come to grief. For outside the form of life there is nothing: no rules, no universals, no sameness, no reality.

What then are we to do: explain moral and mathematical practice only in terms of the natural history of human beings? Can we explain moral and mathematical behaviour only by saying that humans happen to have a constitution such that they tend to react in certain ways on the basis of certain types of training? The sophisticated cognitivist rejects the dilemma which forces us to choose between (moral or mathematical) platonism on the one hand and natural history on the other. The choice looks exhaustive only if we are, confusedly, trying to occupy a vantage point outside our moral and mathematical practices in order to determine what they are really about. From this alleged perspective, we imagine, either there will be something there—moral values, universals—or there will not: no intermediate position seems available. The sophisticated cognitivist tries to deconstruct the dichotomy not by offering something else that we might see from this perspective, but by curing us of the temptation to think that there is any such vantage point to be occupied. The only position from which a statement like '7 + 5 really does equal 12' makes sense is within the context of mathematical practice. So when one says 'It is because 7 + 5 really does equal 12 that humans can be taught to believe that 7 + 5 equals 12', one should not construe oneself as saying anything that disagrees with Wittgenstein. For such a statement is compatible with the belief that it only makes sense to talk of two numbers really equalling another within a form of life. Precisely because there is room for this within a form of life, there is room for a sophisticated cognitivist. He abandons the desire for what he takes to be an unoccupiable vantage point outside our moral and mathematical practices; and he purports to tell us how things really are in the only way in which he can make sense of things really being some way rather than another. Reality, truth, objectivity only make sense within the context of a form of life.

If we wish to ground the objectivity of mathematics in something stronger, the sophisticated cognitivist argues, that could only be because we have not yet been cured of the desire to step outside our form of life and view it 'from sideways on', that is, from an external standpoint. The objectivity of mathematics will only look weak, in need of shoring up, if we fetishistically retain the desire for an external perspective. If we sincerely abandon that desire, if the fever breaks, then we will no longer feel threatened by a collapse into natural history. 7 + 5 really does equal 12, the sophisticated cognitivist tells us, but he also insists that he is not telling us anything we did not already know.

In trying to instill a moral outlook, we will try to get a person to see situations in a certain way: we will appeal to his perceptions of salience, his sense of sympathy, his interests, and indeed to his sense of right and wrong. That we succeed cannot be a matter of his grasping a universal or a rule: it can consist in nothing more than that he comes to see the world the way we do. This does not threaten the objectivity of ethics, the sophisticated cognitivist says, it provides its basis. Within the moral outlook, we see that certain acts—e.g. the killing of innocent humans—really are wrong, and that anyone who thinks otherwise is mistaken. This does not mean that we can convince anyone outside our moral outlook to adopt it: for we've already admitted that outside our moral outlook there is nothing to which we can appeal to commend it. We can only make various appeals to get him to see a situation as we do; if he is not so disposed there is nothing more that can be done. However, this doesn't threaten the objectivity of ethics, it only reveals him to be insensitive. The objectivity of mathematics does not totter every time a child cannot be taught to add. In mathematics, a child's failure to see things our way is taken to be *his* inability. Similarly with ethics: if a man cannot be brought to see that certain acts are right and others wrong, that is his inability. The objectivity of ethics is not threatened even though, in the end, there is nothing we can do other than try to get him to see things the way we do. Ultimately, that is all we can do in any area of discourse and the false belief that we ought to be able to do more in ethics stems from the false belief that we are actually able to do more in other more obviously objective areas of discourse.

4. The sophisticated cognitivist's strategy can thus be outlined as follows. The motivation for non-cognitivism lies in the fact that cognitivism, as traditionally presented, invites criticism. However, traditional cognitivism can be undermined by Wittgensteinian considerations while leaving room for a satisfying form of ethical realism. With traditional cognitivism undermined, the standard motivation for non-cognitivism disappears.

Unfortunately this strategy does not work, for neither the cognitivist nor the non-cognitivist need construe himself as acting upon the stage which the sophisticated cognitivist has set. In so far as a cognitivist *does* wish to act on this stage, he is vulnerable to criticism. Yet neither the cognitivist in ethics nor the platonist in mathematics need conceive of himself as trying to step outside the moral or mathematical realm, as trying to adopt an external point of view. There is no doubt that the cognitivist and the platonist are stepping back from and reflecting on an area of human activity and beliefs. We can easily imagine a people who simply acted according to a moral code, who simply used arithmetic to calculate, and who never thought about the justification of either practice. But when one does step back and ask 'Is this how things really are?' or 'Why do we think that our moral (mathematical) beliefs are true?', this need not be interpreted as a fundamentally illegitimate request to adopt an external standpoint, a non-human perspective. Mathematical and moral practice do not demand self-reflection, but they do invite it. When we, perhaps pre-reflectively, assert, for example, 'Doing that would be wrong', we make a claim that aspires to universality. It is all too human and very much of a piece with moral behaviour to examine critically and test that claim. Since moral claims do have aspirations, and since it is, in part, due to these aspirations that moral claims have significance in our lives, it is natural to ask whether those aspirations are fulfilled. If someone is looking for a Platonic Ought, apprehension of which will force him to act in certain ways, then a Wittgensteinian is correct to criticize him. However, a cognitivist need not be pursuing the unconditioned, trying to stand in space outside of human life, he need only be enquiring whether moral discourse has the status it assigns to itself.

The appeal of cognitivism lies, I think, in the claim that our vocabulary of moral appraisal—e.g. 'right', 'wrong', 'good', 'bad'—can be used without qualification even after reflection on the nature of our moral practices. From earliest childhood a moral outlook is inculcated. We learn moral behaviour long before we can reflect on its status. The original motivation may be a desire for parental approval and fear of punishment, but the moral outlook is soon internalized and comes to supply its own motivation. So by the time we can reflect on moral practices, we cannot be detached observers, impartial consumers wondering which brand to purchase. We already have a moral point of view which embodies a vocabulary that, at least at face value, pretends to universality. When we say that in these circumstances such an action would be the right thing to do, *prima facie* we are claiming that it would be the right thing for anyone in these circumstances, not just for ourselves or for someone who shares our outlook. As we reflect on our moral practices, as we step back, we notice that there are other moral outlooks. There have been and are other moral conceptions that are genuinely incompatible in that they dictate different actions in a given situation, using a vocabulary that pretends to universality. Of course, the different outlooks may conceive of the situation differently—one describing the act as an 'honourable sacrifice to the gods', the other as 'a barbaric murder of an innocent child'—but there may nevertheless be a single question—e.g. 'Shall I kill this child?'—to which the moral outlooks return conflicting answers.

[80]

Cognitivism allows us to retain the full force of our vocabulary of appraisal even after this realization. Those who advocate killing the child are wrong and their own moral outlook a grotesque, perhaps morally culpable, mistake. Of course, cognitivism does not guarantee that our own moral outlook is sacrosanct. Some of our own moral practices may, on reflection, come into question. For example, if we were persuaded by animal liberationists, we would come to see the killing of young animals as murder. The extension of our vocabulary of appraisal would change, not its force. Cognitivism does not legitimize the status quo, but it does hold out the promise that there is some moral outlook, even if it differs from ours, in which the vocabulary of appraisal can legitimately be employed without qualification. Perhaps we are wrong to kill animals, but, if so, cognitivism assures us that we really *are* wrong to do so.

A non-cognitivist may simply be someone who denies this. For example, he may be someone who claims that there are no reasons for action which apply to a person independently of his interests, desires and motives: that there are no 'external' reasons.[5] Further, he recognizes that, as a matter of fact, there have been and are other cultural groups with incompatible moral and social outlooks; and these outlooks do not in any sense appear to be stages in a process of historical convergence. The likely explanation of this phenomenon is, for him, sociological: these groups have developed so as to embody different interests, motives and desires. Thus, he thinks, members of these other groups do not have reason to act as he does. He need not insist on any sharp division between the cognitive and volitional faculties of the mind; he need only see moral discourse as to some extent ideological. The vocabulary of 'right and wrong' is, according to him, more categorical than it should be: it pretends to an objectivity it does not have.

5. The sophisticated cognitivist fails fully to undermine the motivation for non-cognitivism. The question arises as to whether he can put forward a positive conception of ethical cognitivism that we will find attractive and compelling. Mathematics, at least as Wittgenstein construed it, is a practice that combines objectivity, discovery *and* invention.[6] So if ethics can legitimately aspire to its status, perhaps we will have secured all the truth and objectivity that ethics needs.

But how are we to understand the 'creativity' of mathematics and nevertheless legitimately regard it as objective, true and compelling? We can do this, I think, but only by recognizing that there are significant features of mathematical practice to which there are no analogues in ethics. Even in those areas of mathematics that seem to provide a paradigm of truth, objectivity and necessity—for example, the addition of small integers—Wittgenstein claimed that there is only room to talk about truth, objectivity and necessity within the context of a 'form of life': a community that shares perceptions of salience, routes of interest, feelings of naturalness, etc. Of course, this 'agreement in form of life' does not constitute the

5. See Bernard Williams, 'Internal and External Reasons', *Moral Luck* (Cambridge, 1981).
6. See Wiggins, p. 371.

truth of any area of discourse; it provides the space in which one can talk of truth and falsity, necessity and contingency, objectivity and subjectivity.[7] The question is: how can we have this insight and retain a strong sense that arithmetic is a legitimate paradigm of truth, objectivity and necessity? Let us say that a person is *minded in a certain way* if he has the perceptions of salience, routes of interest, feelings of naturalness that constitute being part of a certain form of life.[8] And consider the following claims:

(a) $7 + 5 = 12$. To suppose that any other integer, say 13, is the sum of $7 + 5$ is a mistake.

(b) It is only within the context of our being so minded that $7 + 5 = 12$.

The first claim, (a), is certainly correct. $7 + 5$ equals 12 and anyone who tries to offer a different integer as an answer is in error. After studying the later Wittgenstein, one also wants to say that (b) expresses *some* sort of truth. And yet, how can one believe (b) without that weakening our beliefs in the truth of (a) and the necessity of addition? This is a problem faced by any conventionalist in the philosophy of mathematics. However, I do not think that Wittgenstein was a conventionalist. One can see this when one recognizes that (b) does not express an empirical truth. For if (b) were an empirical truth, then we ought to be able to accept the following counterfactual:

$7 + 5$ would equal something other than 12 if everyone had been other minded.

We cannot make any sense of this counterfactual, for the notion of people being other minded is not something on which we can get any grasp. Our problem is that being minded as we are is not one possibility we can explore among others. We explore what it is to be minded as we are by moving around self-consciously and determining what makes more and less sense. There is no getting a glimpse of what it might be like to be other minded, for as we move toward the outer bounds of our mindedness we verge on incoherence and nonsense.[9]

Explanations, according to Wittgenstein, must come to and end somewhere; and it is the job of philosophy to help us comprehend that which has no explanation.[10] A person who has acquired a basic arithmetical competence will, upon

---

7. The expression is Wittgenstein's. Cf. *Philosophical Investigations,* I.241 (Oxford, 1978).

8. In the next few paragraphs I follow a line of thought that is discussed in more detail in 'Leaving the World Alone', *Journal of Philosophy* (1982).

9. We are, as Thompson Clarke memorably said, 'forced winetasters of the conceivable'. Cf. 'The Legacy of Skepticism', *Journal of Philosophy* (1972), p. 766.

10. See e.g. L. Wittgenstein, *Philosophical Investigations* I.1, 109, 211, 213, 217, 325, 326, 467, 468, 471, 474, 477, 480–485, 496, 497, 516, 599; Zettel (Oxford, 1967) 267, 313–315, 608–611; *On Certainty* (Oxford, 1979), pp. 34, 110, 135, 168, 192, 204, 212, 287, 343–344, 359, 501, 599; *Remarks on the Foundations of Mathematics* (Oxford, 1967), I.34, II.74, 78.

request, add 7 + 5 and get 12 as a result. The full empirical explanation of this action will not quell the sense of puzzlement we feel when, in a Wittgensteinian mood, we wonder why that person acted one way rather than in some radically different way. Our reasons have been stated, we have already given the full empirical explanation, our justifications are spent. And still we want to know: how do we go on? The claim that 'we act in certain ways because we are minded as we are' does seem to do genuine work in making our behaviour comprehensible to ourselves. So it is tempting to see the claim as providing an explanation, as showing why the possibility that constitutes our behaviour was realized rather than some other possibility. We cannot however, make anything of these 'other possibilities': "If someone says 'If our language had not this grammar, it could not express *these* facts'—it should be asked what 'could' means here" (*Investigations* I.497).

Of course, we can certainly *say* 'Imagine a tribe that believed 7 + 5 = 13'; but as we try to *fill out the picture*, to describe in detail how that belief fits in with their other arithmetical and non-arithmetical beliefs, we find that we cannot do it.[11] Either the story will be unable to show how their alleged beliefs and actions fit together or it will stipulate a profound and mysterious blind spot by which their arithmetical belief is inexplicably shielded from the countervailing evidence. Of course it may be heuristically valuable to postulate such a tribe. We can thereby become aware of how our arithmetical beliefs relate to our routes of interest and perceptions of salience. We can also become aware of the outer bounds of our mindedness. For in discovering that we cannot fill out the picture we find that we have passed the outer bounds of our mindedness and lapsed into incoherence.

In the moral case, by contrast, 'form of life' seems to be a sociological expression used to pick out one group among others.[12] We can recognize the existence of other groups who have alternative moral outlooks. Moreover, there still seems to be plenty of room for explanations as to why we have the moral beliefs we do rather than some others. It is hard to see why one should believe that only within the context of a form of life is there room for truth, objectivity and necessity when 'form of life' is being used narrowly, to pick out a subgroup of the community of persons.

6. Another feature of mathematical practice for which there seems to be no analogue in ethics is that one can provide an explanation of the truth and objectivity of mathematics. Of course, the explanation must occur within the context of a form of life, and no such explanation can answer the Wittgensteinian question. 'How do we go on?', but this does not imply that no explanation is possible.

Why are we inclined to think that any area of mathematics contains a body of truths? The reason is, I think, because certain areas of mathematics are applicable

11. See E. J. Craig, 'The Problem of Necessary Truth', in *Meaning, Reference and Necessity*, S. Blackburn (ed.) (Cambridge, 1975).

12. See Charles Taylor 'Understanding and the *Geisteswissenschaften*', in *Wittgenstein: To Follow a Rule;* Bernard Williams, 'Wittgenstein and Idealism', *Moral Luck.*

to the physical world and to our experience of it.[13] There is of course a common belief that mathematical theorems are true irrespective of whether there are any physical instantiations of them. But a study of the shift in perspective that occurred with the discovery of non-Euclidean geometries and the confirmation of relativity theory casts doubt on this belief. For example, Euclid I-32—that a triangle has interior angles equal to two right angles—which was once thought to be an *a priori* truth is no longer even thought to be true. The reason is that it is now believed that physical triangles, if there were any, would not have interior angles equal to two right angles. Euclid I-32 may be a consequence of the Euclidean axioms and thus we can make the limited claim that it is true of triangles in Euclidean space. So while one may believe a theorem true while remaining agnostic on the question, say, of whether there are any physical triangles, one must believe that the theorem truly describes a property a triangle would have.

Those areas of mathematics that we tend to think of as true describe abstract structures that preserve certain structural features that are, generally, to be found in the physical world. It is the nature of this structural preservation that explains the applicability of mathematics. For example, a number n is related to other numbers in ways that are intimately linked to the manner in which n-membered sets of durable physical objects are related to disjoint sets of various cardinality. The union of two disjoint sets of durable physical objects, one seven-membered, one five-membered, is usually a twelve-membered set, and the arithmetical truth '$7 + 5 = 12$' reflects this fact. Of course, physical objects may perish or coalesce—thus the "usually"—and one of the virtues of arithmetic is that in moving to the realm of numbers one can abstract from this possibility.

Since Frege, any philosophy of mathematics that can be labelled 'abstractionist' has been in bad repute. It is worth noting, therefore, that this abstraction does not require any of the Millian inattention which Frege ridiculed nor that arithmetical truths are inductive generalizations from experience. Arithmetic is not the product of induction from experience but of abstraction from experience—abstraction of a wholly legitimate type.

Thus a mathematical statement like '$7 + 5 = 12$' is true in virtue of its applicability to the world: and we can understand how it is applicable because we can see that mathematics preserves structural features that are to be found in the physical world. Of course this structural preservation provides no transcendental guarantee that we will 'stay on the rails' as we continue to do mathematics, nor does it underwrite a *Tractarian* picture of the structure of language mirroring the structure of the world. It is simply an explanation of why mathematical reasoning is applicable to the world.[14] The explanation of the objectivity of mathematics is

13. I discuss this in more detail in 'Aristotle's Philosophy of Mathematics', *Philosophical Review* (1982).

14. McDowell, p. 156, allows that as long as we avoid the mistake of trying to take on a non-human perspective, there is room for a 'satisfying intermediate position, according to which the logical "must" is indeed hard (in the only sense we can give to that idea)'. It is precisely from this intermediate position that an explanation of the truth and objectivity of arithmetic can be given.

not mysterious and one does not have to take up an unoccupiable external position in order to appreciate it.

7. Let us try to fill out the picture of a mathematical case where the features just described do not hold.[15] Imagine that we encounter a new tribe and that the anthropologists on both sides are able to establish successful communication between us. The chief explains to us that he is amazed that we are able to communicate, because a philosopher in his tribe had convinced him that it would be impossible even to interpret a people who had a deviant arithmetic. And yet he finds he can understand our theory of *transfinite* arithmetic, though he is shocked to think that any of us take it seriously. Naturally enough, he would like to persuade us of the folly of our ways:

'Imagine,' the chief begins, 'that you and I have been in existence throughout the previous history of the world. Our physical and mental characteristics change fairly continuously and though our memories are much stronger than ordinary human memories they have the same basic nature: we remember most vividly what has happened in the immediate past and as we move further into the past our memories tend gradually to fade out. (It would take one of your very prodigious psychiatrists to enable us to dredge up our childhood.) Although we have been in existence during the entire previous history of the world, we are not immortal. In fact God has scheduled it so that I must die today and you have been granted another eighty years to live.

'I am inclined to think,' the chief continues, 'that God has granted you eighty *more* years to live than me. And I'm sure if we asked any of your people who had not studied your theory of transfinite arithmetic, this would be their untutored response. However, according to your theory, whether or not my inclination is correct depends upon how long the world has been in existence. If God created the world (and us) a finite number of years ago, then it is certainly true that you will live eighty more years than me. If, however, the previous history of the world extends endlessly into the past then, according to your theory, you will not live any more days than me. For suppose that the previous history of the world can be represented as an ω^*-sequence:

. . .	-3	-2	-1	0
	the day before yesterday	yesterday	today

Then eighty years from today the previous history of the world will also be able to be represented by an ω^*-sequence:

. . .	$-29{,}220$	. . .	-2	-1	0
	today				80 years from today

($29{,}220$ days $= 80$ years). So at the end of our lives you and I will have lived the

---

15. Since it is so unclear what, if anything, the debate between classical and intuitionist mathematicians consists in, it does not provide a helpful example. In fact, I do not think it provides an example of the relevant case at all. Cf. my 'Leaving the World Alone'.

same cardinal number of days: for the days you have lived can be put into a one-one correspondence with the days I have lived. The days of our lives will also be of the same order-type: for the days of our lives can be put into a one-one order-preserving correspondence. Therefore, according to your theory, there is no sense in which you can be said to live longer than me.

'I am amazed that you believers have become used to saying that if there were infinitely many people in the world there would be the same number of right arms as of right arms and left arms. You even become impatient with students who do not quickly accept that there are the same number of odd numbers as of odd and even numbers. I think that you are able to maintain this attitude only because the infinite is so remote from your experience.

'Suppose that today, on the day of my death, I am full of hope and expectation. I feel myself bursting with projects and anticipation. "Why," I might ask God, "must I die today when so much of what I consider important about living my life lies in the future?" It will be small comfort to be told that, however I feel, I am in fact an old man, that I have already lived infinitely many years. For right now I see my life as stretching before me: I feel *deprived* of the life I would lead if I were to live as long as you. In fact I would have died much more happily billions of years ago, during a period when I happened to spend a lot of time dwelling on my past, living in my memories. Then I could have conceived of my life as something I *had* lived.

'Of course we cannot anticipate all the answers God might give. More than likely, He would give no answer at all. Perhaps He might say that long ago, at a time I can no longer remember, I behaved extremely badly and was now to be punished. To my protestations that I have no memory of these deeds, He might respond that He understands personal identity much better than I do and that I am indeed responsible for those bad deeds, whatever confused philosophical ideas I have to the contrary. I still may not be pleased with the fact of my impending death, but I think that I could console myself that my displeasure resulted from my being out of harmony with the universe: there is a reason for my death, even if I cannot fully understand what it is.

'One answer, however, that I hope you will agree is completely unsatisfying is that God saw no difference in what He was granting you and what He was granting me. "After all," He might say, "I am letting you both live exactly the same number of days. It's just that I am going to extinguish one of you today and the other in eighty years. It makes no difference who goes today and who goes in eighty years." That is certainly how it would look if God lived in Cantor's paradise.

'I hope that He does not live there and that He would be moved by the following response. Your theory of the transfinite is nothing more than a mathematical theory. There is no guarantee that it would give an accurate description of our experience of an infinite life. For all we know it is just a consistent mathematical theory that has been tacked onto finite mathematics in such a way that it yields no discordant results at the finite level. The reason that you and I are said to live the same (cardinal and ordinal) number of days is that there is a function which maps the last day of my life onto the last day of yours, the second to the last day of my life onto the second to the last day of yours, and so on. But why should we measure some aspect of our experience on the basis of a function that itself bears no relation to our experience? In terms of how our lives are lived, such a function imposes an irrelevant correlation: it flattens out an important aspect of our experience.

'I would like to suggest that you take seriously the notion of a *relevant correlation*. I cannot give a precise definition of what it is for a function f_r to impose a relevant correlation. But the intuitive, informal idea is that a relevant correlation is one that preserves those aspects of our experience we believe to be important, even when our experience is being characterized mathematically. The relevant correlation for comparing our lives is the identity function: the function that maps any particular day into itself. This function will map the last day of my life onto the 29,220th-to-the-last day of yours, the second-to-last day of my life onto the 29,221st-to-the last day of yours, and so on. This function maps the days of my life into the days of yours. So under this ordering you can be said to live longer than I will. In fact, under this correlation, you can be said to live eighty years longer. That, I submit is precisely how much longer you will live; and that is why this correlation is the relevant one.'

We, as Cantorians, might object that the chief has misdescribed the experience we are to undergo. In particular, we might complain that what is really bothering the chief is that we will live *later* than him.[16] His upset is really that he is to die today and that we will continue to live at a later time when he is not living, even though we are equally worthy in the eyes of God. He is not really upset that we will live *more* days than him.

The chief, however, is not convinced: 'Your objection is not conclusive. For suppose that the eighty years have passed and imagine (perhaps *per impossible*) that we are both somewhere from which we can look back and view our past lives. Of course I sincerely hope that by then I will have been able to adopt a "philosophical" attitude toward the eighty years of experience that you had which I did not. However, I can imagine myself *resenting* the fact that you had been granted eighty more years to live than I. This is not the resentment that you still have life ahead of you whereas I do not; nor need it be a resentment that you have lived later. For by now the question of when it occurred is a matter of relative indifference to me. What continues to irk my residual sense of justice is that you were granted *more* life than me. Such resentment may be pointless, but it is not groundless. (Thus it is like most cases of resentment.)'

16. Cf. e.g. A. N. Prior, 'Thank Goodness That's Over', *Papers in Logic and Ethics* (London, 1976). See also John Perry, 'The Problem of the Essential Indexical', *Noûs*, 1979.

We might try to persuade the chief that what is really bothering him is that we will be alive on days that he is not; he couldn't be upset that we live longer. Again the chief is unmoved. Having already read much of our literature, he quotes to us Homer's account of the twins Castor and Pollux.[17]

> These two live still, though life-creating earth
> embraces them: even in the underworld
> honoured as gods by Zeus, each day in turn
> one comes out alive, the other dies again.

'Were we in such a situation,' says the chief, 'I would not be upset that you were alive on days when I was not. The situation would be fair, because we would each live the same number of days.'

We may object that the chief's example was unfairly chosen, for he has assumed that we have already been in existence for an infinity of time. If we had been, we might say, we could never have 'reached' the present. The chief will have none of this. It is, he says, *he,* who does not believe in the transfinite, who can make this objection, not us. If we believe in the transfinite, then, he thinks, we ought to say that it only takes a transfinite number of days to reach the present from a transfinite number of days ago.[18] Further, the chief might ask us to consider a different example: 'Suppose that life is a vale of tears through which each of us must pass before he earns his Heavenly reward. You have been allowed to begin your Heavenly life of bliss today, while I must soldier on through life's trials and tribulations. Again, I could understand this fate if God thought that we merited such differential treatment. What I could not understand would be God thinking that there was no difference between what He was granting you and what He was granting me. "Relax," he might say, "I have not yet made up my mind when you will go to Heaven. It may be in a hundred years, or a billion, or 10^{27} years. But whenever it is (so long as I don't forget) you will have lived exactly the same number of days as your equally worthy friend." Of course, part of my upset would be that my Job-like existence was still ahead of me. However, this would not account for all of my upset. For I would also be upset that I was being made to live longer than you.'

Nor does the chief think that his argument should be confined to measuring aspects of our lives, leaving all other measurements intact. 'You have measured out your life with coffee spoons,[19] you ought to measure out coffee spoons with your life. There is, of course, a function $f(2n) = n$ which maps the even numbers one-one onto the natural numbers. But the relevant correlation, I think, is again

---

17. Homer, *The Odyssey,* book II, lines 300–304, R. Fitzgerald Jr., tr. (New York, 1961).

18. Cf. Kant, *Critique of Pure Reason,* B435–62. For the next example, I am indebted to Nicholas Denyer.

19. Cf. T. S. Eliot, 'The Love Song of J. Alfred Prufrock', *Collected Poems, 1909–1962* (London, 1975).

the identity function $f(n) = n$. This function maps the even numbers one-one into, but not onto, the natural numbers. Thus one can say that, under this correlation, there are fewer even numbers than there are natural numbers. In fact, one can easily see that for every natural number that is a value of this function, there is another one that is not. This gives sense to the idea that there are half as many even numbers as there are natural numbers.'

It may not come as a surprise that I find the chief's argument rather compelling, perhaps more than you do. For the purposes of the present discussion, what is important is that we recognize that insofar as we see that the chief does have an argument, thus far is our belief in the truth or objectivity of transfinite arithmetic weakened. Unlike someone who just says 'I believe that $7 + 5 = 13$', the chief provides us with a concrete example of an alternative mathematical outlook: one which we could conceivably adopt. There is no reason to think that significantly different approaches to transfinite arithmetic lie beyond the bounds of our minded-ness. Further, one might even suspect that when it comes to applicability, which is, after all, that which grounds the truth and objectivity of elementary arithme-tic,[20] the chief may have sown the seeds of doubt about the value of our own theory. We may wish to say that transfinite arithmetic is more beautiful than the chief's drab alternative, but by moving to aesthetic criticism we acknowledge that we are shifting the focus of the debate away from questions of truth and falsity.

It is not enough simply that there be a group who agree in their judgements of the applicability of a set of concepts. When one learns transfinite arithmetic one learns to apply a set of mathematical concepts that does not make sense outside of this practice. Certain people will be more sensitive in their application of these concepts than others; the temptation is overwhelming to say that they have a clearer perception of the transfinite. It is all right to call such sensitivity 'percep-tion' if one does not thereby think that one has secured the truth and objectivity of the practice. For there may be two practices that are incompatible in the sense that they cannot both be true—for example, the two approaches to transfinite arithme-tic—each of which can be developed with sensitivity by a group whose members agree in their judgements. Indeed, a single person may be considered by both groups to be the most sensitive percipient of their respective outlooks. Thus it is no use for a sophisticated cognitivist to try to rescue the objectivity of ethics by saying that it may be a human invention in the same sense that mathematics is an invention.[21] Transfinite arithmetic is a human invention, and if our moral outlook shares its status, it is not clear that we have secured any degree of moral objec-tivity.

8. If a study of the analogy between ethics and mathematics undermines, rather than supports, sophisticated cognitivism, it also causes problems for sophisticated non-cognitivism. Non-cognitivists tend to be moral relativists, but the traditional

20. Cf. my 'Aristotle's Philosophy of Mathematics', *ibid.*
21. As Wiggins does.

problem with relativism has been to state a version that is both coherent and plausible. Relativism, in vulgar form, is the view that all truth must be truth-in-a-theory, for some theory, and that therefore there can be no truth as to which of two competing theories is correct. Vulgar relativism is incoherent because it uses non-relativistic truth to describe the relativistic situation while denying that there can be any such truth.[22]

Bernard Williams has tried to state the necessary conditions for a more sophisticated, and coherent, form of relativism.[23] First, one must have two theories that are more or less self-contained. Second, the theories must be incompatible, in the minimal sense that there is some yes/no question to which they give conflicting answers. Thus the theories must be comparable to some minimal extent. One theory may describe an event as 'a stone falling due to gravity', the other as 'a bit of earth seeking its natural place', but there must be some question—e.g. 'If the Earth were not in the centre of the universe would this object move in the same direction?'—to which the theories give different answers. Third, the theories must be in *real confrontation:* i.e. there must be a group of people for whom each of the theories is a real option. A theory is a *real option* for a group either if they believe it or if it is possible for the group to believe it. There will, of course, be a vocabulary of appraisal, e.g. 'good—bad', 'true—false', 'right—wrong', in which the options are discussed. In a notional confrontation, by contrast, a group may be aware of both theories and their differences, but at least one of them is not a real option for it.

Relativism, according to Williams, is the doctrine that there is only point in appraising the merits of both theories if they are in real confrontation. 'To stand in only notional confrontation is to lack the relation to our concerns which alone gives any point of substance to the appraisal: the only real questions of appraisal are about real options.' The main virtue of this account, as opposed to vulgar relativism, is that it grants that we can think about alternative theories that may be of concern to us in ways that are not rigidly relativized to our currently held theory. We can explore real options.

It has been objected that such relativism is impossible, on the grounds that successful interpretation of an alternative theory, which is necessary for us to see it as incompatible with our theory, especially in its valuations, demands that the alternative be so similar to our own beliefs and interests that it must be in real confrontation.[24] Any alternative we can understand will be a real option. This objection tries, I think, to get more out of the theory of interpretation than it will plausibly yield. We can understand theories with different valuations that are not real options nor in real confrontation, though the people for whom those theories are real options are recognizably enough like us. For example, Herodotus reports:

22. See Bernard Williams, *Morality: An Introduction to Ethics* (New York, 1972), pp. 20–22.
23. Bernard Williams, 'The Truth in Relativism', in *Moral Luck*.
24. Wiggins, pp. 358–359.

When [Darius] was King of Persia, he summoned the Greeks who happened to be present at his court, and asked them what they would take to eat the dead bodies of their fathers. They replied they would not do it for any money in the world. Later, in the presence of the Greeks, *and through an interpreter, so they could understand what was said,* he asked some Indians of the tribe called Callatiae, who do in fact eat their parents' dead bodies, what they would take to burn them. They uttered a cry of horror and *forbade him to mention such a dreadful thing.*[25]

Nevertheless, Williams' formulation of relativism *is* unacceptable, for it conflates an option and a confrontation. Certainly there can be actual confrontations between two theories in which the adherents of each theory cannot treat the other theory as a real option. For example, the ethical and religious outlooks of fundamental Islam and the West seem to be in actual confrontation: the future of life on the planet is jeopardized by the disagreements, the existence of each theory threatens the ability of the adherents of the other to pass on their beliefs intact to future generations, etc. Yet it is plausible to suppose that, on pain of cultural schizophrenia, adherents of neither theory could have the other as a real option. Part of what it is to be an Islam fundamentalist, I suppose, is to have the Western outlook as an unreal option and *vice versa*. For there to be an actual confrontation between unreal options it seems that the vocabulary of appraisal of each theory must be deployed with full force. It is precisely because one takes the other theory to be false, wicked, absurd etc. that the option is unreal.

Can there be notional confrontations between real options? An examination of the chief's story shows that the answer to the question is affirmative. The chief has, I think, presented us with a real option. His story is sufficiently coherent and his complaint against transfinite arithmetic sufficiently compelling, that one can imagine adopting such an attitude toward the infinite. Yet the confrontation is notional: the chief is not going to persuade a student of the transfinite to abandon his studies. For the student wishes to discover the structure embodied in transfinite set theory and is indifferent to its applicability to human experience. However, in so far as one can see the chief as presenting a real option in but notional confrontation, the language of appraisal loses force. If one could unproblematically label one of the theories 'true', the other 'false', then one of the theories would not be a real option.

This example suggests one plausible characterization of relativism: when two incompatible theories are both real options for their adherents but they are only in notional confrontation. This form of relativism can obtain only in limited areas of thought. A confrontation between real options can be notional only if one admits that there is not much in the dispute: there is no 'fact of the matter' to which one of the theories better approximates. Otherwise the confrontation would be real. Now

25. Herodotus, *The Histories* III.38, A. de Selincourt, Jr. (Harmondsworth, Eng., 1972), my emphasis.

[91]

it seems that the more beliefs about the world we include in our 'theory', the less possible it is for us to treat a confrontation as notional. For we become ever more unable to distance ourselves from our own beliefs to the extent necessary to treat an alternative as a *real* option or as in but notional confrontation.

The extent to which one can be a relativist in this sense with respect to ethical disputes is less than has been suggested. The reason that this form of ethical relativism looks plausible is that there are some disputes, which appear to be ethical, in which unreal options are in notional confrontation. For example, Herodotus' report is shocking precisely because the option is so unreal—there is simply no question of eating our recently deceased loved ones as a token of respect—and yet stepping back a bit from our unreflective beliefs and practices we can see that the confrontation is only notional. We cannot continue to employ the vocabulary of appraisal with full force: we may find their behaviour disgusting, but we do not think it wrong, in the sense that *they* ought to change it. Our ultimate judgment of them is aesthetic, not moral.

Since ethics is concerned with how we should live our lives among others, real options will tend to be in real confrontation. There may be certain limited areas of behaviour, for example questions of etiquette, in which we are able to treat real options as in notional confrontation. As the option encompasses ever wider areas of human behaviour, however, either it will become notional or the confrontation will become real or both. In each case our vocabulary of appraisal will tend to be used with full force. It is precisely because we think that another option is wrong, wicked, absurd, that it is unreal for us; because we believe it wrong, wicked absurd, we are able to sustain ourselves—and our beliefs—through a real confrontation. Given our characterization of relativism, these considerations reveal the *untruth* in ethical relativism.

The relativist, however, may try to strengthen his position. He may grant that because of our makeup we are unable to go on forever treating a confrontation as only notional. Nevertheless, he will insist that '*inside our studies*' we will see that non-cognitivism is true; and we will recognize that, as a matter of empirical fact, there happen to be other groups with incompatible moral outlooks and that the various outlooks can in no sense be thought of as converging. It seems that this relativist can allow there to be real confrontations between options, both real and notional, in which there is ultimately no fact of the matter as to who is right and who is wrong. He will concede that as the option encompasses a wider range of human behaviour we will be ever more unable to treat it as real or as in but notional confrontation. He will concede that 'outside our studies' we will tend to use our vocabulary of appraisal with full force. However, he will also insist that 'inside our studies', sheltered from the onslaught of rival options, we will see that there is no fact of the matter as to who is right and who is wrong.

This form of relativism is, I think, 'untrue' in the Hegelian sense that it is unstable, likely to go under. If it came to be widely believed it would, I think, have serious consequences for our culture. Perhaps some of us are capable of living a life of Humean dissociation, reflectively believing in non-cognitivism but acting as though it were false. However, many of us cannot live such a fragmented life: there is a connection between our actions and our reflective thoughts about the nature of morality.[26]

The belief that this form of relativism is true may produce different effects in different people. In some it may lead to a Machiavellian attempt to impose their own values on others; in some it may encourage a privatized pursuit of individual interests; in some it may lead to ennui and a general undermining of values. Reflective moral agents living in modern liberal society seem to be receptive to the worry that relativism is true; and this may, in Hegelian fashion, lead to a transformation of outlook. The debate between cognitivist and non-cognitivist is not just an abstract one about the (lack of) foundations of ethics; one's ethical outlook may actually be changed by the outcome of the debate. For example, those who are incapable of living a life of Humean dissociation will be unable to have the insight that relativism is true without thereby distancing themselves from the vocabulary of 'right and wrong'. That vocabulary presents itself not as one of approval or disapproval of local behaviour, but as a universal standard against which diverse behaviour can be measured. Perhaps such vocabulary is ideological: perhaps we cannot deploy it sincerely without a false consciousness of our position of judgement. But how can we believe *that* thesis and continue to use the vocabulary sincerely in real confrontations?

In evolutionary terms, it seems that we suffer from a selective disadvantage. Our receptivity to the belief that relativism is true makes us vulnerable in real confrontations—and especially in mortal conflicts—when the other culture tends to be fundamentalist, unreflective and unreceptive to the belief that relativism is true. Even if we accept that some changes of belief may result in an improved moral outlook, it is disheartening to conclude that reflective awareness is selectively disadvantaged by comparison with unreflective single-mindedness.

If this is the end of the story, it is a gloomy one. Since I do not like gloomy endings, I would like to suggest that one way we might continue the story would be to show that this form of relativism is not merely 'untrue' in the Hegelian sense, but actually incoherent. Since we are still very ignorant of the relation between our reflective beliefs—those we entertain 'inside our studies'—and the rest of our beliefs and values, it is not rash to hope that this can be done. If there is a significant analogy between mathematics and ethics, it is that both areas of discourse invite a certain amount of reflection. By reflecting on the mathematical and ethical principles we employ, we become aware of which are worthy of wider application or more general formulation. Moreover, both mathematics and ethics are such remarkable human achievements that they naturally invite wonder as to how they are possible. This is not a desire for a non-human perspective, but a

---

26. This point is beautifully made by Bernard Williams in his T. S. Eliot Lectures, *Shame and Necessity* (London, 1983).

desire for insight into the human perspective. Relativism suggests itself as soon as one reflects on moral practice and notices the existence of different moral practices. But to an observer whose perspective is restricted, the lines of two developing cultures may appear to be parallel even though they are in fact converging. The hope I am expressing, and it is only a hope, is that relativism will come to be seen as a product of an early stage of reflection.

5

The Subjectivity of Values

J. L. Mackie

[handwritten annotation: values are secondary qualities they cannot exist alone in the world]

1. Moral scepticism

There are no objective values. This is a bald statement of the thesis of this essay, but before arguing for it I shall try to clarify and restrict it in ways that may meet some objections and prevent some misunderstanding.

The statement of this thesis is liable to provoke one of three very different reactions. Some will think it not merely false but pernicious; they will see it as a threat to morality and to everything else that is worthwhile, and they will find the presenting of such a thesis in what purports to be a book on ethics paradoxical or even outrageous. Others will regard it as a trivial truth, almost too obvious to be worth mentioning, and certainly too plain to be worth much argument. Others again will say that it is meaningless or empty, that no real issue is raised by the question whether values are or are not part of the fabric of the world. But, precisely because there can be these three different reactions, much more needs to be said.

The claim that values are not objective, are not part of the fabric of the world, is meant to include not only moral goodness, which might be most naturally equated with moral value, but also other things that could be more loosely called moral values or disvalues—rightness and wrongness, duty, obligation, an action's being rotten and contemptible, and so on. It also includes non-moral values, notably aesthetic ones, beauty and various kinds of artistic merit. I shall not discuss these explicitly, but clearly much the same considerations apply to aesthetic and to moral values, and there would be at least some initial implausibility in a view that gave the one a different status from the other.

This essay was originally published as chapter 1 of J. L. Mackie, *Ethics: Inventing Right and Wrong* (London: Penguin Books, 1977), pp. 15–49, copyright © J. L. Mackie, 1977, reprinted by permission of Penguin Books Ltd.

J. L. Mackie

Since it is with moral values that I am primarily concerned, the view I am adopting may be called moral scepticism. But this name is likely to be misunderstood: 'moral scepticism' might also be used as a name for either of two first order views, or perhaps for an incoherent mixture of the two. A moral sceptic might be the sort of person who says 'All this talk of morality is tripe,' who rejects morality and will take no notice of it. Such a person may be literally rejecting all moral judgements; he is more likely to be making moral judgements of his own, expressing a positive moral condemnation of all that conventionally passes for morality; or he may be confusing these two logically incompatible views, and saying that he rejects all morality, while he is in fact rejecting only a particular morality that is current in the society in which he has grown up. But I am not at present concerned with the merits or faults of such a position. These are first order moral views, positive or negative: the person who adopts either of them is taking a certain practical, normative, stand. By contrast, what I am discussing is a second order view, a view about the status of moral values and the nature of moral valuing, about where and how they fit into the world. These first and second order views are not merely distinct but completely independent: one could be a second order moral sceptic without being a first order one, or again the other way round. A man could hold strong moral views, and indeed ones whose content was thoroughly conventional, while believing that they were simply attitudes and policies with regard to conduct that he and other people held. Conversely, a man could reject all established morality while believing it to be an objective truth that it was evil or corrupt.

With another sort of misunderstanding moral scepticism would seem not so much pernicious as absurd. How could anyone deny that there is a difference between a kind action and a cruel one, or that a coward and a brave man behave differently in the face of danger? Of course, this is undeniable; but it is not to the point. The kinds of behaviour to which moral values and disvalues are ascribed are indeed part of the furniture of the world, and so are the natural, descriptive, differences between them; but not, perhaps, their differences in value. It is a hard fact that cruel actions differ from kind ones, and hence that we can learn, as in fact we all do, to distinguish them fairly well in practice, and to use the words 'cruel' and 'kind' with fairly clear descriptive meanings; but is it an equally hard fact that actions which are cruel in such a descriptive sense are to be condemned? The present issue is with regard to the objectivity specifically of value, not with regard to the objectivity of those natural, factual, differences on the basis of which differing values are assigned.

2. Subjectivism

Another name often used, as an alternative to 'moral scepticism', for the view I am discussing is 'subjectivism'. But this too has more than one meaning. Moral

[96]

Subjectivism
— your opinion on the action or you make the decision of it is good or bad

subjectivism too could be a first order, normative, view, namely that everyone really ought to do whatever he thinks he should. This plainly is a (systematic) first order view; on examination it soon ceases to be plausible, but that is beside the point, for it is quite independent of the second order thesis at present under consideration. What is more confusing is that different second order views compete for the name 'subjectivism'. Several of these are doctrines about the meaning of moral terms and moral statements. What is often called moral subjectivism is the doctrine that, for example, 'This action is right' *means* 'I approve of this action', or more generally that moral judgements are equivalent to reports of the speaker's own feelings or attitudes. But the view I am now discussing is to be distinguished in two vital respects from any such doctrine as this. First, what I have called moral scepticism is a negative doctrine, not a positive one: it says what there isn't, not what there is. It says that there do not exist entities or relations of a certain kind, objective values or requirements, which many people have believed to exist. Of course, the moral sceptic cannot leave it at that. If his position is to be at all plausible, he must give some account of how other people have fallen into what he regards as an error, and this account will have to include some positive suggestions about how values fail to be objective, about what has been mistaken for, or has led to false beliefs about, objective values. But this will be a development of his theory, not its core: its core is the negation. Secondly, what I have called moral scepticism is an ontological thesis, not a linguistic or conceptual one. It is not, like the other doctrine often called moral subjectivism, a view about the meanings of moral statements. Again, no doubt, if it is to be at all plausible, it will have to give some account of their meanings. . . . But this too will be a development of the theory, not its core.

It is true that those who have accepted the moral subjectivism which is the doctrine that moral judgements are equivalent to reports of the speaker's own feelings or attitudes have usually presupposed what I am calling moral scepticism. It is because they have assumed that there are no objective values that they have looked elsewhere for an analysis of what moral statements might mean, and have settled upon subjective reports. Indeed, if all our moral statements were such subjective reports, it would follow that, at least so far as we are aware, there are not objective moral values. If we were aware of them, we would say something about them. In this sense this sort of subjectivism entails moral scepticism. But the converse entailment does not hold. The denial that there are objective values does not commit one to any particular view about what moral statements mean, and certainly not to the view that they are equivalent to subjective reports. No doubt if moral values are not objective they are in some very broad sense subjective, and for this reason I would accept 'moral subjectivism' as an alternative name to 'moral scepticism'. But subjectivism in this broad sense must be distinguished from the specific doctrine about meaning referred to above. Neither name is altogether satisfactory: we simply have to guard against the (different) misinterpretations which each may suggest.

just b/c you deny perspective mean value doesn't automatically mean you can give an answer about on values

J. L. Mackie

3. The multiplicity of second order questions

The distinctions drawn in the last two sections rest not only on the well-known and generally recognized difference between first and second order questions, but also on the more controversial claim that there are several kinds of second order moral questions. Those most often mentioned are questions about the meaning and use of ethical terms, or the analysis of ethical concepts. With these go questions about the logic of moral statements: there may be special patterns of moral argument, licensed, perhaps, by aspects of the meanings of moral terms— for example, it may be part of the meaning of moral statements that they are universalizable. But there are also ontological, as contrasted with linguistic or conceptual, questions about the nature and status of goodness or rightness or whatever it is that first order moral statements are distinctively about. These are questions of factual rather than conceptual analysis: the problems of what goodness is cannot be settled conclusively or exhaustively by finding out what the word 'good' means, or what it is conventionally used to say or to do.

Recent philosophy, biased as it has been towards various kinds of linguistic inquiry, has tended to doubt this, but the distinction between conceptual and factual analysis in ethics can be supported by analogies with other areas. The question of what perception is, what goes on when someone perceives something, is not adequately answered by finding out what words like 'see' and 'hear' mean, or what someone is doing in saying 'I perceive . . . ', by analysing, however fully and accurately, any established concept of perception. There is a still closer analogy with colours. Robert Boyle and John Locke called colours 'secondary qualities', meaning that colours as they occur in material things consist simply in patterns of arrangement and movement of minute particles on the surfaces of objects, which make them, as we would now say, reflect light of some frequencies better than others, and so enable these objects to produce colour sensations in us, but that colours as we see them do not literally belong to the surfaces of material things. Whether Boyle and Locke were right about this cannot be settled by finding out how we use colour words and what we mean in using them. Naïve realism about colours might be a correct analysis not only of our pre-scientific colour concepts but also of the conventional meanings of colour words, and even of the meanings with which scientifically sophisticated people use them when they are off their guard, and yet it might not be a correct account of the status of colours.

Error could well result, then, from a failure to distinguish factual from conceptual analysis with regard to colours, from taking an account of the meanings of statements as a full account of what there is. There is a similar and in practice even greater risk of error in moral philosophy. There is another reason, too, why it would be a mistake to concentrate second order ethical discussions on questions of meaning. The more work philosophers have done on meaning, both in ethics and elsewhere, the more complications have come to light. It is by now pretty plain

[98]

that no simple account of the meanings of first order moral statements will be correct, will cover adequately even the standard, conventional, senses of the main moral terms; I think, none the less, that there is a relatively clear-cut issue about the objectivity of moral values which is in danger of being lost among the complications of meaning.

4. Is objectivity a real issue?

It has, however, been doubted whether there is a real issue here. I must concede that it is a rather old-fashioned one. I do not mean merely that it was raised by Hume, who argued that 'The vice entirely escapes you . . . till you turn your reflexion into your own breast,' and before him by Hobbes, and long before that by some of the Greek sophists. I mean rather that it was discussed vigorously in the nineteen thirties and forties, but since then has received much less attention. This is not because it has been solved or because agreement has been reached: instead it seems to have been politely shelved.

But was there ever a genuine problem? R. M. Hare has said that he does not understand what is meant by 'the objectivity of values', and that he has not met anyone who does. We all know how to recognize the activity called 'saying, thinking it to be so, that some act is wrong', and he thinks that it is to this activity that the subjectivist and the objectivist are both alluding, though one calls it 'an attitude of disapproval' and the other 'a moral intuition': these are only different names for the same thing. It is true that if one person says that a certain act is wrong and another that it is not wrong the objectivist will say that they are contradicting one another; but this yields no significant discrimination between objectivism and subjectivism, because the subjectivist too will concede that the second person is negating what the first has said, and Hare sees no difference between contradicting and negating. Again, the objectivist will say that one of the two must be wrong; but Hare argues that to say that the judgement that a certain act is wrong is itself wrong is merely to negate that judgement, and the subjectivist too must negate one or other of the two judgements, so that still no clear difference between objectivism and subjectivism has emerged. He sums up his case thus: 'Think of one world into whose fabric values are objectively built; and think of another in which those values have been annihilated. And remember that in both worlds the people in them go on being concerned about the same things—there is no difference in the "subjective" concern which people have for things, only in their "objective" value. Now I ask, "What is the difference between the states of affairs in these two worlds?" Can any answer be given except "None whatever"?'

Now it is quite true that it is logically possible that the subjective concern, the activity of valuing or of thinking things wrong, should go on in just the same way whether there are objective values or not. But to say this is only to reiterate that

[99]

there is a logical distinction between first and second order ethics: first order judgements are not necessarily affected by the truth or falsity of a second order view. But it does not follow, and it is not true, that there is no difference whatever between these two worlds. In the one there is something that backs up and validates some of the subjective concern which people have for things, in the other there is not. Hare's argument is similar to the positivist claim that there is no difference between a phenomenalist or Berkeleian world in which there are only minds and their ideas and the commonsense realist one in which there are also material things, because it is logically possible that people should have the same experiences in both. If we reject the positivism that would make the dispute between realists and phenomenalists a pseudo-question, we can reject Hare's similarly supported dismissal of the issue of the objectivity of values.

In any case, Hare has minimized the difference between his two worlds by considering only the situation where people already have just such subjective concern; further differences come to light if we consider how subjective concern is acquired or changed. If there were something in the fabric of the world that validated certain kinds of concern, then it would be possible to acquire these merely by finding something out, by letting one's thinking be controlled by how things were. But in the world in which objective values have been annihilated the acquiring of some new subjective concern means the development of something new on the emotive side by the person who acquires it, something that eighteenth-century writers would put under the head of passion or sentiment.

The issue of the objectivity of values needs, however, to be distinguished from others with which it might be confused. To say that there are objective values would not be to say merely that there are some things which are valued by everyone, nor does it entail this. There could be agreement in valuing even if valuing is just something that people do, even if this activity is not further validated. Subjective agreement would give intersubjective values, but intersubjectivity is not objectivity. Nor is objectivity simply universalizability: someone might well be prepared to universalize his prescriptive judgements or approvals— that is, to prescribe and approve in just the same ways in all relevantly similar cases, even ones in which he was involved differently or not at all—and yet he could recognize that such prescribing and approving were his activities, nothing more. Of course if there were objective values they would presumably belong to *kinds* of things or actions or states of affairs, so that the judgements that reported them would be universalizable; but the converse does not hold.

A more subtle distinction needs to be made between objectivism and descriptivism. Descriptivism is again a doctrine about the meanings of ethical terms and statements, namely that their meanings are purely descriptive rather than even partly prescriptive or emotive or evaluative, or that it is not an essential feature of the conventional meaning of moral statements that they have some special illocutionary force, say of commending rather than asserting. It contrasts with the view that commendation is in principle distinguishable from description (however

descriptivism, meanings of ethical terms are purely descriptive rather than prescriptive, emotive or evaluative values have been thought of as being both prescriptive + objective

difficult they may be to separate in practice) and that moral statements have it as at least part of their meaning that they are commendatory and hence in some uses intrinsically action-guiding. But descriptive meaning neither entails nor is entailed by objectivity. Berkeley's subjective idealism about material objects would be quite compatible with the admission that material object statements have purely descriptive meaning. Conversely, the main tradition of European moral philosophy from Plato onwards has combined the view that moral values are objective with the recognition that moral judgements are partly prescriptive or directive or action-guiding. Values themselves have been seen as at once prescriptive and objective. In Plato's theory the Forms, and in particular the Form of the Good, are eternal, extra-mental, realities. They are a very central structural element in the fabric of the world. But it is held also that just knowing them or 'seeing' them will not merely tell men what to do but will ensure that they do it, overruling any contrary inclinations. The philosopher-kings in the *Republic* can, Plato thinks, be trusted with unchecked power because their education will have given them knowledge of the Forms. Being acquainted with the Forms of the Good and Justice and Beauty and the rest they will, by this knowledge alone, without any further motivation, be impelled to pursue and promote these ideals. Similarly, Kant believes that pure reason can by itself be practical, though he does not pretend to be able to explain how it can be so. Again, Sidgwick argues that if there is to be a science of ethics—and he assumes that there can be, indeed he defines ethics as 'the science of conduct'—what ought to be 'must in another sense have objective existence: it must be an object of knowledge and as such the same for all minds'; but he says that the affirmations of this science 'are also precepts', and he speaks of happiness as 'an end *absolutely* prescribed by reason'. Since many philosophers have thus held that values are objectively prescriptive, it is clear that the ontological doctrine of objectivism must be distinguished from descriptivism, a theory about meaning.

other generation like Plato thinked the Form of the God as being an external primary quality

But perhaps when Hare says that he does not understand what is meant by 'the objectivity of values' he means that he cannot understand how values could be objective, he cannot frame for himself any clear, detailed, picture of what it would be like for values to be part of the fabric of the world. This would be a much more plausible claim; as we have seen, even Kant hints at a similar difficulty. Indeed, even Plato warns us that it is only through difficult studies spread over many years that one can approach the knowledge of the Forms. The difficulty of seeing how values could be objective is a fairly strong reason for thinking that they are not so; this point will be taken up in Section 9 but it is not a good reason for saying that this is not a real issue.

I believe that as well as being a real issue it is an important one. It clearly matters for general philosophy. It would make a radical difference to our metaphysics if we had to find room for objective values—perhaps something like Plato's Forms—somewhere in our picture of the world. It would similarly make a difference to our epistemology if it had to explain how such objective values are or

mackie believes it would definitely alter things if values were to be objective

[101]

can be known, and to our philosophical psychology if we had to allow such knowledge, or Kant's pure practical reason, to direct choices and actions. Less obviously, how this issue is settled will affect the possibility of certain kinds of moral argument. For example, Sidgwick considers a discussion between an egoist and a utilitarian, and points out that if the egoist claims that his happiness or pleasure is objectively desirable or good, the utilitarian can argue that the egoist's happiness 'cannot be more objectively desirable or more a good than the similar happiness of any other person: the mere fact . . . that *he is he* can have nothing to do with its objective desirability or goodness'. In other words, if ethics is built on the concept of objective goodness, then egoism as a first order system or method of ethics can be refuted, whereas if it is assumed that goodness is only subjective it cannot. But Sidgwick correctly stresses what a number of other philosophers have missed, that this argument against egoism would require the objectivity specifically of goodness: the objectivity of what ought to be or of what it is rational to do would not be enough. If the egoist claimed that it was objectively rational, or obligatory upon him, to seek his own happiness, a similar argument about the irrelevance of the fact that he is he would lead only to the conclusion that it was objectively rational or obligatory for each other person to seek *his* own happiness, that is, to a universalized form of egoism, not to the refutation of egoism. And of course insisting on the universalizability of moral judgements, as opposed to the objectivity of goodness, would yield only the same result.

5. Standards of evaluation

One way of stating the thesis that there are no objective values is to say that value statements cannot be either true or false. But this formulation, too, lends itself to misinterpretation. For there are certain kinds of value statements which undoubtedly can be true or false, even if, in the sense I intend, there are no objective values. Evaluations of many sorts are commonly made in relation to agreed and assumed standards. The classing of wool, the grading of apples, the awarding of prizes at sheepdog trials, flower shows, skating and diving championships, and even the marking of examination papers are carried out in relation to standards of quality or merit which are peculiar to each particular subject-matter or type of contest, which may be explicitly laid down but which, even if they are nowhere explicitly stated, are fairly well understood and agreed by those who are recognized as judges or experts in each particular field. Given any sufficiently determinate standards, it will be an objective issue, a matter of truth and falsehood, how well any particular specimen measures up to those standards. Comparative judgements in particular will be capable of truth and falsehood: it will be a factual question whether this sheepdog has performed better than that one.

The subjectivist about values, then, is not denying that there can be objective evaluations relative to standards, and these are as possible in the aesthetic and

[handwritten: Justice + injustice can be objective i.e. True or false]

moral fields as in any of those just mentioned. More than this, there is an objective distinction which applies in many such fields, and yet would itself be regarded as a peculiarly moral one: the distinction between justice and injustice. In one important sense of the word it is a paradigm case of injustice if a court declares someone to be guilty of an offence of which it knows him to be innocent. More generally, a finding is unjust if it is at variance with what the relevant law and the facts together require, and particularly if it is known by the court to be so. More generally still, any award of marks, prizes, or the like is unjust if it is at variance with the agreed standards for the contest in question: if one diver's performance in fact measures up better to the accepted standards for diving than another's, it will be unjust if the latter is awarded higher marks or the prize. In this way the justice or injustice of decisions relative to standards can be a thoroughly objective matter, though there may still be a subjective element in the interpretation or application of standards. But the statement that a certain decision is thus just or unjust will not be objectively prescriptive: in so far as it can be simply true it leaves open the question whether there is any objective requirement to do what is just and to refrain from what is unjust, and equally leaves open the practical decision to act in either way.

Recognizing the objectivity of justice in relation to standards, and of evaluative judgements relative to standards, then, merely shifts the question of the objectivity of values back to the standards themselves. The subjectivist may try to make his point by insisting that there is no objective validity about the choice of standards. Yet he would clearly be wrong if he said that the choice of even the most basic standards in any field was completely arbitrary. The standards used in sheepdog trials clearly bear some relation to the work that sheepdogs are kept to do, the standards for grading apples bear some relation to what people generally want in or like about apples, and so on. On the other hand, standards are not as a rule strictly validated by such purposes. The appropriateness of standards is neither fully determinate nor totally indeterminate in relation to independently specifiable aims or desires. But however determinate it is, the objective appropriateness of standards in relation to aims or desires is no more of a threat to the denial of objective values than is the objectivity of evaluation relative to standards. In fact it is logically no different from the objectivity of goodness relative to desires. Something may be called good simply in so far as it satisfies or is such as to satisfy a certain desire; but the objectivity of such relations of satisfaction does not consitute in our sense an objective value.

[handwritten margin notes: relation then of values to standards — Subj? No obj validity about the choice of standards]

6. Hypothetical and categorical imperatives

We may make this issue clearer by referring to Kant's distinction between hypothetical and categorical imperatives, though what he called imperatives are more naturally expressed as 'ought'-statements than in the imperative mood. 'If you want X, do Y' (or 'You ought to do Y') will be a hypothetical imperative if it is

[103]

based on the supposed fact that Y is, in the circumstances, the only (or the best) available means to X, that is, on a causal relation between Y and X. The reason for doing Y lies in its causal connection with the desired end, X; the oughtness is contingent upon the desire. But 'You ought to do Y' will be a categorical imperative if you ought to do Y irrespective of any such desire for any end to which Y would contribute, if the oughtness is not thus contingent upon any desire. But this distinction needs to be handled with some care. An 'ought'-statement is not in this sense hypothetical merely because it incorporates a conditional clause. 'If you promised to do Y, you ought to do Y' is not a hypothetical imperative merely on account of the stated if-clause; what is meant may be either a hypothetical or a categorical imperative, depending upon the implied reason for keeping the supposed promise. If this rests upon some such further unstated conditional as 'If you want to be trusted another time', then it is a hypothetical imperative; if not, it is categorical. Even a desire of the agent's can figure in the antecedent of what, though conditional in grammatical form, is still in Kant's sense of a categorical imperative. 'If you are strongly attracted sexually to young children you ought not to go in for school teaching' is not, in virtue of what it explicitly says, a hypothetical imperative: the avoidance of school teaching is not being offered as a means to the satisfaction of the desires in question. Of course, it could still be a hypothetical imperative, if the implied reason were a prudential one; but it could also be a categorical imperative, a moral requirement where the reason for the recommended action (strictly, avoidance) does not rest upon that action's being a means to the satisfaction of any desire that the agent is supposed to have. Not every conditional ought-statement or command, then, is a hypothetical imperative; equally, not every non-conditional one is a categorical imperative. An appropriate if-clause may be left unstated. Indeed, a simple command in the imperative mood, say a parade-ground order, which might seem most literally to qualify for the title of a categorical imperative, will hardly ever be one in the sense we need here. The implied reason for complying with such an order will almost always be some desire of the person addressed, perhaps simply the desire to keep out of trouble. If so, such an apparently categorical order will be in our sense a hypothetical imperative. Again, an imperative remains hypothetical even if we change the 'if' to 'since': the fact that the desire for X is actually present does not alter the fact that the reason for doing Y is contingent upon the desire for X by way of Y's being a means to X. In Kant's own treatment, while imperatives of skill relate to desires which an agent may or may not have, imperatives of prudence relate to the desire for happiness which, Kant assumes, everyone has. So construed, imperatives of prudence are no less hypothetical than imperatives of skill, no less contingent upon desires that the agent has at the time the imperatives are addressed to him. But if we think rather of a counsel of prudence as being related to the agent's future welfare, to the satisfaction of desires that he does not yet have—not even to a present desire that his future desires should be satisfied— then a counsel of prudence is a categorical imperative, different indeed from a moral one, but analogous to it.

A categorical imperative, then, would express a reason for acting which was unconditional in the sense of not being contingent upon any present desire of the agent to whose satisfaction the recommended action would contribute as a means—or more directly: 'You ought to dance', if the implied reason is just that you want to dance or like dancing, is still a hypothetical imperative. Now Kant himself held that moral judgements are categorical imperatives, or perhaps are all applications of one categorical imperative, and it can plausibly be maintained at least that many moral judgements contain a categorically imperative element. So far as ethics is concerned, my thesis that there are no objective values is specifically the denial that any such categorically imperative element is objectively valid. The objective values which I am denying would be action-directing absolutely, not contingently (in the way indicated) upon the agent's desires and inclinations.

Another way of trying to clarify this issue is to refer to moral reasoning or moral arguments. In practice, of course, such reasoning is seldom fully explicit: but let us suppose that we could make explicit the reasoning that supports some evaluative conclusion, where this conclusion has some action-guiding force that is not contingent upon desires or purposes or chosen ends. Then what I am saying is that somewhere in the input to this argument—perhaps in one or more of the premisses, perhaps in some part of the form of the argument—there will be something which cannot be objectively validated—some premiss which is not capable of being simply true, or some form of argument which is not valid as a matter of general logic, whose authority or cogency is not objective, but is constituted by our choosing or deciding to think in a certain way.

7. The claim to objectivity

If I have succeeded in specifying precisely enough the moral values whose objectivity I am denying, my thesis may now seem to be trivially true. Of course, some will say, valuing, preferring, choosing, recommending, rejecting, condemning, and so on, are human activities, and there is no need to look for values that are prior to and logically independent of all such activities. There may be widespread agreement in valuing, and particular value-judgements are not in general arbitrary or isolated: they typically cohere with others, or can be criticized if they do not, reasons can be given for them, and so on: but if all that the subjectivist is maintaining is that desires, ends, purposes, and the like figure somewhere in the system of reasons, and that no ends or purposes are objective as opposed to being merely intersubjective, then this may be conceded without much fuss.

But I do not think that this should be conceded so easily. As I have said, the main tradition of European moral philosophy includes the contrary claim, that there are objective values of just the sort I have denied. I have referred already to Plato, Kant, and Sidgwick. Kant in particular holds that the categorical imperative

[105]

is not only categorical and imperative but objectively so: though a rational being gives the moral law to himself, the law that he thus makes is determinate and necessary. Aristotle begins the *Nicomachean Ethics* by saying that the good is that at which all things aim, and that ethics is part of a science which he calls 'politics', whose goal is not knowledge but practice; yet he does not doubt that there can be *knowledge* of what is the good for man, nor, once he has identified this as well-being or happiness, *eudaimonia,* that it can be known, rationally determined, in what happiness consists; and it is plain that he thinks that this happiness is intrinsically desirable, not good simply because it is desired. The rationalist Samuel Clarke holds that

> these eternal and necessary differences of things make it *fit and reasonable* for creatures so to act . . . even separate from the consideration of these rules being the *positive will* or *command of God;* and also antecedent to any respect or regard, expectation or apprehension, of any *particular private and personal advantage or disadvantage, reward or punishment,* either present or future . . .

Even the sentimentalist Hutcheson defines moral goodness as 'some quality apprehended in actions, which procures approbation . . . ', while saying that the moral sense by which we perceive virtue and vice has been given to us (by the Author of nature) to direct our actions. Hume indeed was on the other side, but he is still a witness to the dominance of the objectivist tradition, since he claims that when we 'see that the distinction of vice and virtue is not founded merely on the relations of objects, nor is perceiv'd by reason', this 'wou'd subvert all the vulgar systems of morality'. And Richard Price insists that right and wrong are 'real characters of actions', not 'qualities of our minds', and are perceived by the understanding; he criticizes the notion of moral sense on the ground that it would make virtue an affair of taste, and moral right and wrong 'nothing in the objects themselves'; he rejects Hutcheson's view because (perhaps mistakenly) he sees it as collapsing into Hume's.

But this objectivism about values is not only a feature of the philosophical tradition. It has also a firm basis in ordinary thought, and even in the meanings of moral terms. No doubt it was an extravagance for Moore to say that 'good' is the name of a non-natural quality, but it would not be so far wrong to say that in moral contexts it is used as if it were the name of a supposed non-natural quality, where the description 'non-natural' leaves room for the peculiar evaluative, prescriptive, intrinsically action-guiding aspects of this supposed quality. This point can be illustrated by reflection on the conflicts and swings of opinion in recent years between non-cognitivist and naturalist views about the central, basic, meanings of ethical terms. If we reject the view that it is the function of such terms to introduce objective values into discourse about conduct and choices of action, there seem to be two main alternative types of account. One (which has importantly different subdivisions) is that they conventionally express either attitudes which the speaker purports to adopt towards whatever it is that he characterizes morally, or prescriptions or recommendations, subject perhaps to the logical constraint of universalizability. Different views of this type share the central thesis that ethical terms have, at least partly and primarily, some sort of non-cognitive, non-descriptive, meaning. Views of the other type hold that they are descriptive in meaning, but descriptive of natural features, partly of such features as everyone, even the non-cognitivist, would recognize as distinguishing kind actions from cruel ones, courage from cowardice, politeness from rudeness, and so on, and partly (though these two overlap) of relations between the actions and some human wants, satisfactions, and the like. I believe that views of both these types capture part of the truth. Each approach can account for the fact that moral judgements are action-guiding or practical. Yet each gains much of its plausibility from the felt inadequacy of the other. It is a very natural reaction to any non-cognitive analysis of ethical terms to protest that there is more to ethics than this, something more external to the maker of moral judgements, more authoritative over both him and those of or to whom he speaks, and this reaction is likely to persist even when full allowance has been made for the logical, formal, constraints of full-blooded prescriptivity and universalizability. Ethics, we are inclined to believe, is more a matter of knowledge and less a matter of decision than any non-cognitive analysis allows. And of course naturalism satisfies this demand. It will not be a matter of choice or decision whether an action is cruel or unjust or imprudent or whether it is likely to produce more distress than pleasure. But in satisfying this demand, it introduces a converse deficiency. On a naturalist analysis, moral judgements can be practical, but their practicality is wholly relative to desires or possible satisfactions of the person or persons whose actions are to be guided; but moral judgements seem to say more than this. This view leaves out the categorical quality of moral requirements. In fact both naturalist and non-cognitive analyses leave out the apparent authority of ethics, the one by excluding the categorically imperative aspect, the other the claim to objective validity or truth. The ordinary user of moral language means to say something about whatever it is that he characterizes morally, for example a possible action, as it is in itself, or would be if it were realized, and not about, or even simply expressive of, his, or anyone else's, attitude or relation to it. But the something he wants to say is not purely descriptive, certainly not inert, but something that involves a call for action or for the refraining from action, and one that is absolute, not contingent upon any desire or preference or policy or choice, his own or anyone else's. Someone in a state of moral perplexity, wondering whether it would be wrong for him to engage, say, in research related to bacteriological warfare, wants to arrive at some judgement about this concrete case, his doing this work at this time in these actual circumstances; his relevant characteristics will be part of the subject of the judgement, but no relation between him and the proposed action will be part of the predicate. The question is not, for example, whether he really wants to do this work, whether it will satisfy or dissatisfy him, whether he will in the long run have a pro-attitude towards it, or even whether this is an action of a sort that he can happily and sincerely recommend in all relevantly similar cases. Nor is he even wondering just whether to recommend such action in all relevantly similar cases. He wants to know whether this course of action would be wrong in itself. Something like this is the everyday objectivist concept of which talk about non-natural qualities is a philosopher's reconstruction.

The prevalence of this tendency to objectify values—and not only moral ones—is confirmed by a pattern of thinking that we find in existentialists and those influenced by them. The denial of objective values can carry with it an extreme emotional reaction, a feeling that nothing matters at all, that life has lost its purpose. Of course this does not follow; the lack of objective values is not a good reason for abandoning subjective concern or for ceasing to want anything. But the abandonment of a belief in objective values can cause, at least temporarily, a decay of subjective concern and sense of purpose. That it does so is evidence that the people in whom this reaction occurs have been tending to objectify their concerns and purposes, have been giving them a fictitious external authority. A claim to objectivity has been so strongly associated with their subjective concerns and purposes that the collapse of the former seems to undermine the latter as well.

This view, that conceptual analysis would reveal a claim to objectivity, is sometimes dramatically confirmed by philosophers who are officially on the other side. Bertrand Russell, for example, says that 'ethical propositions should be expressed in the optative mood, not in the indicative'; he defends himself effectively against the charge of inconsistency in both holding ultimate ethical valuations to be subjective and expressing emphatic opinions on ethical questions. Yet at the end he admits:

> Certainly there *seems* to be something more. Suppose, for example, that some one were to advocate the introduction of bull-fighting in this country. In opposing the proposal, I should *feel,* not only that I was expressing my desires, but that my desires in the matter are *right,* whatever that may mean. As a matter of argument, I can, I think, show that I am not guilty of any logical inconsistency in holding to the above interpretation of ethics and at the same time expressing strong ethical preferences. But in feeling I am not satisfied.

But he concludes, reasonably enough, with the remark: 'I can only say that, while my own opinions as to ethics do not satisfy me, other people's satisfy me still less.'

I conclude, then, that ordinary moral judgements include a claim to objectivity, an assumption that there are objective values in just the sense in which I am concerned to deny this. And I do not think it is going too far to say that this assumption has been incorporated in the basic, conventional, meanings of moral terms. Any analysis of the meanings of moral terms which omits this claim to objective, intrinsic, prescriptivity is to that extent incomplete; and this is true of any non-cognitive analysis, any naturalist one, and any combination of the two.

If second order ethics were confined, then, to linguistic and conceptual analysis, it ought to conclude that moral values at least are objective: that they are so is part of what our ordinary moral statements mean: the traditional moral concepts of the ordinary man as well as of the main line of western philosophers are concepts of objective value. But it is precisely for this reason that linguistic and conceptual analysis is not enough. The claim to objectivity, however ingrained in our language and thought, is not self-validating. It can and should be questioned. But the denial of objective values will have to be put forward not as the result of an analytic approach, but as an 'error theory', a theory that although most people in making moral judgements implicitly claim, among other things, to be pointing to something objectively prescriptive, these claims are all false. It is this that makes the name 'moral scepticism' appropriate.

But since this is an error theory, since it goes against assumptions ingrained in our thought and built into some of the ways in which language is used, since it conflicts with what is sometimes called common sense, it needs very solid support. It is not something we can accept lightly or casually and then quietly pass on. If we are to adopt this view, we must argue explicitly for it. Traditionally it has been supported by arguments of two main kinds, which I shall call the argument from relativity and the argument from queerness, but these can, as I shall show, be supplemented in several ways.

8. The argument from relativity

The argument from relativity has as its premiss the well-known variation in moral codes from one society to another and from one period to another, and also the differences in moral beliefs between different groups and classes within a complex community. Such variation is in itself merely a truth of descriptive morality, a fact of anthropology which entails neither first order nor second order ethical views. Yet it may indirectly support second order subjectivism: radical differences between first order moral judgements make it difficult to treat those judgements as apprehensions of objective truths. But it is not the mere occurrence of disagreements that tells against the objectivity of values. Disagreement on questions in history or biology or cosmology does not show that there are no objective issues in these fields for investigators to disagree about. But such scientific disagreement results from speculative inferences or explanatory hypotheses based on inadequate evidence, and it is hardly plausible to interpret moral disagreement in the same way. Disagreement about moral codes seems to reflect people's adherence to and participation in different ways of life. The causal connection seems to be mainly that way round: it is that people approve of monogamy because they participate in a monogamous way of life rather than that they participate in a monogamous way of life because they approve of monogamy. Of course, the standards may be an idealization of the way of life from which they arise: the monogamy in which people participate may be less complete, less rigid, than that of which it leads them to approve. This is not to say that moral judgements are purely conventional. Of course there have been and are moral heretics and moral reformers, people who have turned against the established rules and practices of their own communities for moral reasons, and often for moral reasons that we would endorse. But this can usually be understood as the extension, in ways which, though new and unconventional, seemed to them to be required for inconsistency, of rules to which they already adhered as arising out of an existing way of life. In short, the argument from relativity has some force simply because the actual variations in the moral codes are more readily explained by the hypothesis that they reflect ways of life than by the hypothesis that they express perceptions, most of them seriously inadequate and badly distorted, of objective values.

But there is a well-known counter to this argument from relativity, namely to say that the items for which objective validity is in the first place to be claimed are not specific moral rules or codes but very general basic principles which are recognized at least implicitly to some extent in all society—such principles as provide the foundations of what Sidgwick has called different methods of ethics: the principle of universalizability, perhaps, or the rule that one ought to conform to the specific rules of any way of life in which one takes part, from which one profits, and on which one relies, or some utilitarian principle of doing what tends, or seems likely, to promote the general happiness. It is easy to show that such general principles, married with differing concrete circumstances, different existing social patterns or different preferences, will beget different specific moral rules; and there is some plausibility in the claim that the specific rules thus generated will vary from community to community or from group to group in close agreement with the actual variations in accepted codes.

The argument from relativity can be only partly countered in this way. To take this line the moral objectivist has to say that it is only in these principles that the objective moral character attaches immediately to its descriptively specified ground or subject: other moral judgments are objectively valid or true, but only derivatively and contingently—if things had been otherwise, quite different sorts of actions would have been right. And despite the prominence in recent philosophical ethics of universalization, utilitarian principles, and the like, these are very far from constituting the whole of what is actually affirmed as basic in ordinary moral thought. Much of this is concerned rather with what Hare calls 'ideals' or, less kindly, 'fanaticism'. That is, people judge that some things are good or right, and others are bad or wrong, not because—or at any rate not only because—they exemplify some general principle for which widespread implicit acceptance could be claimed, but because something about those things arouses certain responses immediately in them, though they would arouse radically and irresolvably different responses in others. 'Moral sense' or 'intuition' is an initially more plausible description of what supplies many of our basic moral judgements than 'reason'. With regard to all these starting points of moral thinking the argument from relativity remains in full force.

9. The argument from queerness

Even more important, however, and certainly more generally applicable, is the argument from queerness. This has two parts, one metaphysical, the other epistemological. If there were objective values, then they would be entities or qualities or relations of a very strange sort, utterly different from anything else in the universe. Correspondingly, if we were aware of them, it would have to be by some special faculty of moral perception or intuition, utterly different from our ordinary ways of knowing everything else. These points were recognized by Moore when he spoke of non-natural qualities, and by the intuitionists in their talk about a 'faculty of moral intuition'. Intuitionism has long been out of favour, and it is indeed easy to point out its implausibilities. What is not so often stressed, but is more important, is that the central thesis of intuitionism is one to which any objectivist view of values is in the end committed: intuitionism merely makes unpalatably plain what other forms of objectivism wrap up. Of course the suggestion that moral judgements are made or moral problems solved by just sitting down and having an ethical intuition is a travesty of actual moral thinking. But, however complex the real process, it will require (if it is to yield authoritatively prescriptive conclusions) some input of this distinctive sort, either premises or forms of argument or both. When we ask the awkward question, how we can be aware of this authoritative prescriptivity, of the truth of these distinctively ethical premises or of the cogency of this distinctively ethical pattern of reasoning, none of our ordinary accounts of sensory perception or introspection or the framing and confirming of explanatory hypotheses or inference or logical construction or conceptual analysis, or any combination of these, will provide a satisfactory answer; 'a special sort of intuition' is a lame answer, but it is the one to which the clear-headed objectivist is compelled to resort.

Indeed, the best move for the moral objectivist is not to evade this issue, but to look for companions in guilt. For example, Richard Price argues that it is not moral knowledge alone that such an empiricism as those of Locke and Hume is unable to account for, but also our knowledge and even our ideas of essence, number, identity, diversity, solidity, inertia, substance, the necessary existence and infinite extension of time and space, necessity and possibility in general, power, and causation. If the understanding, which Price defines as the faculty within us that discerns truth, is also a source of new simple ideas of so many other sorts, may it not also be a power of immediately perceiving right and wrong, which yet are real characters of actions?

This is an important counter to the argument from queerness. The only adequate reply to it would be to show how, on empiricist foundations, we can construct an account of the ideas and beliefs and knowledge that we have of all these matters. I cannot even begin to do that here, though I have undertaken some parts of the task elsewhere. I can only state my belief that satisfactory accounts of most of these can be given in empirical terms. If some supposed metaphysical necessities or essences resist such treatment, then they too should be included, along with objective values, among the targets of the argument from queerness.

This queerness does not consist simply in the fact that ethical statements are 'unverifiable'. Although logical positivism with its verifiability theory of descriptive meaning gave an impetus to non-cognitive accounts of ethics, it is not only logical positivists but also empiricists of a much more liberal sort who should find objective values hard to accommodate. Indeed, I would not only reject the verifiability principle but also deny the conclusion commonly drawn from it, that moral judgements lack descriptive meaning. The assertion that there are objective values or intrinsically prescriptive entities or features of some kind, which ordinary moral judgements presuppose, is, I hold, not meaningless but false.

Plato's Forms give a dramatic picture of what objective values would have to be. The Form of the Good is such that knowledge of it provides the knower with both a direction and an overriding motive, something's being good both tells the person who knows this to pursue it and makes him pursue it. An objective good would be sought by anyone who was acquainted with it, not because of any contingent fact that this person, or every person, is so constituted that he desires this end, but just because the end has to-be-pursuedness somehow built into it. Similarly, if there were objective principles of right and wrong, any wrong (possible) course of action would have not-to-be-doneness somehow built into it. Or we should have something like Clarke's necessary relations of fitness between situations and actions, so that a situation would have a demand for such-and-such an action somehow built into it.

The need for an argument of this sort can be brought out by reflection on Hume's argument that 'reason'—in which at this stage he includes all sorts of knowing as well as reasoning—can never be an 'influencing motive of the will'. Someone might object that Hume has argued unfairly from the lack of influencing power (not contingent upon desires) in ordinary objects of knowledge and ordinary reasoning, and might maintain that values differ from natural objects precisely in their power, when known, automatically to influence the will. To this Hume could, and would need to, reply that this objection involves the postulating of value-entities or value-features of quite a different order from anything else with which we are acquainted, and of a corresponding faculty with which to detect them. That is, he would have to supplement his explicit argument with what I have called the argument from queerness.

Another way of bringing out this queerness is to ask, about anything that is supposed to have some objective moral quality, how this is linked with its natural features. What is the connection between the natural fact that an action is a piece of deliberate cruelty—say, causing pain just for fun—and the moral fact that it is wrong? It cannot be an entailment, a logical or semantic necessity. Yet it is not merely that the two features occur together. The wrongness must somehow be 'consequential' or 'supervenient'; it is wrong because it is a piece of deliberate cruelty. But just what *in the world* is signified by this 'because'? And how do we know the relation that it signifies, if this is something more than such actions being socially condemned, and condemned by us too, perhaps through our having absorbed attitudes from our social environment? It is not even sufficient to postulate a faculty which 'sees' the wrongness: something must be postulated which can see at once the natural features that constitute the cruelty, and the wrongness, and the mysterious consequential link between the two. Alternatively, the intuition required might be the perception that wrongness is a higher order property belonging to certain natural properties; but what is this belonging of properties to other properties, and how can we discern it? How much simpler and more comprehensible the situation would be if we could replace the moral quality with some sort of subjective response which could be causally related to the detection of the natural features on which the supposed quality is said to be consequential.

It may be thought that the argument from queerness is given an unfair start if we thus relate it to what are admittedly among the wilder products of philosophical fancy—Platonic Forms, non-natural qualities, self-evident relations of fitness, faculties of intuition, and the like. Is it equally forceful if applied to the terms in which everyday moral judgements are more likely to be expressed—though still, as has been argued in Section 7, with a claim to objectivity—'you must do this', 'you can't do that', 'obligation', 'unjust', 'rotten', 'disgraceful', 'mean', or talk about good reasons for or against possible actions? Admittedly not; but that is because the objective prescriptivity, the element a claim for whose authoritativeness is embedded in ordinary moral thought and language, is not yet isolated in these forms of speech, but is presented along with relations to desires and feelings, reasoning about the means to desired ends, interpersonal demands, the injustice which consists in the violation of what are in the context the accepted standards of merit, the psychological constituents of meanness, and so on. There is nothing queer about any of these, and under cover of them the claim for moral authority may pass unnoticed. But if I am right in arguing that it is ordinarily there, and is therefore very likely to be incorporated almost automatically in philosophical accounts of ethics which systematize our ordinary thought even in such apparently innocent terms as these, it needs to be examined, and for this purpose it needs to be isolated and exposed as it is by the less cautious philosophical reconstructions.

10. Patterns of objectification

Considerations of these kinds suggest that it is in the end less paradoxical to reject than to retain the common-sense belief in the objectivity of moral values, provided that we can explain how this belief, if it is false, has become established and is so resistant to criticisms. This proviso is not difficult to satisfy.

On a subjectivist view, the supposedly objective values will be based in fact upon attitudes which the person has who takes himself to be recognizing and responding to those values. If we admit what Hume calls the mind's 'propensity to spread itself on external objects', we can understand the supposed objectivity of moral qualities as arising from what we can call the projection or objectification of moral attitudes. This would be analogous to what is called the 'pathetic fallacy', the tendency to read our feelings into their objects. If a fungus, say, fills us with disgust, we may be inclined to ascribe to the fungus itself a non-natural quality of foulness. But in moral contexts there is more than this propensity at work. Moral attitudes themselves are at least partly social in origin: socially established—and socially necessary—patterns of behaviour put pressure on individuals, and each individual tends to internalize these pressures and to join in requiring these patterns of behaviour of himself and of others. The attitudes that are objectified into moral values have indeed an external source, though not the one assigned to them by the belief in their absolute authority. Moreover, there are motives that would support objectification. We need morality to regulate interpersonal relations, to control some of the ways in which people behave towards one another, often in opposition to contrary inclinations. We therefore want our moral judgements to be authoritative for other agents as well as for ourselves: objective validity would give them the authority required. Aesthetic values are logically in the same position as moral ones; much the same metaphysical and epistemological considerations apply to them. But aesthetic values are less strongly objectified than moral ones; their subjective status, and an 'error theory' with regard to such claims to objectivity as are incorporated in aesthetic judgements, will be more readily accepted, just because the motives for their objectification are less compelling.

But it would be misleading to think of the objectification of moral values as primarily the projection of feelings, as in the pathetic fallacy. More important are wants and demands. As Hobbes says, 'whatsoever is the object of any man's Appetite or Desire, that is it, which he for his part calleth *Good*'; and certainly both the adjective 'good' and the noun 'goods' are used in non-moral contexts of things because they are such as to satisfy desires. We get the notion of something's being objectively good, or having intrinsic value, by reversing the direction of dependence here, by making the desire depend upon the goodness, instead of the goodness on the desire. And this is aided by the fact that the desired thing will indeed have features that make it desired, that enable it to arouse a desire or that make it such as to satisfy some desire that is already there. It is fairly easy to confuse the way in which a thing's desirability is indeed objective with its having in our sense objective value. The fact that the word 'good' serves as one of our main moral terms is a trace of this pattern of objectification.

Similarly related uses of words are covered by the distinction between hypothetical and categorical imperatives. The statement that someone 'ought to' or, more strongly, 'must' do such-and-such may be backed up explicitly or implicitly by reference to what he wants or to what his purposes and objects are. Again, there may be a reference to the purposes of someone else, perhaps the speaker: 'You must do this'—'Why?'—'Because I want such-and-such'. The moral categorical imperative which could be expressed in the same words can be seen as resulting from the suppression of the conditional clause in a hypothetical imperative without its being replaced by any such reference to the speaker's wants. The action in question is still required in something like the way in which it would be if it were appropriately related to a want, but it is no longer admitted that there is any contingent want upon which its being required depends. Again this move can be understood when we remember that at least our central and basic moral judgements represent social demands, where the source of the demand is indeterminate and diffuse. Whose demands or wants are in question, the agent's, or the speaker's, or those of an indefinite multitude of other people? All of these in a way, but there are advantages in not specifying them precisely. The speaker is expressing demands which he makes as a member of a community, which he has developed in and by participation in a joint way of life; also, what is required of this particular agent would be required of any other in a relevantly similar situation; but the agent too is expected to have internalized the relevant demands, to act as if the ends for which the action is required were his own. By suppressing any explicit reference to demands and making the imperatives categorical we facilitate conceptual moves from one such demand relation to another. The moral uses of such words as 'must' and 'ought' and 'should', all of which are used also to express hypothetical imperatives, are traces of this pattern of objectification.

It may be objected that this explanation links normative ethics too closely with descriptive morality, with the mores or socially enforced patterns of behaviour that anthropologists record. But it can hardly be denied that moral thinking starts from the enforcement of social codes. Of course it is not confined to that. But even when moral judgements are detached from the mores of any actual society they are liable to be framed with reference to an ideal community of moral agents, such as Kant's kingdom of ends, which but for the need to give God a special place in it would have been better called a commonwealth of ends.

Another way of explaining the objectification of moral values is to say that ethics is a system of law from which the legislator has been removed. This might have been derived either from the positive law of a state or from a supposed system of divine law. There can be no doubt that some features of modern European moral concepts are traceable to the theological ethics of Christianity. The stress on quasi-imperative notions, on what ought to be done or on what is wrong in a sense that is close to that of 'forbidden', are surely relics of divine commands. Admittedly, the central ethical concepts for Plato and Aristotle also are in a broad sense prescriptive or intrinsically action-guiding, but in concentrating rather on 'good' than on 'ought' they show that their moral thought is an objectification of the desired and the satisfying rather than of the commanded. Elizabeth Anscombe has argued that modern, non-Aristotelian, concepts of *moral* obligation, *moral* duty, of what is *morally* right and wrong, and of the *moral* sense of 'ought' are survivals outside the framework of thought that made them really intelligible, namely the belief in divine law. She infers that 'ought' has 'become a word of mere mesmeric force', with only a 'delusive appearance of content', and that we would do better to discard such terms and concepts altogether, and go back to Aristotelian ones.

There is much to be said for this view. But while we can explain some distinctive features of modern moral philosophy in this way, it would be a mistake to see the whole problem of the claim to objective prescriptivity as merely local and unnecessary, as a post-operative complication of a society from which a dominant system of theistic belief has recently been rather hastily excised. As Cudworth and Clarke and Price, for example, show, even those who still admit divine commands, or the positive law of God, may believe moral values to have an independent objective but still action-guiding authority. Responding to Plato's *Euthyphro* dilemma, they believe that God commands what he commands because it is in itself good or right, not that it is good or right merely because and in that he commands it. Otherwise God himself could not be called good. Price asks, 'What can be more preposterous, than to make the Deity nothing but will; and to exalt this on the ruins of all his attributes?' The apparent objectivity of moral value is a widespread phenomenon which has more than one source: the persistence of a belief in something like divine law when the belief in the divine legislator has faded out is only one factor among others. There are several different patterns of objectification, all of which have left characteristic traces in our actual moral concepts and moral language.

11. The general goal of human life

The argument of the preceding sections is meant to apply quite generally to moral thought, but the terms in which it has been stated are largely those of the Kantian and post-Kantian tradition of English moral philosophy. To those who are more familiar with another tradition, which runs through Aristotle and Aquinas, it may seem wide of the mark. For them, the fundamental notion is that of the good for man, or the general end or goal of human life, or perhaps of a set of basic goods or primary human purposes. Moral reasoning consists partly in achieving a more adequate understanding of this basic goal (or set of goals), partly in working out the best way of pursuing and realizing it. But this approach is open to two radically different interpretations. According to one, to say that something is the good for man or the general goal of human life is just to say that this is what men in fact pursue or will find ultimately satisfying, or perhaps that it is something which, if postulated as an implicit goal, enables us to make sense of actual human strivings and to detect a coherent pattern in what would otherwise seem to be a chaotic jumble of conflicting purposes. According to the other interpretation, to say that something is the good for man or the general goal of human life is to say that this is man's proper end, that this is what he ought to be striving after, whether he in fact is or not. On the first interpretation we have a descriptive statement, on the second a normative or evaluative or prescriptive one. But this approach tends to combine the two interpretations, or to slide from one to the other, and to borrow support for what are in effect claims of the second sort from the plausibility of statements of the first sort.

I have no quarrel with this notion interpreted in the first way. I would only insert a warning that there may well be more diversity even of fundamental purposes, more variation in what different human beings will find ultimately satisfying, than the terminology of '*the* good for man' would suggest. Nor indeed, have I any quarrel with the second, prescriptive, interpretation, provided that it is recognized as subjectively prescriptive, that the speaker is here putting forward his own demands or proposals, or those of some movement that he represents, though no doubt linking these demands or proposals with what he takes to be already in the first, descriptive, sense fundamental human goals. . . . But if it is claimed that something is objectively the right or proper goal of human life, then this is tantamount to the assertion of something that is objectively categorically imperative, and comes fairly within the scope of our previous arguments. Indeed, the running together of what I have here called the two interpretations is yet another pattern of objectification: a claim to objective prescriptivity is constructed by combining the normative element in the second interpretation with the objectivity allowed by the first, by the statement that such and such are fundamentally pursued or ultimately satisfying human goals. The argument from relativity still applies: the radical diversity of the goals that men actually pursue and find satisfying makes it implausible to construe such pursuits as resulting from an imperfect grasp of a unitary true good. So too does the argument from queerness; we can still ask what this objectively prescriptive rightness of the true goal can be, and how this is linked on the one hand with the descriptive features of this goal and on the other with the fact that it is *to some extent* an actual goal of human striving.

To meet these difficulties, the objectivist may have recourse to the purpose of God: the true purpose of human life is fixed by what God intended (or, intends) men to do and to be. Actual human strivings and satisfactions have some relation to this true end because God made men for this end and made them such as to pursue it—but only *some* relation, because of the inevitable imperfection of created beings.

I concede that if the requisite theological doctrine could be defended, a kind of objective ethical prescriptivity could be thus introduced. Since I think that theism cannot be defended, I do not regard this as any threat to my argument. . . .

12. Conclusion

I have maintained that there is a real issue about the status of values, including moral values. Moral scepticism, the denial of objective moral values, is not to be confused with any one of several first order normative views, or with any linguistic or conceptual analysis. Indeed, ordinary moral judgements involve a claim to objectivity which both non-cognitive and naturalist analyses fail to capture. Moral scepticism must, therefore, take the form of an error theory, admitting that a belief in objective values is built into ordinary moral thought and language, but holding that this ingrained belief is false. As such, it needs arguments to support it against 'common sense'. But solid arguments can be found. The considerations that favour moral scepticism are: first, the relativity or variability of some important starting points of moral thinking and their apparent dependence on actual ways of life; secondly, the metaphysical peculiarity of the supported objective values, in that they would have to be intrinsically action-guiding and motivating; thirdly, the problem of how such values could be consequential or supervenient upon natural features; fourthly, the corresponding epistemological difficulty of accounting for our knowledge of value entities or features and of their links with the features on which they would be consequential; fifthly, the possibility of explaining, in terms of several different patterns of objectification, traces of which remain in moral language and moral concepts, how even if there were no such objective values people not only might have come to suppose that there are but also might persist firmly in that belief. These five points sum up the case for moral scepticism; but of almost equal importance are the preliminary removal of misunderstandings that often prevent this thesis from being considered fairly and explicitly, and the isolation of those items about which the moral sceptic is sceptical from many associated qualities and relations whose objective status is not in dispute.

6

Ethics and Observation

Gilbert Harman

1. The basic issue

Can moral principles be tested and confirmed in the way scientific principles can? Consider the principle that, if you are given a choice between five people alive and one dead or five people dead and one alive, you should always choose to have five people alive and one dead rather than the other way round. We can easily imagine examples that appear to confirm this principle. Here is one:

> You are a doctor in a hospital's emergency room when six accident victims are brought in. All six are in danger of dying but one is much worse off than the others. You can just barely save that person if you devote all of your resources to him and let the others die. Alternatively, you can save the other five if you are willing to ignore the most seriously injured person.

It would seem that in this case you, the doctor, would be right to save the five and let the other person die. So this example, taken by itself, confirms the principle under consideration. Next, consider the following case.

> You have five patients in the hospital who are dying, each in need of a separate organ. One needs a kidney, another a lung, a third a heart, and so forth. You can save all five if you take a single healthy person and remove his heart, lungs, kidneys, and so forth, to distribute to these five patients. Just such a healthy person is in room 306. He is in the hospital for routine tests. Having seen his test results, you know that he is perfectly healthy and of the right tissue compatibility. If you do nothing, he will survive without incident; the other patients will die, however. The other five patients can be saved only if the person in Room 306 is cut up and his organs distributed. In that case, there would be one dead but five saved.

This essay was originally published as chapter 1 of Gilbert Harman, *The Nature of Morality* (New York: Oxford University Press), pp. 3–10.

The principle in question tells us that you should cut up the patient in Room 306. But in this case, surely you must not sacrifice this innocent bystander, even to save the five other patients. Here a moral principle has been tested and disconfirmed in what may seem to be a surprising way.

This, of course, was a "thought experiment." We did not really compare a hypothesis with the world. We compared an explicit principle with our feelings about certain imagined examples. In the same way, a physicist performs thought experiments in order to compare explicit hypotheses with his "sense" of what should happen in certain situations, a "sense" that he has acquired as a result of his long working familiarity with current theory. But scientific hypotheses can also be tested in real experiments, out in the world.

Can moral principles be tested in the same way, out in the world? You can observe someone do something, but can you ever perceive the rightness or wrongness of what he does? If you round a corner and see a group of young hoodlums pour gasoline on a cat and ignite it, you do not need to *conclude* that what they are doing is wrong; you do not need to figure anything out; you can *see* that it is wrong. But is your reaction due to the actual wrongness of what you see or is it simply a reflection of your moral "sense," a "sense" that you have acquired perhaps as a result of your moral upbringing?

2. Observation

The issue is complicated. There are no pure observations. Observations are always "theory laden." What you perceive depends to some extent on the theory you hold, consciously or unconsciously. You see some children pour gasoline on a cat and ignite it. To really see that, you have to possess a great deal of knowledge, know about a considerable number of objects, know about people: that people pass through the life stages infant, baby, child, adolescent, adult. You must know what flesh and blood animals are, and in particular, cats. You must have some idea of life. You must know what gasoline is, what burning is, and much more. In one sense, what you "see" is a pattern of light on your retina, a shifting array of splotches, although even that is theory, and you could never adequately describe what you see in that sense. In another sense, you see what you do because of the theories you hold. Change those theories and you would see something else, given the same pattern of light.

Similarly, if you hold a moral view, whether it is held consciously or unconsciously, you will be able to perceive rightness or wrongness, goodness or badness, justice or injustice. There is no difference in this respect between moral propositions and other theoretical propositions. If there is a difference, it must be found elsewhere.

Observation depends on theory because observation involves forming a belief as a fairly direct result of perceiving something; you can form a belief only if you understand the relevant concepts and a concept is what it is by virtue of its role in some theory or system of beliefs. To recognize a child as a child is to employ, consciously or unconsciously, a concept that is defined by its place in a framework of the stages of human life. Similarly burning is an empty concept apart from its theoretical connections to the concepts of heat, destruction, smoke, and fire.

Moral concepts—Right and Wrong, Good and Bad, Justice and Injustice— also have a place in your theory or system of beliefs and are the concepts they are because of their context. If we say that observation has occurred whenever an opinion is a direct result of perception, we must allow that there is moral observation, because such an opinion can be a moral opinion as easily as any other sort. In this sense, observation may be used to confirm or disconfirm moral theories. The observational opinions that, in this sense, you find yourself with can be in either agreement or conflict with your consciously explicit moral principles. When they are in conflict, you must choose between your explicit theory and observation. In ethics, as in science, you sometimes opt for theory, and say that you made an error in observation or were biased or whatever, or you sometimes opt for observation, and modify your theory.

In other words, in both science and ethics, general principles are invoked to explain particular cases and, therefore, in both science and ethics, the general principles you accept can be tested by appealing to particular judgments that certain things are right or wrong, just or unjust, and so forth; and these judgments are analogous to direct perceptual judgments about facts.

3. Observational evidence

Nevertheless, observation plays a role in science that it does not seem to play in ethics. The difference is that you need to make assumptions about certain physical facts to explain the occurrence of the observations that support a scientific theory, but you do not seem to need to make assumptions about any moral facts to explain the occurrence of the so-called moral observations I have been talking about. In the moral case, it would seem that you need only make assumptions about the psychology or moral sensibility of the person making the moral observation. In the scientific case, theory is tested against the world.

The point is subtle but important. Consider a physicist making an observation to test a scientific theory. Seeing a vapor trail in a cloud chamber, he thinks, "There goes a proton." Let us suppose that this is an observation in the relevant sense, namely, an immediate judgment made in response to the situation without any conscious reasoning having taken place. Let us also suppose that his observation confirms his theory, a theory that helps give meaning to the very term "proton" as it occurs in his observational judgment. Such a confirmation rests on inferring an explanation. He can count his making the observation as confirming evidence for his theory only to the extent that it is reasonable to explain his making the observation by assuming that, not only is he in a certain psychological "set," given the theory he accepts and his beliefs about the experimental apparatus, but furthermore, there really was a proton going through the cloud chamber, causing the vapor trail, which he saw as a proton. (This is evidence for the theory to the extent that the theory can explain the proton's being there better than competing theories can.) But, if his having made that observation could have been equally well explained by his psychological set alone, without the need for any assumption about a proton, then the observation would not have been evidence for the existence of that proton and therefore would not have been evidence for the theory. His making the observation supports the theory only because, in order to explain his making the observation, it is reasonable to assume something about the world over and above the assumptions made about the observer's psychology. In particular, it is reasonable to assume that there was a proton going through the cloud chamber, causing the vapor trail.

Compare this case with one in which you make a moral judgment immediately and without conscious reasoning, say, that the children are wrong to set the cat on fire or that the doctor would be wrong to cut up one healthy patient to save five dying patients. In order to explain your making the first of these judgments, it would be reasonable to assume, perhaps, that the children really are pouring gasoline on a cat and you are seeing them do it. But, in neither case is there any obvious reason to assume anything about "moral facts," such as that it really is wrong to set the cat on fire or to cut up the patient in Room 306. Indeed, an assumption about moral facts would seem to be totally irrelevant to the explanation of your making the judgment you make. It would seem that all we need assume is that you have certain more or less well articulated moral principles that are reflected in the judgments you make, based on your moral sensibility. It seems to be completely irrelevant to our explanation whether your intuitive immediate judgment is true or false.

The observation of an event can provide observational evidence for or against a scientific theory in the sense that the truth of that observation can be relevant to a reasonable explanation of why that observation was made. A moral observation does not seem, in the same sense, to be observational evidence for or against any moral theory, since the truth or falsity of the moral observation seems to be completely irrelevant to any reasonable explanation of why that observation was made. The fact that an observation of an event was made at the time it was made is evidence not only about the observer but also about the physical facts. The fact that you made a particular moral observation when you did does not seem to be evidence about moral facts, only evidence about you and your moral sensibility. Facts about protons can affect what you observe, since a proton passing through the cloud chamber can cause a vapor trail that reflects light to your eye in a way that, given your scientific training and psychological set, leads you to judge that what you see is a proton. But there does not seem to be any way in which the actual rightness or wrongness of a given situation can have any effect on your perceptual apparatus. In this respect, ethics seems to differ from science.

In considering whether moral principles can help explain observations, it is therefore important to note an ambiguity in the word "observation." You see the children set the cat on fire and immediately think, "That's wrong." In one sense, your observation is that what the children are doing is wrong. In another sense, your observation is your thinking that thought. Moral principles might explain observations in the first sense but not in the second sense. Certain moral principles might help to explain why it was *wrong* of the children to set the cat on fire, but moral principles seem to be of no help in explaining *your thinking* that that is wrong. In the first sense of "observation," moral principles can be tested by observation—"That this act is wrong is evidence that causing unnecessary suffering is wrong." But in the second sense of "observation," moral principles cannot clearly be tested by observation, since they do not appear to help explain observations in this second sense of "observation." Moral principles do not seem to help explain your observing what you observe.

Of course, if you are already given the moral principle that it is wrong to cause unnecessary suffering, you can take your seeing the children setting the cat on fire as observational evidence that they are doing something wrong. Similarly, you can suppose that your seeing the vapor trail is observational evidence that a proton is going through the cloud chamber, if you are given the relevant physical theory. But there is an important apparent difference between the two cases. In the scientific case, your making that observation is itself evidence for the physical theory because the physical theory explains the proton, which explains the trail, which explains your observation. In the moral case, your making your observation does not seem to be evidence for the relevant moral principle because that principle does not seem to help explain your observation. The explanatory chain from principle to observation seems to be broken in morality. The moral principle may "explain" why it is wrong for the children to set the cat on fire. But the wrongness of that act does not appear to help explain the act, which you observe, itself. The explanatory chain appears to be broken in such a way that neither the moral principle nor the wrongness of the act can help explain why you observe what you observe.

A qualification may seem to be needed here. Perhaps the children perversely set the cat on fire simply "because it is wrong." Here it may seem at first that the actual wrongness of the act does help explain why they do it and therefore indirectly helps explain why you observe what you observe just as a physical theory, by explaining why the proton is producing a vapor trail, indirectly helps explain why the observer observes what he observes. But on reflection we must agree that this is probably an illusion. What explains the children's act is not clearly the actual wrongness of the act but, rather, their belief that the act is wrong. The actual rightness or wrongness of their act seems to have nothing to do with why they do it.

Observational evidence plays a part in science it does not appear to play in ethics, because scientific principles can be justified ultimately by their role in explaining observations, in the second sense of observation—by their explanatory role. Apparently, moral principles cannot be justified in the same way. It appears to be true that there can be no explanatory chain between moral principles and particular observings in the way that there can be such a chain between scientific principles and particular observings. Conceived as an explanatory theory, morality, unlike science, seems to be cut off from observation.

Not that every legitimate scientific hypothesis is susceptible to direct observational testing. Certain hypotheses about "black holes" in space cannot be directly tested, for example, because no signal is emitted from within a black hole. The connection with observation in such a case is indirect. And there are many similar examples. Nevertheless, seen in the large, there is the apparent difference between science and ethics we have noted. The scientific realm is accessible to observation in a way the moral realm is not.

4. Ethics and mathematics

Perhaps ethics is to be compared, not with physics, but with mathematics. Perhaps such a moral principle as "You ought to keep your promises" is confirmed or disconfirmed in the way (whatever it is) in which such a mathematical principle as "$5 + 7 = 12$" is. Observation does not seem to play the role in mathematics it plays in physics. We do not and cannot perceive numbers, for example, since we cannot be in causal contact with them. We do not even understand what it would be like to be in causal contact with the number 12, say. Relations among numbers cannot have any more of an effect on our perceptual apparatus than moral facts can.

Observation, however, *is* relevant to mathematics. In explaining the observations that support a physical theory, scientists typically appeal to mathematical principles. On the other hand, one never seems to need to appeal in this way to moral principles. Since an observation is evidence for what best explains it, and since mathematics often figures in the explanations of scientific observations, there is indirect observational evidence for mathematics. There does not seem to be observational evidence, even indirectly, for basic moral principles. In explaining why certain observations have been made, we never seem to use purely moral assumptions. In this respect, then, ethics appears to differ not only from physics but also from mathematics.

II

MORAL
REALISM

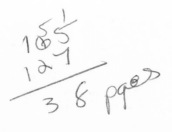

7

Truth, Invention, and the Meaning of Life

David Wiggins

Nul n'est besoin d'espérer pour entreprendre, ni de réussir pour persévérer.

William the Silent

Eternal survival after death completely fails to accomplish the purpose for which it has always been intended. Or is some riddle solved by my surviving for ever? Is not this eternal life as much of a riddle as our present life?

Wittgenstein

1. Even now, in an age not much given to mysticism, there are people who ask 'What is the meaning of life?' Not a few of them make the simple 'unphilosophical' assumption that there is something to be known here. (One might say that they are 'cognitivists' with regard to this sort of question.) And most of these same people make the equally unguarded assumption that the whole issue of life's meaning presupposes some positive answer to the question whether it can be plainly and straight-forwardly *true* that this or that thing or activity or pursuit is good, has value, or is worth something. And then, what is even harder, they suppose that questions like that of life's meaning must be among the central questions of moral philosophy.

The question of life's having a meaning and the question of truth are not at the centre of moral philosophy as we now have it. The second is normally settled by something bordering on stipulation,[1] and the first is under suspicion of belonging in the same class as 'What is the greatest good of the greatest number?' or 'What is the will?' or 'What holds the world up?' This is the class of questions not in good order, or best not answered just as they stand.

A lecture delivered on November 24, 1976. First published in *Proceedings of the British Academy* LXII (1976), and reprinted here, with some editorial changes, by permission of the British Academy. The version given here supersedes that reprinted as Essay III in *Needs, Values, Truth* (Oxford, 1987).

1. Cp. Essay IV, Wiggins, *Needs, Values, Truth*, pp. 139–184. In 1976, at the time of speaking, the remark stood in less need of qualification than it does now.

David Wiggins

If there is a semantical crux about this sort of occurrence of the word 'meaning', then all logical priority attaches to it; and no reasonable person could pretend that a perfectly straight-forward purport attaches to the idea of life's meaning something. But logical priority is not everything; and, most notably, the order of logical priority is not always or necessarily the same as the order of discovery. Someone who was very perplexed or very persistent would be well within his rights to insist that, where a question has been asked as often as this one has, a philosopher must make what he can of it: and that, if the sense really is obscure, then he must find what significance the effort to frame an answer is apt to *force* upon the question.

In what follows, I try to explore the possibility that the question of truth and the question of life's meaning are among the most fundamental questions of moral philosophy. The outcome of the attempt may perhaps indicate that, unless we want to continue to think of moral philosophy as the casuistry of emergencies, these questions and the other questions that they bring to our attention are a better focus for ethics and meta-ethics than the textbook problem 'What [under this or that or the other circumstance] shall I do?' My finding will be that the question of life's meaning does, as the untheoretical suppose, lead into the question of truth— and conversely. Towards the end I shall also claim to uncover the possibility that philosophy has put happiness in the place that should have been occupied in moral philosophy by meaning. This is a purely theoretical claim, but if it is correct, it is not without consequences; and if (as some say) weariness and dissatisfaction have issued from the direct pursuit of happiness as such, then it is not without all explanatory power.

2. I have spoken in favour of the direct approach, but it is impossible to reach out to the perplexity for which the question of meaning is felt to stand without first recording the sense that, during relatively recent times, there has been some shift in the way the question of life's meaning is seen, and in the kind of answer it is felt to require. Here is an answer made almost exactly two hundred years ago, two years before the death of Voltaire:

> We live in this world to compel ourselves industriously to enlighten one another by means of reasoning and to apply ourselves always to carrying forward the sciences and the arts. (W. A. Mozart to Padre Martini: letter of 4 December 1776.)[2]

What we envy here is the specificity and the certainty of purpose. But, even as we feel envy, it is likely that we want to rejoice in our freedom to disbelieve in that which provided the contingent foundation of the specificity and certainty. I make this remark, not because I think that we ought to believe in what Mozart and Padre

---

2. Compare the composer's choice of expression on the occasion of his father's birthday anniversary in 1777: 'I wish you as many years as are needed to have nothing left to do in music.'

Martini believed in, but in outright opposition to the hope that some relatively painless accommodation can be made between the freedom and the certainty. The foundation of what we envy was the now (I think) almost unattainable conviction that there exists a God whose purpose ordains certain specific duties for all men, and appoints particular men to particular roles or vocations.

That conviction was not only fallible: there are many who would say that it was positively dangerous—and that the risk it carried was that, if the conviction were false, then one might prove to have thrown one's life away. It is true that in the cases we are considering, 'throwing one's life away' seems utterly the wrong thing to say of the risk carried by the conviction. It seems wrong even for the aspects of these men's lives that were intimately conditioned by the belief in God. But if one doubts that God exists, then it is one form of the problem of meaning to justify not wanting to speak here of throwing a life away. It is a terrible thing to try to live a life without believing in *anything*. But surely that doesn't mean that just—*you need some kind of focus or direction to hold on to* any old set of concerns and beliefs will do, provided one *could* live a life by them. Surely if any old set would do, that is the same as life's being meaningless.

If we envy the certainty of the 1776 answer, then most likely this is only one of several differences that we see between our own situation and the situation of those who lived before the point at which Darwin's theory of evolution so confined the scope of the religious imagination. History has not yet carried us to the point where it is impossible for a description of such differences to count as exaggerated. But they are formidable. And, for the sake of the clarity of what is to come, I must pause to express open dissent from two comments that might be made about them.

First, someone more interested in theory than in what it was like to be alive then and what it is like now may try to diminish the differences that we sense, by arguing from the accessibility to both eighteenth and twentieth centuries of a core notion of God, a notion that he may say persists in the concept of God championed by modern theologians. To this use of their ideas I object that, whatever gap it is which lies between 1776 and 1976, such notions as *God as the ground of our being* cannot bridge it. For recourse to these exemplifies a tendency towards an *a priori* conception of God which, even if the eighteenth century had had it, most of the men of that age would have hastened to amplify with a more hazardous *a posteriori* conception. Faith in God conceived *a posteriori* was precisely the cost of the particularity and definiteness of the certainty that we envy.

The other thing someone might say is that, in one crucial respect, our situation is not different from a late Enlightenment situation, because there is a conceptually determined need in which the eighteenth century stood and in which we stand equally. This, it might be said, is the need for commitment. In the eighteenth-century case, this extra thing was commitment to submission to God's purpose. We shall come in §4 to what these theorists think it is in our case. Faced however with this second comment, one might well wonder how someone could get to the point of recognizing or even suspecting that it was God's purpose that he should be a composer (say) and yet be indifferent to that. Surely no extra anything, over and above some suspicion that this or that is God's purpose, is required to create the concern we should expect to find that the suspicion itself would have implanted in him. On the other hand, if this extra thing were supplied, then it would bring too much. For the commitment to submission seems to exclude rebellion; and rebellion against what is taken as God's purpose has never been excluded by the religious attitude as such.

What then are the similarities and the differences between the eighteenth-century orientation and our own orientation upon the meaning of life? It seems that the similarities that persist will hold between the conceptual scheme with which they in that century confronted the world of everyday experience and the scheme with which we, in spite of our thoroughgoing acceptance of natural science, confront it: and the dissimilarities will relate to the specificity and particularity of the focus of the various concerns in which their world-view involved them and our world-view involves us. For us there is less specificity and much less focus.

If this is still a dark statement, it is surely not so dark as to obscure the relationship between this difference between them and us and a cognate difference that will have signalled its presence and importance so soon as I prepared to approach the divide between the eighteenth and twentieth centuries by reference to the purposive or practical certainty of individual human agents. Unless we are Marxists, we are much more resistant in the second half of the twentieth century than the eighteenth or nineteenth century knew how to be against attempts to locate the meaning of human life or human history in mystical or metaphysical conceptions—in the emancipation of mankind, or progress, or the onward advance of Absolute Spirit. It is not that we have lost interest in emancipation or progress themselves. But, whether temporarily or permanently, we have more or less abandoned the idea that the importance of emancipation or progress (or a correct conception of spiritual advance) is that these are marks by which our minute speck in the universe can distinguish itself as the spiritual focus of the cosmos. Perhaps that is what makes the question of the meaning we can find in life so difficult and so desolate for us.

With these bare and inadequate historical assertions, however, the time is come to go straight to a modern philosophical account of the matter. There are not very many to choose from.

3. The account I have taken is that given in Chapter 18 of Richard Taylor's book *Good and Evil*—an account rightly singled out for praise by the analytical philosopher who reviewed the book for the *Philosophical Review*.[3]

Taylor's approach to the question whether life has any meaning is first to 'bring

3. See Richard Taylor, *Good and Evil* (New York: Macmillan, 1970). The review was by Judith Jarvis Thomson, *Philosophical Review* vol. 81, 1973, p. 113.

to our minds a clear image of meaningless existence', and then determine what would need to be inserted into the meaningless existence so depicted in order to make it not meaningless. Taylor writes:

> A perfect image of meaninglessness of the kind we are seeking is found in the ancient myth of Sisyphus. Sisyphus, it will be remembered, betrayed divine secrets to mortals, and for this he was condemned by the gods to roll a stone to the top of the hill, the stone then immediately to roll back down, again to be pushed to the top by Sisyphus, to roll down once more, and so on again and again, *forever*.

Two ways are then mentioned in which this meaninglessness could be alleviated or removed. First: — *reason for removal*

> . . . if we supposed that these stones . . . were assembled [by Sisyphus] at the top of the hill . . . in a beautiful and enduring temple, then . . . his labours would have a point, something would come of them all . . . — *give his labour a point & beautiful temple being built*

That is one way. But Taylor is not in the end disposed to place much reliance in this species of meaning, being more impressed by a second mode of enrichment.

> Suppose that the gods, as an afterthought, waxed perversely merciful by implanting in [Sisyphus] a strange and irrational impulse . . . to roll stones. . . . To make this more graphic, suppose they accomplish this by implanting in him some substance that has this effect on his character and drives. . . . This little afterthought of the gods . . . was . . . merciful. For they have by this device managed to give Sisyphus precisely what he wants—by making him want precisely what they inflict on him. However it may appear to us, Sisyphus' . . . life is now filled with mission and meaning, and he seems to himself to have been given an entry to heaven. . . . The *only* thing that has happened is this: Sisyphus has been reconciled to [his existence]. . . . He has been led to embrace it. Not, however, by reason or persuasion, but by nothing more rational than the potency of a new substance in his veins. . . .

So much for meaninglessness, and two ways of alleviating it. Meaninglessness, Taylor says,

> is essentially endless pointlessness, and meaningfulness is therefore the opposite. Activity, and even long drawn out and repetitive activity, has a meaning if it has some significant culmination, some more or less lasting end that can be considered to have been the direction and purpose of the activity.

That is the temple-building option, of course.

> But the descriptions so far also provide something else; namely, the suggestion of how an existence that is objectively meaningless, in this sense, can nevertheless acquire a meaning for him whose existence it is.

This 'something else' is the option of implanting in Sisyphus the impulse to push what he has to push. Here Taylor turns aside to compare, in point of meaningless-ness or meaningfulness, the condition of Sisyphus and the lives of various animals, working from the lower to the higher animals—cannibalistic blind-worms, the cicada, migratory birds, and so on up to ourselves. His verdict is that the point of any living thing's life is evidently nothing but life itself.

> This life of the world thus presents itself to our eyes as a vast machine, feeding on itself, running on and on forever to nothing. And we are part of that life. To be sure, we are not just the same, but the differences are not so great as we like to think; many are merely invented and none really cancels meaninglessness. . . . We are conscious of our activity. Our goals, whether in any significant sense we choose them or not, are things of which we are at least partly aware and can . . . appraise. . . . Men have a history, as other animals do not. [Still] . . . if we think that, unlike Sisyphus', [our] labours do have a point, that they culminate in something lasting and, independently of our own deep interests in them, very worthwhile, then we simply have not considered the thing closely enough. . . . For [Sisyphus' temple] to make any difference it had to be a temple that would at least endure, adding beauty to the world for the remainder of time. Our achievements . . . , those that do last, like the sand-swept pyramids, soon become mere curiosities, while around them the rest of mankind continues its perpetual toting of rocks, only to see them roll down. . . .

Here is a point that obsesses the author. Paragraph upon paragraph is devoted to describing the lamentable but undoubted impermanence (futility *sub specie aeter-nitatis*) of the architectural or built monuments of human labour. It is not entirely clear that the same effect could have been contrived if the gradual accumulation of scientific understanding or the multiplication of the sublime utterances of litera-ture or music had been brought into the argument. What is clear is that Taylor is committed to a strong preference for the second method of enriching Sisyphus' life—that is the compulsion caused by the substance put into Sisyphus' veins. For as for the first method, and temple-building for the sake of the temple,

> Suppose . . . that after ages of dreadful toil, all directed at this final result [Sisyphus] did at last complete his temple, [so] that now he could say his work was done, and he could rest and forever enjoy the result. Now what? What picture now presents itself to our minds? It is precisely the picture of infinite boredom! Of Sisyphus doing nothing ever again, but contemplating what he has already wrought and can no longer add anything to, and contemplating it for eternity! Now in this picture we have a meaning for Sisyphus' existence, a point for his prodigious labour, because we have put it there; yet, at the same time, that which is really worthwhile seems to have slipped away entirely.

The final reckoning would appear to be this: (a) a lasting end or *telos* could constitute a purpose for the work; but (b) there is no permanence; and (c), even if there were such permanence, its point would be effectively negated by boredom with the outcome of the work. And so we are thrown inexorably into the arms of the other and second sort of meaning.

> We can reintroduce what has been resolutely pushed aside in an effort to view our lives and human existence with objectivity; namely, our own wills, our deep interest in what we find ourselves doing. . . . Even the glow worms . . . whose cycles of existence over the millions of years seem so pointless when looked at by us, will seem utterly different to us if we can somehow try to view their existence from within. . . . If the philosopher is apt to see in this a pattern similar to the unending cycles of the existence of Sisyphus, and to despair, then it is indeed because the meaning and point he is seeking is not there—but mercifully so. The meaning of life is from within us, it is not bestowed from without, and it far exceeds in its beauty and permanence any heaven of which men have ever dreamed or yearned for.

4. Connoisseurs of twentieth-century ethical theory in its Anglo-Saxon and Continental variants will not be slow to see the affinities of this account. Practitioners of the first of these kinds are sometimes singled out for their failure to say anything about such questions as the meaning of life. But, if the affinities are as strong as I think, then, notwithstanding Taylor's philosophical distance from his contemporaries, what we have just unearthed has a strong claim to be their secret doctrine of the meaning of life.

Consider first the sharp supposedly unproblematic distinction, reinforced by the myth as told and retold here, between what we discover already there in the world—the facts, including the gods' enforcement of their sentence—and what is invented or, by thinking or willing, somehow *put into* or *spread onto* the factual world—namely the values.[4] Nobody who knows the philosophical literature on value will be surprised by Taylor's variant on the myth. . . . Here, however, at the point where the magic stuff is to be injected into the veins of Sisyphus, I must digress for the sake of what is to come, in order to explain the deliberate way in which I shall use the word 'value'.

I propose that we distinguish between *valuations* (typically recorded by such forms as '*x* is good', 'bad', 'beautiful', 'ugly', 'ignoble', 'brave', 'just', 'mischievous', 'malicious', 'worthy', 'honest', 'corrupt', 'disgusting', 'amusing', 'diverting', 'boring', *etc.*—no restrictions at all on the category of *x*) and *directive* or *deliberative* (or *practical*) *judgements* (e.g. 'I must ψ', 'I ought to ψ', 'it would be best, all things considered, for me to ψ', *etc.*).[5] It is true that between these there is an important no-man's-land (comprising, e.g., general judgments of the strongly deprecatory or commendatory kind about vices and virtues, and

4. On the differences between discovery and invention, and on some abuses of the distinction, see William Kneale, 'The Idea of Invention', *Proceedings of the British Academy* vol. 39, 1955.

5. Note that this is not a distinction whose rationale is originally *founded* in a difference in the motivating force of judgments of the two classes, even if such a difference may be forthcoming from the distinction. In both cases, the thinking that *p* is arguably derivative from the *finding* that *p*. (Cp. Essay V, §11, Wiggins, *Needs, Values, Truth*, pp. 202–204.)

David Wiggins

general or particular statements about actions that it is ignoble or inhuman or unspeakably wicked to do or not to do).[6] But the fact that many other kinds of judgment lie between pure valuations and pure directives is no objection; and it does nothing to obstruct the discrimination I shall seek to effect in due course between the fact-value distinction and the is-ought or is-must distinction. The unavailability of any well-grounded notion of the factual that will make the fact-value distinction an exclusive distinction can only promote our interest in the possibility of there being some *oughts* or *musts* that will not count as a case of an *is*. If we then conceive of a distinction between *is* and *must* as corresponding to the distinction between appreciation and decision and at the same time emancipate ourselves from a limited and absurd idea of what *is,* then there can be a new verisimilitude in our several accounts of all these things.[7]

Having proposed a usage of the word 'value' to be adhered to throughout this paper, I return now to Sisyphus and the body of doctrine that is illustrated by Taylor's version of his story. At one moment, we are told, Sisyphus sees his task as utterly futile and degrading: a moment later, supposedly without any initiating change in his cognitive appreciation, he sees his whole life as infinitely rewarding. Before the digression I was going to say that there is only one philosophy of value that can even attempt to accommodate this possibility.

Consider next Taylor's account of the escape from meaninglessness—or what he might equally well have followed the Existentialists in calling *absurdity.* Taylor's mode of escape is simply a variation on the habitual philosophical reaction to the perception of the real or supposed meaninglessness of human existence. As a method for escape it is co-ordinate with every other proposal that is known, suicide (always one recognized way), scorn or defiance (Albert Camus), resignation or drift (certain orientally influenced positions), various kinds of commitment (R. M. Hare and J.-P. Sartre), and what may be the most recently enlisted member of this équipe, which is irony.[8]

Again, few readers of *Freedom and Reason* will fail to recognize in Sisyphus, after the injection of the gods' substance into his veins, a Mark I, stone-rolling model of R. M. Hare's further elaborated, rationally impregnable 'fanatic'.[9] As for the mysterious substance itself, surely this is some extra oomph, injected afterwards *ad libitum,* that will enable Sisyphus' factual judgments about stone-rolling to take on 'evaluative meaning'.

6. For some purposes, judgments that philosophers describe as judgments of *prima facie* obligation (better *pro tanto* obligation) might almost or without excessive distortion be assimilated to valuational judgments.

7. See below, §6 and §10. In the language of note 20, ad fin, my own view is that the fact-value distinction is not like a *bat/elephant* distinction, but like an *animal/elephant* distinction. On the other hand, if §11 is right, then the *is/ought* distinction is like either a *bat/elephant* or a *mammal/carnivore* distinction. For the latter possibility, see the diagram illustrating the overlap of concept extensions.

8. See Thomas Nagel, 'The Absurd' in *Journal of Philosophy* vol. 68, 1971.

9. R. M. Hare, *Freedom and Reason* (Oxford University Press, 1963). I mean that Sisyphus is the *stuff* of which the fanatic is made.

Finally, nor has nineteenth- or twentieth-century Utilitarianism anythir from Taylor's style of fable-telling. For the *locus* or origin of all value firmly confined within the familiar area of psychological states conceived in independence of what they are directed to.[10]

In order to have a name, I shall call Taylor's and all similar accounts non-cognitive accounts of the meaning of life. This choice of name is not inappropriate if it helps to signal the association of these accounts with a long-standing philo-sophical tendency to strive for descriptions of the human condition by which will and intellect-cum-perception are kept separate and innocent of all insider transac-tions. The intellect supplies uncontaminated factual perception, deduction, and means-end reasoning. Ends are supplied (in this picture) by feeling or will, which are not conceived either as percipient or as determinants in any interesting way of perception.

What I shall argue next is that, in spite of the well-tried familiarity of these ideas, the non-cognitive account depends for its whole plausibility upon abandon-ing at the level of theory the inner perspective that it commends as the only possible perspective upon life's meaning. This is a kind of incoherence, and one that casts some doubt upon the distinction of the inside and the outside view-points. I also believe that, once we break down the supposed distinction between the inner or participative and the outer, supposedly objective viewpoints, there will be a route by which we can advance (though not to anything like the particularity of the moral certainty that we began by envying).

5. Where the non-cognitive account essentially depends on the existence and availability of the inner view, it is a question of capital importance whether the non-cognitivist's account of the inner view makes such sense of our condition as it actually has for us from the inside.

The first ground for suspecting distortion is that, if the non-cognitive view is put in the way Taylor puts it, then it seems to make too little difference to the meaningfulness of life how well or badly our strivings are apt to turn out. Stone rolling for its own sake, and stone-rolling for successful temple building, and stone-rolling for temple building that will be frustrated—all seem to come to much the same thing. I object that that is not how it feels to most people. No doubt there are 'committed' individuals like William the Silent or the doctor in Camus' *La Peste* who will constitute exceptions to my claim. But in general, the larger the obstacles that nature or other people put in our way, and the more truly hopeless the prospect, the less point most of us will feel anything has. 'Where there is no hope, there is no endeavour' as Samuel Johnson observed. In the end point is partly dependent on expectation of outcome; and expectation is dependent on past outcomes. So point is not independent of outcome.

10. For efforts in the direction of a better account of some of these states, see below §6 and Essay V, *Needs, Values, Truth*.

David Wiggins

The non-cognitivist may make two replies here. The first is that, in so far as the outcome is conceived by the agent as independent of the activity, the activity itself is merely instrumental and must lead back to other activities that are their own outcome. And these he will say are what matter. But in opposition to this,

(a) I shall show in due course how activities that can be regarded as 'their own goals' typically depend on valuations that non-cognitivism makes bad sense of (§6 below);

(b) I shall question whether all activities that have a goal independent of the activity itself are perceived by their agents as only derivatively meaningful (§13 below).

The non-cognitivists' second reply will be directed against the objection that he makes it matter too little how well or badly our strivings turn out. Is it not a point on *his* side that the emptier and worse worlds where one imagines everything having even less point than it has now are worlds where the will itself will falter? To this I say Yes, I hear the reply. But if the non-cognitive view was to make the sense of our condition that we attribute to it, then something needed to be written into the non-cognitive account about what kinds of object will engage with the will as important. And it is still unclear at this stage how much room can be found within non-cognitivism for the will's own distinctions between good and bad reasons for caring about anything as important. Objectively speaking (once 'we disengage our wills'), any reason is as good or as bad as any other reason, it seems to say. For on the non-cognitive account, life is *objectively* meaningless. So, by the non-cognitivist's lights, it must appear that whatever the will chooses to treat as a good reason to engage itself is, for the will, a good reason. But the will itself, taking the inner view, picks and chooses, deliberates, weighs, and tests its own concerns. It craves objective reasons; and often it could not go forward unless it thought it had them. The extension of the concept *objective* is quite different on the inner view from the extension assigned to it by the outer view. And the rationale for determining the extension is different also.

There is here an incoherence. To avoid it without flying in the face of what we think we know already about the difference between meaning and meaninglessness, the disagreement between the inner and the outer views must be softened somehow. The trouble is that, if we want to preserve any of the distinctive emphases of Taylor's and similar accounts, then we are bound to find that, for purposes of the validation of any given concern, the non-cognitive view always readdresses the problem to the inner perspective *without itself adopting that perspective*. It cannot adopt the inner perspective because, according to the picture that the non-cognitivist paints of these things, the inner view has to be unaware of the outer one, and has to enjoy essentially illusory notions of objectivity, importance, and significance: whereas the outer view has to hold that life is objectively meaningless. The non-cognitivist mitigates the outrageousness of so

[136]

categorical a denial of meaning as the outer view issues by pointing to the availability of the participant perspective. But the most that he can do is to point to it. Otherwise the theorist is himself engulfed by a view that he must maintain to be false.

So much for the first distortion I claim to find in non-cognitivism and certain inconclusive defences of that approach. There is also a second distortion.

To us there seems to be an important difference between the life of the cannibalistic blindworms that Taylor describes and the life of (say) a basking seal or a dolphin at play, creatures that are conscious, can rest without sleeping, can adjust the end to the means as well as the means to the end, and can take in far more about the world than they have the immediate or instrumental need to take in. There also seems to us to be a difference, a different difference, between the life of seals or dolphins and the life of human beings living in communities with a history. And there is even a third difference, which as participants we insist upon, between the life of someone who contributes something to a society with a continuing history and a life lived on the plan of a southern pig-breeder who (in the economics textbooks, if not in real life) buys more land to grow more corn to feed more hogs to buy more land, to grow more corn to feed more hogs . . . The practical concerns of this man are at once regressive and circular. And we are keenly interested, on the inner view, in the difference between these concerns and non-circular practical reasonings or life plans.

For the inner view, this difference undoubtedly exists. If the outside view is right to commend the inside view, then the outside view must pay some heed to the differences that the inner view perceives. But it can accord them no importance that is commensurate with the weight that the non-cognitive theory of life's meaning thrusts upon the inner view. 'The differences are merely invented,' Taylor has to say, 'and none really cancels the kind of meaninglessness we found in Sisyphus'.

To the participant it may seem that it is far harder to explain what is so good about buying more land to grow more corn to feed more hogs to buy more land to grow more corn to feed more hogs . . . than it is to explain what is good about digging a ditch with a man whom one likes, or helping the same man to talk or drink the sun down the sky. It might seem to a participant that the explanation of the second sort of thing, so far from having nowhere to go but round and round in circles, fans out into a whole arborescence of concerns; that, unlike any known explanation of what is so good about breeding hogs to buy more land to breed more hogs . . . , it can be pursued backwards and outwards to take in more and more of the concerns of a whole human life. But on the non-cognitive view of the inner view there is no way to make these differences stick. They count for so little that it is a mystery why the non-cognitivist doesn't simply say: life is meaningless; and that's all there is to it. If only he would make that pronouncement, we should know where we were.

But why do the differences just mentioned count for so little for the non-

Handwritten margin notes: Taylor claims no diff w/ various animals (who rest) we think animals of conscious are quite different. + further diff between animals & humans. + Thirdly a diff between people w/in a society & those who contribute a lot & those who bond.

Handwritten note at bottom: gives the examples

cognitivist? Because they all arise from subjective or anthropocentric considerations, and what is subjective or anthropocentric is not by the standards of the outer view objective. (Taylor insists that to determine whether something matters, we have to view it 'independently of our own deep interest'.) I shall come back to this when I reconstruct the non-cognitive view; but let me point out immediately the *prima facie* implausibility of the idea that the distinction between objectivity and non-objectivity (which appears to have to do with the existence of publicly accepted and rationally criticizable standards of argument, or of ratiocination towards truth) should coincide with the distinction between the anthropocentric and the non-anthropocentric (which concerns orientation towards human interests or a human point of view). The distinctions are not without conceptual links, but the *prima facie* appearance is that a matter that is anthropocentric may be either more objective or less objective, or (at the limit) *merely* subjective.[11] This is how things will appear until we have an argument to prove rigorously the mutual coincidence of independently plausible accounts of the anthropocentric/non-anthropocentric distinction, the non-objective/objective distinction, and the subjective/non-subjective distinction.[12]

✳ The third and last distortion of experience I find in Taylor's presentation of non-cognitivism I shall try to convey by an anecdote. Two or three years ago, when I went to see some film at the Academy Cinema, the second feature of the evening was a documentary film about creatures fathoms down on the ocean-bottom. When it was over, I turned to my companion and asked, 'What is it about these films that makes one feel so utterly desolate?' Her reply was: 'apart from the fact that so much of the film was about monstrous predators eating one another, the unnerving thing was that nothing down there ever seemed to *rest*.' As for play, disinterested curiosity, or merely contemplating, she could have added, these seemed inconceivable.

At least about the film we had just seen, these were just the points that needed to be made—untrammelled by all pseudo-philosophical inhibitions, which are irrelevant in any case to the 'inner' or participant perspective. And the thought the film leads to is this. If we can project upon a form of life nothing but the pursuit of life itself, if we find there no non-instrumental concerns and no interest in the world considered as lasting longer than the animal in question will need the world to last

---

11. For an independent account of the subjective, see Wiggins, *Needs, Values, Truth*, p. 208.

12. A similar observation needs to be entered about all the other distinctions that are in the offing here—the distinctions between the neutral and the committed, the neutral and the biased, the descriptive and the prescriptive, the descriptive and the evaluative, the quantifiable and the unquantifiable, the absolute and the relative, the scientific and the unscientific, the not essentially contestable and the essentially contestable, the verifiable or falsifiable and the neither verifiable nor falsifiable, the factual and the normative. . . . In common parlance, and in sociology and economics—even among political scientists who should know better—these distinctions are used almost interchangeably. But they are different, and they are separately interesting. Each of these contrasts has its own rationale. An account of all of them would be a contribution not only to philosophy but to life.

[138]

in order to sustain the animal's own life; then the form of life must be to some considerable extent alien to us.[13] Any adequate description of the point we can attach to our form of life must do more than treat our appetitive states in would-be isolation from their relation to the things they are directed at.

For purposes of his eventual philosophical destination, Richard Taylor had to forge an intimate and very direct link between contemplation, permanence, and boredom. But, at least on the inner view, the connection between these things is at once extremely complex and relatively indirect.[14] And, once one has seen the final destination towards which it is Taylor's design to move the whole discussion, then one sees in a new light his obsession with monuments. Surely these are his hostages for the objects of psychological states in general; and all such objects are due to be in some sense discredited. (Discredited on the outer view, or accorded a stultifyingly indiscriminate tolerance on the outer account of the inner view.) And one comprehends all too well Taylor's sour grapes insistence on the impermanence of monuments—as if by this he could reduce to nil the philosophical (as opposed, he might say, to subjective) importance of all the objects of psychological states, longings, lookings, reverings, contemplatings, or whatever.

6. Leaving many questions still dangling, I shall conclude discussion of the outer account of the inner perspective with a general difficulty, and a suggestion.

There is a tendency, in Utilitarian writings and in the writings of economists,[15] to locate all ultimate or intrinsic value in human appetitive states.[16] They are contrasted (as we also see Taylor contrasting them for his purposes) with every-

13. Here, I think, or in this neighbourhood, lies the explanation of the profound unease that some people feel at the systematic and unrelenting exploitation of nature and animals which is represented by factory farming, by intensive livestock rearing, or by the mindless spoliation of non-renewable resources. This condemnation of evil will never be understood till it is distinguished by its detractors from its frequent, natural, but only contingent concomitant—the absolute prohibition of all killing not done in self-defense.

14. On permanence, *cf.* Ludwig Wittgenstein, *Tractatus Logico-Philosophicus* 6.4312 quoted *ad init.*; F. P. Ramsey, 'Is there anything to discuss?', *Foundations of Mathematics and Other Essays* (London, 1931): "I apply my perspective not merely to space but also to time. In time the world will cool and everything will die; but that is a long time off still and its percent value at compound discount is almost nothing. Nor is the present less valuable because the future will be blank."

15. *Cf.* Wilfred Beckerman, *New Statesman*, June 21, 1974, p. 880: "The second, and real question is: at what rate should we use up resources in order to maximise the welfare of human beings. . . . Throughout existence man has made use of the environment, and the only valid question for those who attach—as I do (in accordance with God's first injunction to Adam)—*complete and absolute priority to human welfare* is what rate of use provides the maximum welfare for humans, including future generations."

I quote this relatively guarded specimen to illustrate the hazards of making too easy a distinction between human welfare on the one side and the environment on the other. But it also illustrates the purely ornamental role which has devolved upon the Hebrew scriptures. They constitute matter for the literary decoration of sentiments formed and apprehended by quite different methods of divination. It is irrelevant for instance that the world-view given voice in the first chapters of *Genesis* is perceptibly more complicated than the one Beckerman expresses.

16. Or in the case of vegetarian utilitarian writings, to locate all ultimate value in conscious animal appetitive states.

thing else in the world. According to this sort of view, the value of anything that is not a psychological state derives from the psychological state or states for which it is an actual or potential object. See here what Bentham says in *An Introduction to the Principles of Morals and Legislation:*

> Strictly speaking, nothing can be said to be good or bad, but either in itself, which is the case only with pain or pleasure; or on account of its effects; which is the case only with things that are the causes or preventives of pain and pleasure.

One has only to put the matter like this, however, to be troubled by a curious instability. Since nothing at all can count for the outer view as inherently or intrinsically good, the doctrine must belong to the inner or inside view. But, as experienced, the inner view too will reject this view of value. For, adopting that inner view,[17] and supposing with Bentham that certain conscious states are good in themselves, we must take these states as they appear to the inner view. But then one cannot say without radical misconception that these states are all that is intrinsically valuable. For (a) many of these conscious states have intentional objects; (b) many of the conscious states in which intrinsic value supposedly resides are strivings *after* objects that are not states, or are contemplations *of* objects that are not themselves states; and (c) it is of the essence of these conscious states, experienced as strivings or contemplations or whatever, to accord to their intentional objects a non-instrumental value. For from the inside of lived experience, and by the scale of value that that imposes, the shape of an archway or the sound of the lapping of the sea against the shore at some place at some time may appear to be of an altogether different order of importance from the satisfaction that some human being once had from his breakfast.[18]

The participant, with the going concepts of the objective and the worth while, descries certain external properties in things and states of affairs. And the presence there of these properties is what invests them with importance in his eyes. The one thing that properties cannot be, at least for him, is mere projections resulting from a certain kind of efficacy in the causation of satisfaction. For no appetitive or aesthetic or contemplative state can see its own object as having a value that is derivative in the special way that is required by the thesis that all non-instrumental value resides in human states of satisfaction. But, if that is right, then

17. Perhaps some one individual man's inner view. For here and only here could it be held to be perfectly or fully obvious that the special goodness in themselves of certain of his pleasurable states is something simply above or beyond argument for him. Beyond that point—notwithstanding utilitarian explanations of the superfluity of argument on something so allegedly evident—it is less obvious to him.

18. This feature of experience is of course lamented by thinkers who seek to make moral philosophy out of ((‘formal value theory’ + moral earnestness) + some values of the theorist's own, generalized and thereby tested) + applications. But the feature is part of what is given in the phenomenology of some of the very same ‘satisfaction’ experiences that are the starting-point of the utilitarians themselves. And there is nothing to take fright at in this feature of them, inconsistent though it is with absurd slogans of the literally absolute priority of human welfare.

the outer view cannot rely for its credibility upon the meaning that the inner view perceives in something. To see itself and its object in the alien manner of the outer view, the state as experienced would have to be prepared to suppose that it, the state, could just as well have lighted on any other object (even any other kind of object), provided only that the requisite attitudes could have been induced. But in this conception of such states we are entitled to complain that nothing remains that we can recognize, or that the inner perspective will not instantly disown.[19]

I promised to conclude the critique of non-cognitivism with a suggestion about values. It is this: no attempt to make sense of the human condition can make sense of it if it treats the objects of psychological states as unequal partners or derivative elements in the conceptual structure of values and states and their objects. This is far worse than Aristotle's opposite error:

We desire the object because it seems good to us, rather than the object's seeming good to us because we desire it. *Metaphysics,* 1072a29 (*cf. NE* 1175a).

Spinoza appears to have taken this sentence as it stood and deliberately negated it (*Ethics,* part III, proposition 9, note). But maybe it is the beginning of real wisdom to see that we may have to side against both Aristotle and Spinoza here and ask: 'Why should the *because* not hold both ways round?'' Surely an adequate account of these matters will have to treat psychological states and their objects as equal and reciprocal partners, and is likely to need to see the identifications of the states and of the properties under which the states subsume their objects as interdependent. (If these interdependencies are fatal to the distinction of inner and outer, we are already in a position to be grateful for that.)

Surely it can be true both that we desire *x* because we think *x* good, and that *x* is good because (it is such that) we desire *x*. It does not count against the point I am making that the explanation of the 'because' is different in each direction. Nor does it count against the particular anti-non-cognitivist position that is now emerging in opposition to non-cognitivism that the second 'because' might have to be explained in some such way as this: such desiring by human beings directed

---

19. An example will make these claims clearer perhaps. A man comes at dead of night to a hotel in a place where he has never been before. In the morning he stumbles out from his darkened room and, following the scent of coffee out of doors, he finds a sunlit terrace looking out across a valley on to a range of blue mountains in the half-distance. The sight of them—a veritable vale of Tempe—entrances him. In marvelling at the valley and mountains he thinks only how overwhelmingly beautiful they are. The value of the state depends on the value attributed to the object. But the theory I oppose says all non-instrumental value resides here in the man's own state, and in the like states of others who are actually so affected by the mountains. The more numerous such states are, the greater, presumably, the theory holds, is the 'realized' value of the mountains. The theory says that the whole actual value of the beauty of the valley and mountains is dependent upon arranging for the full exploitation of the capacity of these things to produce such states in human beings. (Exploitation now begun and duly recorded in Paul Jennings's Wordsworthian emendation: 'I wandered lonely as a crowd.') What I am saying about the theory is simply that it is untrue to the actual experience of the object-directed states that are the starting-point of that theory.

in this way is one part of what is required for there to be such a thing as the perspective from which the non-instrumental goodness of *x* is there to be perceived.

There is an analogy for this suggestion. We may see a pillar-box as red because it is red. But also pillar-boxes, painted as they are, *count* as red only because there actually exists a perceptual apparatus (*e.g.* our own) that discriminates, and learns on the direct basis of experience to group together, all and only the actually red things. Not every sentient animal that sees a red postbox sees it as red. But this in no way impugns the idea that redness is an external, monadic property of a postbox. 'Red postbox' is not short for 'red to human beings postbox'. Red is not a relational property. (It is certainly not relational in the way in which 'father of' is relational, or 'moves' is relational on a Leibniz-Mach view of space.) All the same, it is in one interesting sense a *relative* property. For the category of colour is an anthropocentric category. The category corresponds to an interest that can only take root in creatures with something approaching our own sensory apparatus.

Philosophy has dwelt nearly exclusively on differences between 'good' and 'red' or 'yellow'. I have long marvelled at this.[20] For there resides in the

20 Without of course wishing to deny the difference that good is 'attributive' to a marked degree, whereas colour words are scarcely attributive at all. I think that, in these familiar discussions, philosophers have misdescribed the undoubted fact that, because there is no standing interest to which yellowness answers, 'yellow' is not such as to be *cut out* (by virtue of standing for what it stands for) to commend a thing or evaluate it favourably. But, surely, if there were such a standing interest, 'yellow' would be at least as well suited to commend as 'sharp' or 'beautiful' or even 'just' are.

Against the suggestion that axiological predicates are a species of predicate not clearly marked off from the factual, there is a trick the non-cognitivist always plays and he ought not to be allowed to play. He picks himself a 'central case' of a descriptive predicate, and a 'central case' of a valuational predicate. Then he remarks how very different the predicates he has picked are. But what on earth can that show? Nobody thinks you could prove a bat was not an animal by contrasting some bat (a paradigm case of a bat) with some elephant (a paradigm case of an animal). Nothing can come clear from such procedures in advance of explanation of the point of the contrast. In the present case the point of the factual/non-factual distinction has not been explained; and it has to be explained without begging the question in favour of the non-cognitivist, who picked the quarrel in the first place. What was the nature or rationale of the difference which was by these means to have been demonstrated? Till it is explained there must remain all of the following possibilities:

It would be unfair to say there have been no attempts at all to elucidate the point of the fact-value contrast as exclusive. Wittgenstein tried (quite unsuccessfully, I think) to explain it in his 'Lecture on Ethics', *Philosophical Review* vol. 74, 1965, p. 6. And prescriptivists explain it as exclusive by reference to the link they allege holds between evaluation and action. But, although there is some such link between deliberative judgment and action, the required link does not hold between evaluation and action. That was one part of the point of the contrast I proposed at the beginning of §4.

combined objectivity and anthropocentricity of colour a striking analogy to il-
luminate not only the externality that human beings attribute to the properties by
whose ascription they evaluate things, people, and actions, but also the way in
which the quality *by* which the thing qualifies as good and the desire *for* the thing
are equals—are, 'made for one another' so to speak. Compare the way in which
the quality by which a thing counts as funny and the mental set that is presupposed
to being amused by it are made for one another.

7. The time has come to sort out the non-cognitive theory to accommodate these
findings and expel contradiction. But it is just possible that I have not convinced
you that any sorting out is necessary, and that you have found more coherent than I
have allowed it to be the non-cognitivist's use of the idea of perspective, and of
different and incompatible perspectives.

Perspective is not a form of illusion, distortion, or delusion. All the different
perspectives of a single array of objects are perfectly consistent with one another.
Given a set of perspectives, we can recover, if only they be reliably collected, a
unified true account of the shape, spatial relations, and relative dimensions of the
objects in the array. If we forget these platitudes then we may think it is much
more harmless than it really is that the so-called outer and inner perspectives
should straightforwardly contradict one another. There is nothing whatever in the
idea of a perspective to license this scandalous idea—no more than the truism that
two perspectives may include or exclude different aspects will create the licence to
think that the participant and external views, as the non-cognitivist has described
them, may unproblematically conflict over whether a certain activity or pursuit is
really (or objectively) worth while or not[21]

There are several different reasons then why the non-cognitivist theory must be
redeployed if it is to continue to be taken seriously. The traditional twentieth-
century way of amending the theory to secure its self-consistency would have
been *meta-ethics,* conceived as an axiologically neutral branch of 'logic'. Meta-
ethics is not as neutral as was supposed, or so I shall claim. But it may be that it is
still the best way for us to understand ourselves better, and we should make the
experiment.

Let us take the language of practice or morals as an object language. Call it L.
The theorist's duty is then to discover, and to explain in the meta-language which
is his own language, both a *formal theory* and a more discursive *informal theory* of
L-utterances, not least L-utterances concerning what is worth while or a good

[21] Still less does the language of perspective license the supposition that the philosopher who
answers the question of the meaning of life could make a virtue out of committing himself to neither, or
neither and both perspectives. Where on earth is *he* looking at things from? Or does he think of himself
as one who mysteriously somehow looks at everything from no perspective at all? For the closest
approximation he could coherently conceive of attaining to this aspiration, see §10 below.

L = object language

[143]

thing to do with one's life. In place of philosophical analysis, scarcely as yet a successful philosophical genre, let him concentrate his attention on the informal elucidation of the assertibility predicate in its application to various types of moral judgment and seek to determine its approximation there to genuine truth.

What does this involve? First, and this is the humble formal task that is presupposed to his more distinctively ethical aspirations, the theorist needs to be able to say, or assume that someone can say, what each of the sentences of the object language means. To achieve this, a procedure is needed for parsing L-sentences into their primitive semantic components, and an axiom is required for each primitive component accounting for its particular contribution to assertion conditions. Then, given any L-sentence(s), the axioms can be deployed to derive a pairing of (s) with an assertion condition p, the pairing being stated in the metalanguage by a theorem in the form:

$$s \text{ is assertible if and only if } p.$$

Moral philosophy as we now know it makes many sophisticated claims about meanings, some of them very hard to assess, both about the 'analysis' of particular concepts and about meaning in general. Compared with everything that would be involved in making those assessments, what we are assuming here is minimal. What we are assuming is only that the discursive or informal comments that the moral theorist hopes to make about the status of this, that, or the other judgment in L will presuppose that such a biconditional can be constructed for each sentence of L. These assertion conditions give the meanings of the judgments he wants to comment upon. All that has been contended so far is that, if no such principled understanding of what they mean may be thought of as obtainable, then (whatever other treasures he possesses) he cannot even count on the first thing.

I speak of *assertion* conditions as that by which meaning is given, and not yet of *truth* conditions, but only because within this meta-ethical framework the non-cognitivist's most distinctive non-formal thesis is likely to be the denial that the assertibility of a value judgment or of a deliberative judgment can amount to anything as objective as we suppose truth to be. To do justice to this denial of his, the aim must be to leave undecided *pro tempore*—as Dummett in one way and Davidson and McDowell in another have shown to be possible—the relationship of truth and assertibility.[22] In this way we arrange matters so that it can turn out—

---

22. See M. A. E. Dummett, *Frege: Philosophy of Language* (London: Duckworth, 1973), and John McDowell, 'Bivalence and Verificationism' in *Truth and Meaning: Essays in Semantics* (Oxford University Press, 1972), edited by Gareth Evans and John McDowell. McDowell shows how we can build up an independent account of what a semantic predicate F will have to be like if the sentences of an object language are to be interpreted by means of equivalences which will say what the object language sentences mean. His way of showing that it can be a *discovery*, so to speak, that it is the truth predicate which fulfils the requirements on F is prefigured at p. 210 of Donald Davidson, 'Truth and Meaning', *Synthèse* vol. 17, 1967.

as it does for empirical or scientific utterances—that truth assertibility. But it is not theoretically excluded that, for ce ments, assertibility should fall short of truth. The matter is metaethics and the informal theory that is built around the it. I come now to this informal theory.

Adapting Tarski's so-called 'Convention T' to the p_ theory, we may say now that the meta-language has a materially adequate tion of the predicate 'assertible' just in case it has as consequences all sentences obtained from the schema '*s* is assertible if and only if *p*' by substituting for '*s*' a name of any sentence of L and substituting for '*p*' the translation or interpretation of this sentence in the meta-language.[23] So if the ethical theorist is to erect a theory of objectivity, subjectivity, relativism, or whatever upon these foundations, then the next thing we need to say some more about is how a theory of L-assertibility is to be constrained in order to ensure that the sentence used on the right-hand side of any particular equivalence that is entailed by the theory of assertibility should indeed interpret or translate the sentence mentioned on the left. What is *interpretation* or *translation* in this context? If we can supply this constraint then, as a bonus, we shall understand far better the respective roles of participant and theorist and what assertibility would have to amount to.

It seems obvious that the only way to by-pass Tarski's explicit use of the word 'translation' is by reference to what Davidson has called radical interpretation.[24] A promising proposal is this. Rewrite convention T to state that the meta-language possesses an empirically correct definition of 'assertible' just in case the semantical axioms, in terms of which the definition of assertibility is given, all taken together, entail a set Σ of equivalences '*s* is assertible just in case *p*', one equivalence for each sentence of L, with the following overall property: a theorist who employs the condition *p* with which each sentence *s* is mated in a Σ-equivalence, and who employs the equivalence to interpret utterances of *s*, is in the best position he can be to make the *best possible overall sense* there is to be made of L-speakers. This goal sets a real constraint—witness the fact that the theorist may test his theory, try it out as a way of making sense of his subjects, even as he constructs it. By 'making sense of them' would be meant ascribing to the speakers of L, on the strength of their linguistic and other actions, an intelligible collection

---

[23] See p. 187 of A. Tarski, 'The Concept of Truth in Formalized Languages', in *Logic, Semantics, and Metamathematics* (Oxford University Press, 1956). For my present doubts that there is anything to be gained, except an expository point, by the fabrication of a predicate of semantic assessment that is independent of 'true' in the fashion that 'assertible' might seem to promise to be, see Essay IV, §18, *Needs, Values, Truth*, pp. 176–8.

[24] See D. Davidson, 'Radical Interpretation', *Dialectica* vol. 27, 1973. The original problem is of course Quine's. See W. V. Quine, *Word and Object* (Cambridge, Mass: MIT Press, 1960). Davidson's own conception has been progressively refined by many philosophers, notably by Richard Grandy, Donald Davidson, Christopher Peacocke, Gareth Evans, and John McDowell. See also Essay IV, pp. 139–84, esp. pp. 141–7, Wiggins, *Needs, Values, Truth*.

beliefs, needs, and concerns. That is a collection that diminishes to the bare minimum the need for the interpreter to ascribe inexplicable error or inexplicable irrationality to them.[25] By 'interpreting an utterance of *s*' is intended here: saying what it is that *s* is used to say.

This general description is intended to pass muster for the interpretation of a totally alien language. But now suppose that we envisage the object-language and meta-language both being English. Then we can turn radical interpretation to advantage in order to envisage ourselves as occupying simultaneously the roles of theorist or interpreter and subject or participant. That will be to envisage ourselves as engaged in an attempt to understand ourselves.

Whether we think of things in this way or not, it is very important to note how essentially similar are the positions of the linguistic theorist and his subjects. The role of the theorist is only to *supplement,* for theoretical purposes, the existing understanding of L-speakers. It is true that, subject to the constraint upon which the whole exercise of interpretation itself rests—namely sufficient agreement in beliefs, concerns, and conceptions of what is rational and what is not—the theorist need not have exactly the same beliefs as his subjects. But the descriptions of the world that are available to him are essentially the same sorts of description as those available to his subjects. He uses the very same sort of sentence to describe the conditions under which *s* is assertible as the sentence *s* itself: and the meta-language is at no descriptive distance from the object-language. If the theorist believes his own semantic theory, then he is committed to be ready to put his mind where his mouth is at least once for each sentence *s* of the object-language, in a statement of assertion conditions for *s* in which he himself uses either *s* or a faithful translation of *s*. It follows that the possibility simply does not exist for the theorist to stand off entirely from the language of his subjects or from the viewpoint that gives this its sense. He has to begin at least by embracing—or by making as if to embrace—the very same commitments and world-view as the

25. See Richard Grandy, 'Reference, Meaning, and Belief', *Journal of Philosophy* vol. 70, 1973: John McDowell, *op. cit.* The requirement that we diminish to the minimum the theoretical need to postulate inexplicable error or irrationality is a precondition of trying to project any interpretation at all upon alien speakers. It was phrased by Davidson in another way, and called by him the requirement of charity. The replacement given here is closer to what has been dubbed by Richard Grandy the requirement of *humanity.* The further alterations reflect the belief that philosophy must desist from the systematic destruction of the sense of the word 'want', and that what Davidson calls 'primary reasons' must be diversified to embrace a wider and more diverse class of affective states than *desire.* (For a little more on these points, see now my *Sameness and Substance* (Oxford: Blackwell, 1980), Longer Note 6.36.)

Note that, even though we must for purposes of radical interpretation project upon L-speakers our own notions of rationality (and there is no proof they are the sole possible), and even though we take all the advantage we can of the fact that the speakers of the object-language are like us in being human, there is no guarantee that there must be a unique best theory of the assertibility conditions of their utterances. It has not been excluded that there might be significant disagreement between interpreters who have made equally good overall sense of the shared life of speakers of L, but at some points rejected one another's interpretations of L.

ordinary speakers of the object language. This is not to say that having understood them, he cannot back off from that world-view. What will require careful statement is how he is to do so.

8. [Even if this is a disappointment to those who have supposed that by means of metaethics, the theorist of value could move straight to a position of complete neutrality, it faces us in the right direction for the reconstitution of the non-cognitive theory.] In fact the framework I have been proposing precisely enables him to register his own distinctive point. He can do so in at least two distinctive ways. The first accepts and the second actually requires that framework.

First, using the language of his subjects but thinking (as a moralist like a Swift or an Aristophanes should, or as any moral theorist may) a bit harder than the generality of his subjects, he may try to make them look at themselves; and he may prompt them to see their own pursuits and concerns in unaccustomed ways. There is an optical metaphor that is much more useful here than that of perspective. Staying within the participant perspective, what the theorist may do is *lower the level of optical resolution*. Suppressing irrelevancies and trivialities he may perceive, and then persuade others to perceive, the capriciousness of some of the discriminations we unthinkingly engage in; or the obtuseness of some of the assimilations that we are content with. Again, rather differently but placing the non-cognitivist closer within reach of his own hobby-horse, he may direct the attention of his audience to what Aurel Kolnai called 'the incongruities of ordinary practice.'[26] Here Kolnai alluded to the irremovable disproportion between how heroic is the effort that it is biologically instinct in us to put into the pursuit of certain of our concerns, and how 'finite, limited, transient, perishable, tiny, tenuous' we ourselves and our goods and satisfactions all are. To lower the level of resolution, not down to the point where human concerns themselves are invisible—we shall come to that—but to the point where both the disproportion and its terms are manifest, is a precondition of human (as opposed to merely animal) resilience, of humour, of sense of proportion, of sanity even. It is the traditional function of the moralist who is a participant and of the satirist (who may want not to be). But this way of seeing is not the seeing of the total meaninglessness that Taylor spoke of. Nor, in the existentialist philosopher's highly technical sense, is it the perception of absurdity. For the participant perspective can contain together both the perception of incongruity and a nice appreciation of the limited but not necessarily infinitesimal importance of this or that particular object or concern. (It is not perfectly plain what Kolnai thought about the affinity of existentialist absurdity and incongruity—for his manuscript is unrevised—but, if Kolnai had doubted the compatibility of the perceptions of incongruity and importance, I think I could have convinced him by making the

26. 'The Utopian Mind' (unpublished typescript), p. 77.

following entirely Kolnaistic point. The disproportion between our effort and our transience is a fugitive quantity. It begins to disappear as soon as one is properly impressed by it. For it is only to us or our kind that our own past or future efforts can seem heroic.)

So much then for the non-cognitivist's first way of making his chief point. It will lead to nothing radical enough for him. The second way to make his point is to abstract it from the long sequence of preposterous attempts at traditional philosophical analysis of *good, ought, right, etc.* in terms of pleasure or feeling or approval . . . , and to transform it into an informal observation concerning the similarity or difference between the status of assertibility enjoyed by evaluative judgments and practical judgments, on the one hand, and the status of plain, paradigmatic, or canonical truth enjoyed by (for example) historical or geographical judgments on the other hand.[27]

What then is plain truth? Well, for purposes of the comparison, perhaps it will be good enough to characterize it by what may be called the truisms of plain truth. These truisms I take to be (1) the primacy of truth as a dimension for the assessment of judgments; (2) the answerability of truth to evidenced argument that will under favourable conditions converge upon agreement whose proper explanation involves that very truth; and (3) the independence of truth both from our will and from our own limited means of recognizing the presence or absence of the property in a statement. (2) and (3) together suggest the truism (4) that every truth is true in virtue of something. We shall expect further (5) that every plain truth is compatible with every other plain truth. Finally, a putative further truism (6*) requires the complete determinacy of truth and of all questions whose answers can aspire to that status.[28]

Does the assertibility of evaluative judgments and/or deliberative judgments come up to this standard? If we press this question within the framework just proposed, the non-cognitivist's distinctive doctrine becomes the contention that the answer is *no*. The question can be pressed from a point that is well within reach. We do not need to pretend to be outside our own conceptual scheme, or at a point that ought to have been both inaccessible and unthinkable.[29] The question is one we can pursue by working with informal elucidations of truth and assertibility that can be fruitfully constrained by the project of radical interpretation.[30] And as regards the apparent incoherence of Taylor's non-cognitivism, we can supersede the separate outer and inner perspectives by a common perspective that is accessible to both theorist and participant. Suppose it is asserted that this, that, or the

---

27. See in this connexion Essay IV, Wiggins, *Needs, Values, Truth*, pp. 139–184.

28. These formulations are superseded by the statement of the marks of truth given in §5 of Essay IV (Wiggins, pp. 147–148). (6*) does not survive there, for reasons that emerge in §10 below.

29. Compare the manner in which we could ascertain from within the space that we occupy certain of the geometrical properties of that very space: e.g., discover whether all equilateral triangles we encounter, of whatever size, are in fact similar triangles. If not, then the space is non-Euclidean.

30. Cp. Essay IV, §4 (Wiggins, pp. 146–147).

other thing is worth doing, and that the assertion is made on the best sort of grounds known to participant or theorist. Or suppose that a man dies declaring that his life has been marvellously worth while. The non-cognitive theory is first and foremost a theory not about the meaning but about the *status* of those remarks: that their assertibility is not plain truth and reflects no fact of the matter. What is more, this is precisely the suspicion that sometimes troubles and perplexes the un-theoretical participant who is moved to ask the questions from which we began this inquiry. Finally, let it be noticed, and put down to the credit of the framework being commended, that within this it was entirely predictable that this sort of question would be there to be asked.

9. The non-cognitivist's answer to the question can now be considered under two separate heads, value judgments (strict valuations) in general (this §) and deliberative judgments in general (§11).

For the non-cognitive critique of the assertibility predicate as it applies to value judgments I propose to employ a formulation given by Bernard Williams in 'The Truth in Relativism', *Proceedings of the Aristotelian Society* (1974/5).

Relativism will be true, Williams says, just in case there are or can be systems of beliefs S_1, and S_2 such that:

(1) S_1 and S_2 are distinct and to some extent self-contained;

(2) Adherents of S_1 can understand adherents of S_2;

(3) S_1 and S_2 exclude one another—by (a) being comparable and (b) returning divergent yes/no answers to at least one question identifying some action or object type which is the locus of disagreement under some agreed description;

(4) S_1 and S_2 do not (for us here now, say) stand in real confrontation because, whichever of S_1 and S_2 is ours, the question of whether the other one is right lacks the relation to our concerns 'which alone gives any point or substance to appraisal: the only real questions of appraisal [being] about real options' (p. 255). 'For we recognize that there can be many systems S which have insufficient relation to our concerns for our [own] judgments to have any grip on them.'

If this is right then the non-cognitivist critique of valuations comes to this. Their mere assertibility as such lacks one of the truistic properties of plain truth: for an assertible valuation may fail even under favourable conditions to command agreement (*cf.* truism (2) §8). Again, there is nothing in the assertibility property itself to guarantee that all one by one assertible evaluations are *jointly* assertible (*cf.* truism (5)). Nor is it clear that where there is disagreement there is always something or other at issue (*cf.* truism (4)). For truth on the other hand we expect and demand all of this.

The participant will find this disturbing, even discouraging. But is Williams

[149]

right about the compatibility of his four conditions?[31] He mentions among other things undifferentiated judgments of 'right' and 'wrong', 'ought' and 'ought not'. Here, where the point of agreement of disagreements or opting one way or another lies close to action, and radical interpretation is correspondingly less problematical, I think he is on strong ground. We can make good sense of conditions (2) and (3) being satisfied together. We can easily imagine condition (4) being satisfied. But for valuations in the strict and delimited sense, such as 'brave', 'dishonest', 'ignoble', 'just,' 'malicious', 'priggish', there is a real difficulty. The comparability condition (3) requires that radical interpretation be possible. But radical interpretation requires the projection by one person upon another of a collection of beliefs, desires, and concerns that differ from the interpreter's own only in a fashion that the interpreter can describe and, to some extent, explain: and the remoter the link between the word to be interpreted and action, and (which is different) the more special the flavour of the word, the more detailed and delicate the projection that has to be possible to anchor interpretation. Evaluations raise both of these problems at once. (And one of the several factors that make the link between strict valuations and action so remote is something that Williams himself has prominently insisted upon in other connections—the plurality, mutual irreducibility, and incommensurability of goods.) The more feasible interpretation is here, the smaller must be the distance between the concerns of interpreter and subject.[32] But then the harder condition (4) is to satisfy.

In the theoretical framework of radical interpretation we shall suddenly see the point of Wittgenstein's dictum (*Philosophical Investigations*, §242) 'If language is to be a means of communication there must be agreement not only in definitions but, queer as this may sound, agreement in judgments also.'[33]

10. The difficulty the non-cognitivist is having in pressing his claim at this point is scarcely a straightforward vindication of cognitivism. If the case for the coincidence of truth and assertibility in evaluative judgments is made in the terms of §9, then truth itself is in danger of coming in the process to seem a fairly parochial thing. Is it not disquieting to be driven to the conclusion that the more

---

31. Both for Williams's purposes and for ours—which is the status of the assertibility concept as it applies to value judgments, and then as it applies (§11) to deliberative or practical judgments—we have to be able to convert a relativism such as this, concerning as it does overall systems S_1 and S_2, into a relativism concerning this or that particular judgment or class of judgments identifiable and reidentifiable across S_1 and S_2. Williams requires this in order that disagreement shall be focused. I require it in order to see whether it is possible to distinguish judgments in S_1 or S_2 whose assertibility conditions coincide with plain truth from other judgments where this is dubious. For more on all these questions, see *Needs, Values, Truth*, Essays II, IV, V, Postcript §3.

32. There are valuations which are so specific, and so special in their point, that interpretation requires interpreter and subject to have in some area of concern the very same interests and the same precise focus. But specificity is only one part of the problem.

33. *Cf.* §241 and the rest of §242. For the interpretation I should offer of the rest of this passage, see *Needs, Values, Truth*, p. 350. *Cf.* also Wittgenstein op. cit. p. 223 (*passim*): 'If a lion could talk, we couldn't understand him.'

idiosyncratic the customs of a people, the more inscrutable their form of life, and the more special and difficult their language to interpret, the smaller the problem of the truth status of their evaluations?

It would be natural for someone perplexed by the question of the meaning of life to insist at this point that we shall not have found what it takes for individual lives to have the meaning we attribute to them unless we link meaning with rationality. He will say that the threat of relativism does not depend on Williams's condition (3) in Section VIII being satisfied. The threat is rather that, contrary to the tenor of §5, the reasons that impress us as good reasons have no foundation in reason at all. Or as Hume states the point in a famous passage of the First Appendix to the *Inquiry concerning the Principles of Morals:*

> It appears evident that the ultimate ends of human actions can never, in any case, be accounted for by *reason,* but recommend themselves entirely to the sentiments and affections of mankind, without any dependence on the intellectual faculties. Ask a man *why he uses exercise;* he will answer, *because he desires to keep his health.* If you then enquire, *why he desires health,* he will readily reply, *because sickness is painful.* If you push your enquiries farther, and desire a reason *why he hates pain,* it is impossible he can ever give any. This is an ultimate end, and is never referred to any other object.
>
> Perhaps to your second question, *why he desires health,* he may also reply, that *it is necessary for the exercise of his calling.* If you ask, *why he is anxious on that head,* he will answer, *because he desires to get money.* If you demand *Why? It is the instrument of pleasure,* says he. And beyond this it is an absurdity to ask for a reason. It is impossible there can be a progress *in infinitum;* and that one thing can always be a reason why another is desired. Something must be desirable on its own account, and because of its immediate accord or agreement with human sentiment and affection.

Not only is it pointless to hope to discover a *rational* foundation in human sentiment and affection. It is not even as if human sentiment and affection will *effectively* determine the difference between the worth while and the not worth while. Each culture, and each generation in each culture, confronts the world in a different way and reacts to it in a different way.

This scepticism pointedly ignores all the claims I made earlier, in §5. Rallying to their support, I ask: What does this scepticism show about our own judgments of significance or importance? After all there is no such thing as a rational creature of no particular neuro-physiological formation or a rational man of no particular historical formation. And even if, inconceivably, there were such, why should we care about what this creature would find compelling? It is not in this make-believe context that we are called upon to mount a critique of our own conceptions of the objective, the true, and the worth while.

So much seems to hang on this, but the reply comes so close to a simple repetition of the words of the relativist whom it is meant to challenge, that there is no alternative but to illustrate what happens when we do try to think of rationality

David Wiggins

in the absolute impersonal or cosmic fashion that it seems our interlocutor requires.

It is interesting that, so far as rationality in theoretical beliefs is concerned, it is by no means impossible for us to conceive of thinking in the impersonal way. Suppose we take a Peircean view of Science as discovering that which is destined, the world being what it is, to be ultimately agreed by all who investigate.[34] Let 'all' mean 'all actual or possible intelligent beings competent, whatever their conceptual scheme, to look for the fundamental explanatory principles of the world'. Then think of all these theories gradually converging through isomorphism towards identity. Cosmic rationality in belief will then consist in conforming one's beliefs so far as possible to the truths that are destined to survive in this process of convergence.[35]

Perhaps this is all make-believe. (I don't think it is entirely.) But the important thing is that, if we identify properties across all theories that converge upon what are destined to be agreed upon (by us or any other determined natural researchers) as the fundamental principles of nature, then the only non-logical, non-mathematical predicates we shall not discard from the language of rational belief are those which, in one guise or another, will always pull their weight in all explanatorily adequate theories of the world. As a result, and corresponding to predicates fit and not fit so to survive, we shall have a wonderful contrast between the primary qualities of nature and all other qualities. We can then make for ourselves a fact-value distinction that has a real and definite point. We can say that no value predicate stands for any real primary quality, and that the real properties of the world, the properties which inhere in the world *however it is viewed,* are the primary qualities.[36]

This is a very stark view. It expresses what was an important element of truth in

34. *Cf.* C. S. Peirce: 'How to Make Our Ideas Clear,' *Popular Science Monthly* vol. 12, 1878, pp. 286–302: ''Different minds may set out with the most antagonistic views, but the progress of investigations carries them by a force outside themselves to one and the same conclusion. This activity of thought by which we are carried, not where we wish but to a foreordained goal, is like the operation of destiny. No modification of the point of view taken, no selection of other facts for study, no natural bent of mind even, can enable a man to escape the predestinate opinion. This great law is embodied in the conception of truth and reality. The opinion which is fated to be ultimately agreed to by all who investigate is what we mean by the truth, and the object represented in this opinion is the real. That is the way I would explain reality.''

35. Inasmuch as there is a reality which dictates the way a scientific theory has to be in order that what happens in the world be explained by the theory, the difficulties of radical interpretation, attempted against the background of the truth about the world and the unwaveringly constant desire of speakers of the language to understand the material world, are at their slightest.

36. One should talk here also of the fundamental physical constants. *Cf.* B. A. W. Russell, *Human Knowledge* (London, 1948), p. 41: ''These constants appear in the fundamental equations of physics . . . it should be observed that we are much more certain of the importance of these constants than we are of this or that interpretation of them. Planck's constant, in its brief history since 1900, has been represented in various ways, but its numerical value has not been affected . . . Electrons may disappear completely from modern physics but e [charge] and m [mass] are pretty certain to survive. In a sense it may be said that the discovery and measurement of these constants is what is most solid in modern physics.''

the 'external' perspective. Seeing the world in this way, one sees no meaning in anything.[37] But it is evidently absurd to try to reduce the sharpness of the viewpoint by saying that meaning can be introduced into the world thus seen by the addition of human commitment. Commitment to what? This Peircean conceptual scheme *articulates* nothing that it is humanly possible to care about. It does not even have the expressive resources to pick out the extensions of ordinary predicates like 'red', 'chair', 'earthquake', 'person', 'famine' . . . For none of these has a strong claim to be a factual predicate by the scientific criterion. The distinction of fact and value we reach here, at the very limit of our understanding of scientific understanding, cannot be congruent with what the non-cognitivists intended as their distinction. It is as dubious as ever that there is anything for them to have intended. Starting out with the idea that value properties are mental projections, they have discovered that, if value properties are mental projections, then, except for the primary qualities, all properties are mental projections.

We come now to practical rationality for all conceivable rational agents. (Cosmically valid practical rationality.) The idea here would be, I suppose, that to be serious about objective reasons, or why anything matters, one must try to ascend closer to the viewpoint of an impersonal intelligence;[38] and that the properties of such an intelligence should be determinable *a priori*. A great deal of time and effort has been channelled into this effort. It might have been expected that the outcome would be the transformation of the bareness of our conception of an impersonal intelligence into the conception of an impersonal intelligence of great bareness. What was not so plainly to be expected was that the most elementary part of the subject should immediately collide—as it has—with a simple and (within the discipline thus *a priori* conceived) unanswerable paradox—the so-called 'Prisoner's Dilemma'.[39] What underlies the paradox (or the

37. Cf. Tolstoy, *Anna Karenina*, Penguin, p. 820: '[Levin was] stricken with horror not so much at death as at life, without the least conception of its origin, its purpose, its reason, its nature. The organism, its decay, the indestructibility of matter, the law of the conservation of energy, evolution were the terms that had superseded those of his early faith.' This is a description of what might pass as one stage in the transition we have envisaged as completed.

38. Or of a human being surviving, Nagel, *op. cit.*, p. 720, 'with that detached amazement which comes from watching an ant struggle up a heap of sand'. Cf. p. 722, 'the philosophical judgment [of absurdity] contrasts the pretensions of life with a larger context in which *no* standards can be discovered, rather than with a context from which alternative overriding standards may be applied'.

39. Cf. R. C. Jeffrey, *The Logic of Decision* (New York: McGraw-Hill, 1965), pp. 11–12. I take this as a 'paradox' in the following sense: a general principle of decision-theoretic prudence, generalizable to any agent whatever caught in the relevant circumstances, will lead in a wide variety of applications to what must be agreed by everybody to be a situation which is worse than it might have been for each participant if he had not acted on the generalizable principle—or if there had been another generalizable decision-theoretic principle to recommend (which there is not).

To say what is said in the text is not of course to 'solve' the paradox. It cannot be solved on the terms given. But it could only be accounted a real paradox if there were some antecedent grounds to suppose that it *should* have been possible to construct an *a priori* theory of rationality or prudence such that 'rational (A)' is incompatible with 'rational (not-A)', and such that that rationality is definable both independently of morality and ideals of agency and in such a way as to have independent leverage in these ancient disputes. (Cf. Plato, *Republic*, 445a.) Nothing in the derivation of the paradox excludes a

idea that there *is* here some paradox) is the supposition that it is simply obvious that an *a priori* theory of rational action ought to be possible—that some cosmic peg must exist on which we can fasten a set of concerns clearly and unproblematically identified *independently* of all ideals of agency and rationality themselves. First you have a set of concerns; then you think of a way by which they may best be achieved. That was the picture. But, in a new guise, it was nothing other than the absurd idea that all deliberation is really of means.[40]

11. I conclude that there is no such thing as a pure *a priori* theory of rationality conceived in isolation from what it is for us as we are to have a reason: and that even if there were such a thing, it would always have been irrelevant to the problem of finding a meaning in life, or seeing anything as worth while. What we need is to define non-cognitivist relativism in a way that is innocent of all dependence on a contrast between our rationality and some purer rationality yet restates the point we find in Richard Taylor.

It now says: Perhaps all strict valuations of the more specific and interesting kind have the interesting property that the interpretation of the value predicate itself presupposes a shared viewpoint, and a set of concerns common between interpreter and subject. Let it be admitted that this does an exclusive fact-value distinction no good. If someone insists, then there is nothing to prevent him from exploiting the collapse of that distinction in order to redescribe in terms of a shift or wandering of the 'value-focus' all the profound changes in valuation that have occurred in history, when the Greek world became the Christian world, or the Christian world the Renaissance world. He may elect to say with Nicolai Hartmann, as John Findlay reports him, that these changes were all by-products of an intense consciousness of new values, whose swimming into focus pushed out the old: that such newly apprehended values were not really new, only hitherto ignored.[41]

---

competing demonstration, on a different basis altogether, that the opposite course of action is also reasonable.

For an illuminating account of some of the asymmetries it is rational to expect between an *a priori* theory of belief and an *a priori* theory of practical reasonableness, see Ronald de Sousa, 'The Good and the True', *Mind* vol. 83, 1974.

40. That practically all interesting deliberation relates to ends and their practical specification in the light of actually or potentially available constituents, and that the place of means-ends reasoning is subordinate in practical reason, is argued by A. T. Kolnai, 'Deliberation is of Ends', *Proceedings of the Aristotelian Society* (1962). See also Essay VI, Wiggins, *Need, Values, Truth*, for a divergent interpretation of Aristotle's thought on this point, but an account similar to Kolnai's, of the problem itself.

41. See J. N. Findlay, *Axiological Ethics* (London: Macmillan, 1970). *Cf.* William James, *Talks to Teachers on Psychology: and to Students on some of Life's Ideals* (Longman, Green & Co., 1899), p. 299: "In this solid and tridimensional sense, so to call it, those philosophers are right who contend that the world is a standing thing with no progress, no real history. The changing conditions of history touch only the surface of the show. The altered equilibriums and redistributions only diversify our opportunities and open chances to us for new ideals. But with each new ideal that comes into life, the chance for a life based on some old ideal will vanish; and he would needs be a presumptuous calculator who should with confidence say that the total sum of significance is positively and absolutely greater at any one epoch than at any other of the world.''

All this the non-cognitivist may let pass as harmless, however eccentrically expressed; and may in less colourful language himself assert. He may even allow *totidem verbis* that, just as the world cannot be prised by us away from our manner of conceiving it, so our manner of conceiving it cannot be prised apart from our concerns themselves.[42] (It is also open to him to assert the compatibility of anthropocentricity with the only thing that there is for us to mean by objectivity, and to concede that the differences between higher and lower forms of life are not fictitious. They are even objective, he will say, if you use the word 'objective' like that.) But here he will stick. Where he will not back down from Taylor's original position is in the denial that the differences are simply discovered. Rather they are invented. Not only are some of them invented in the strictest and most straightforward sense. All of them depend for their significance upon a framework that is a free construct.

Here at last we approach the distinctive nucleus of non-cognitivism (married, without the consent of either, to Williams's relativism). What the new position will say is that, in so far as anything matters, and in so far as human life has the meaning we think it has, that possibility is rooted in something that is arbitrary, contingent, unreasoned, objectively non-defensible—and not one whit the less arbitrary, contingent and indefensible by virtue of the fact that the unconstrained inventive processes underlying it have been gradual, unconscious, and communal. Our form of life—or that in our form of life which gives individual lives a meaning—is not something that we as a species have ever (as they are apt to say) found or discovered. It is not something that we can criticize or regulate or adjust with an eye to what is true or correct or reasonable. Even within the going enterprise of existing concerns and deliberations, it would be a sad illusion to suppose that the judgment that this or that is worthwhile, or that life is worth living (or worth leaving), would be simply and plainly true. That sort of *terra firma* is simply not to be had.

The doctrine thus reconstructed from the assets of bankrupted or naïve non-cognitivism I shall call the doctrine of cognitive underdetermination. Unlike the positions it descends from, this position does not contradict itself. It is consistent with its own rationale. It can be explained without entering at all into the difficulties and ineffabilities of cultural relativism. It can even be stated in a manner innocent of the common confusion between the idea of human culture's being a construct and the idea of the assessment (truth-value) of a culturally conditioned judgment's being a construct. (A sort of sense-reference confusion.)

Suppose someone says: 'For me it is neither here nor there that I cannot prise my way of seeing the world apart from my concerns. This does nothing to answer my complaint that there is not *enough* meaning in the world. My life doesn't add up. Nothing matters sufficiently to me. My concerns themselves are too unimportant, too scattered, and too disparate.' Equally devastatingly to the naïve cognitiv-

---

42. Cf. A. J. Ayer, *The Central Questions of Philosophy* (London: Macmillan, 1974), p. 235: 'we have seen that the world cannot be prised away from our manner of conceiving it'.

ism that the doctrine of cognitive underdetermination bids us abandon, another may say that he finds that the objects of his concern beckon to him too insistently, too cruelly beguilingly, from too many different directions. 'I have learned that I cannot strive after all of these objects, or minister even to most of the concerns that stand behind them. To follow more than a minute subset is to be doomed to be frustrated in all. The mere validity—if it were valid—of the total set from which I am to choose one subset would provide no guarantee at all that any subset I can actually have will *add up* to anything that means anything to me.'

It is the cognitive underdeterminationist's role to comment here that things can never add up for the complainant who finds too frustratingly much, or for the complainant who finds too inanely little, unless each one supplies something extra, some conception of his own, to make sense of things *for himself.*

The problem of living a life, he may say, is to realize or respect a long and incomplete or open-ended list of concerns which are always at the limit conflicting. The claims of all true beliefs (about how the world is) are reconcilable. Everything true must be consistent with everything else that is true (*cf.* truism (5) of §8). But not all the claims of all rational concerns or even of all moral concerns (that the world *be* thus or so) need be actually reconcilable. When we judge that this is what we must do now,[43] or that that is what we'd better do, or that our life must now take one direction rather than another direction, we are not fitting truths (or even probabilities) into a pattern where a discrepancy proves that we have mistaken a falsehood for a truth.[44] Often we have to make a practical choice that another rational agent might understand through and through, not fault or even disagree with, but (as Winch has stressed)[45] make differently himself; whereas, if there is disagreement over what is factually true and two rational men have come to different conclusions, then we think it has to be theoretically possible to uncover some discrepancy in their respective views of the evidence. In matters of fact, we suppose that, if two opposing answers to a yes/no question are equally good, then they might as well have been equally bad. But in matters of practice, we are grateful for the existence of alternative answers. The choice between them is then up to us. Here is our freedom. But here too is the bareness of the world we inhabit. If there were practical truth it would have to violate the third and fifth

43. I have put '*must*', because *must* and *must not*, unlike *ought* and *ought not,* are genuine contraries.

44. See B. A. O. Williams, 'Consistency and Realism', *Proceedings of the Aristotelian Society Supplementary Volume,* 1966 and *cf.* J. N. Findlay, *op. cit.,* pp. 74–5: "What is good [Hartmann tells us] necessarily lies in a large number of incompatible directions, and it is intrinsically impossible that all of these should be followed out into realisation. One cannot, for example, achieve pure simplicity and variegated richness in the same thing or occasion, and yet both incontestably make claims upon us . . . in practice we sacrifice one good to another, or we make compromises and accommodations . . . such practical accommodations necessarily override the claims of certain values and everywhere consummate something that in some respect [ideally] ought not to be . . . a man [ideally should] be as wise as a serpent and gentle as a dove, but that does not mean that . . . it is *possible* for him to be both of them."

45. Peter Winch, 'The Universalizability of Moral Judgements', *Monist* vol. 49, 1965.

truisms of truth ((3) and (5) of §8 above). In living a life there is no truth, and there is nothing very like plain truth, for us to aim at. Anybody who supposes that the assertibility of 'I must do this' or the assertibility of 'This is the way for me to live, not that' consists in their plain truth is simply deluded.

Aristotle wrote (*NE* 1094a23): 'Will not knowledge of the good have a great influence on life? Shall we not, like archers who have a mark to aim at, be more likely to hit upon the right thing?' But in reality there is no such thing as *The Good*, no such thing as knowledge of it, and nothing fixed independently of ourselves to aim at. Or that is what is implied by the thesis of cognitive underdetermination.

12. If there is any common ground to be discovered in modern literature and one broad stream of modern philosophy it is here. What philosophers, even philosophers of objectivist formation, have constantly stressed is the absence of the unique solutions and unique determinations of the practical that naïve cognitivism would have predicted.[46] They have thus supplied the theoretical basis for what modern writers (not excluding modern writers who have believed in God) have felt rather as a void in our experience of the apprehension of value, and have expressed not so much in terms of the plurality and mutual irreducibility of goods as in terms of the need for an organizing focus or meaning or purpose that we ourselves *bring* to life. The mind is not only a receptor: it is a projector.[47]

At the end of *Anna Karenina* Levin says to himself: 'I shall still lose my temper with Ivan the coachman, I shall still embark on useless discussions and expressing my opinions inopportunely; there will still be the same wall between the sanctuary of my inmost soul and other people, even my wife . . . but my life now, my whole life, independently of anything that can happen to me, every minute of it is no longer meaningless as it was before, but has a positive meaning of goodness with which I have the power to invest it.'

However remote such declarations may appear from the language of the non-cognitivist philosopher, this need for autonomous making or investing of which Levin speaks is one part of what, in my presentation of him, the non-cognitive philosopher means by cognitive underdetermination. The familiar idea is that we do not discover a meaning for life or strictly find one: we have to make do with

---

46. The plurality and mutual irreducibility of things good has been stressed by F. Brentano (*Origins of Our Knowledge of Right and Wrong*, see especially para. 32); by N. Hartmann (see J. Findlay, *op. cit.*); by Isaiah Berlin, see, for instance, *Four Essays on Liberty* (Oxford University Press, 1969), Introduction p. xlix; by A. T. Kolnai and B. A. O. Williams (*op cit.*). See also Leszek Kolakowski, 'In Praise of Inconsistency' in *Marxism and Beyond* (London, 1969); Stuart Hampshire, *Morality and Pessimism* (Cambridge University Press, 1972); and Essay VII, Wiggins, *Needs, Values, Truth*, pp. 239–67.

47. For the seed of this idea in Plotinus' theory of cognition and for its transplantation and subsequent growth, see M. H. Abrams, *The Mirror and the Lamp* (Oxford University Press, 1953), Plotinus, *Ennead*, IV. 6.2–3: 'The mind affirms something not contained within impression: this is the characteristic of a power—within its allotted sphere to act.' 'The mind gives radiance to the objects of sense out of its own store.'

an artifact or construct or projection—something as it were invented.[48] And, whereas discovery is answerable to truth, invention and construction are not. From this he concludes that a limited and low-grade objectivity is the very best one could hope for in predications of meaning or significance.

The non-cognitivist takes two steps here and the assessment of the second step concerning objectivity depends markedly on the notion of truth that is employed at the first. What is this notion, we need to know, and to what extent does the cognitivist's position depend upon a naïve and precritical understanding of it? Give or take a little—subtract perhaps the more indeterminate among subjunctive conditionals—the precritical notion of truth covers empirical judgments fairly well. But it consorts less well with conceptions of truth or assertibility defended in mathematics by mathematical intuitionists or mathematical constructivists. It is well worth remarking that, for someone who wanted to combine objectivity with a doctrine of qualified cognitivism or of underdetermination, there might be no better model than Wittgenstein's normative conception of the objectivity of mathematics; and no better exemplar than Wittgenstein's extended description of how a continuing cumulative process of making or constructing can amount to the creation of a shared form of life that is constitutive of rationality itself, furnishing proofs that are not compulsions but procedures to guide our conceptions, explaining without explaining away our sense that sometimes we have no alternative but to infer this from that.[49]

Perhaps this is a million miles from ethics. Or perhaps Wittgenstein's philosophy of mathematics is completely unsuccessful. But if the subject-matter of moral philosophy had any of the features that Wittgenstein attributed to the sort of subject-matter he thought he was treating, then the issue whether the assertibility of practical judgments was truth, and did or did not sufficiently approximate to the truth of statements universally agreed to be factual, might become less important.[50] We could measure the distance, assess its importance, and think how to live with what we discover in this way. (Is there an independent case for tamper-

---

48. For a remarkable expression of the non-cognitivist's principal point and some others, see Aldous Huxley, *Do As You Will* (London, 1929), p. 101: "The purpose of life, outside the mere continuance of living (already a most noble and beautiful end), is the purpose we put into it. Its meaning is whatever we may choose to call the meaning. Life is not a crossword puzzle, with an answer settled in advance and a prize for the ingenious person who noses it out. The riddle of the universe has as many answers as the universe has living inhabitants. Each answer is a working hypothesis, in terms of which the answerer experiments with reality. The best answers are those which permit the answerer to live most fully, the worst are those which condemn him to partial or complete death. . . . Every man has an inalienable right to the major premiss of his philosophy of life." If anything need be added to this, presumably it is only that, concerning what 'living most fully' is for each man, the final authority must be the man himself. There is something right with this; but there is something wrong with it too.

49. *Cf.* L. Wittgenstein, *Remarks on the Foundation of Mathematics* (Oxford: Blackwell, 1956), III–30.

50. There is a cheap victory to be won even here of course. For it has proved much easier to achieve convergence or reflective equilibrium within our culture about the value of, say, civil liberty than about how exactly printing extra bank-notes will act upon conditions of economic recession. But this is not the point I am making.

ing in certain ways with the received truisms of truth? Or should we leave them to define an ideal that practical judgment must fall short of? How important really is the shortfall?)

Of course, if practical judgments were candidates to be accounted simply true, then what made them true, unlike valuations,[51] could not be the world itself, whatever that is.[52] But, saying what they say, the world is not really what they purport to characterize. (Compare what Wittgenstein, whether rightly or wrongly, wanted to say about statements of arithmetic.) In the assertibility (or truth) of mathematical statements we see what perhaps we can never see in the assertibility of empirical (such as geographical or historical) statements: the compossibility of objectivity, discovery, *and* invention.[53]

If we combine Wittgenstein's conception of mathematics with the constructivist or intuitionist views that are its cousins, then we find an illuminating similarity. One cannot get more out of the enterprise of making than one has in one way or another put there. ('What if someone were to reply to a question: "So far there is no such thing as an answer to this question"?' *Remarks on the Foundations of Mathematics, IV. 9.*) And at any given moment one will have put less than everything into it. So however many determinations have been made, we never have a reason to think we have reached a point where no more decisions or determinations will be needed. No general or unrestricted affirmation is possible of the law of excluded middle. But then anyone who wishes to defend the truth status for practical judgments is released from claiming that every practical question already has an answer. For reasons both independent of the practical and helpful to its pretensions, we may doubt how mandatory it ever was to enter into the system of ideas and preconceptions that issues in such declarations as truism (6*) of §8 above.

I shall break off from these large questions with two points of comparison and contrast.

(i) It seems that in the sphere of the practical we may know for certain that there

---

51. Note that the distinction proposed at §4 between evaluation and practical judgment is observed both here and throughout this essay.

52. Everything would be the wrong way round. *Cf.* B. A. O. Williams, 'Consistency and Realism' (*op. cit.*, n. 44), p. 19: "the line on one side of which consistency plays its peculiarly significant role is the line between the theoretical and the practical, the line between discourse which (to use a now familiar formula) has to fit the world, and discourse which the world must fit. With discourse that is practical in these terms, we can see why . . . consistency . . . should admit of exception and should be connected with coherence notions of a less logical character." This whole passage suggests something important, not only about statements of what ideally should be, but also about deliberative judgments,—namely that the exigencies of having to decide what to believe are markedly dissimilar from the exigencies of having to decide how to act. What the argument does *not* show is that the only truth there could be in a practical judgment is a peculiar truth which transposes the onus of match on to the world. (Still less that, if one rejects that idea, then the onus of match would be from the sentence or its annexed action to an *ideal* world.) Williams has illuminatingly glossed (1) precisely why truth in a practical judgement would not be like that; (2) the reasons why 'Ought (A)' and 'Ought (not-A)' are actually consistent; and (3) why 'must (A)' (which *is* consistent with 'must (not-A)') is only strictly assertible or true if A is the unique thing you must here do.

53. See *Needs, Values, Truth*, p. 350.

exist absolutely undecidable questions—*e.g.*, cases where the situation is so
calamitous or the choices so insupportable that nothing will count as *the* morally
correct, right answer. In mathematics, on the other hand, it appears to be an
undecidable question even how much sense attaches to the idea of an absolutely
undecidable question. This is a potentially important discrepancy between the two
subject matters. If we insist upon the actuality of some absolute undecidability in
the practical sphere, then we shall burst the bounds of ordinary, plain truth. To
negate the law of excluded middle is to import a contradiction into the intuitionist
logic which our comparison makes the natural choice for practical judgments. The
denial of '((A would be right) or not (A would be right))' contradicts the intuitionist theorem '(not (not (p or not p)))'.

 If a man makes an arithmetical mistake he may collide with a brick wall or
miss a train. He may bankrupt himself. For each calculation there is some risk,
and for each risk a clear mark of the worst's having befallen us. There is nothing
so definite with practical judgments. But surely it is begging the question to
require it. Equally, it is begging the question to shrug this off without another
word.

13. Let us review what has been found, before trying to advance further.
However rarely or frequently practical judgments can attain to truth or not, and
whatever is the extent and importance of cognitive underdetermination, we have
found no overwhelming reason to deny all objectivity even to practical judgments.
That practical questions sometimes have more than one answer, and that there is
not always an ordering of better or worse answers, is no reason to conclude that
good and bad answers cannot be argumentatively distinguished.

It is either false or senseless to deny that what valuational predicates stand for
are properties in a world. It is neither here nor there that these value properties are
not primary qualities, provided that they be objectively discriminable and can
impinge upon practical appreciation and judgment. No extant argument shows
that they cannot.

Individual human lives can have more or less point in a manner partially
dependent upon the disposition in the world of these value properties. The naïve
non-cognitivist has sometimes given the impression that the way we give point to
our lives is as if by blindfolding ourselves and attaching to something—anything—some free floating commitment, a commitment that is itself sustained by
the mere fact of our animal life. But that was a mistake. There is no question here
of blindfolding. And that is not what is said or implied by the reconstructed
doctrine of cognitive underdetermination.

In as much as invention and discovery are distinguishable, and in so far as either
of these ideas properly belongs here, life's having a point may depend as much
upon something contributed by the person whose life it is as it depends upon
something discovered. Or it may depend upon what the owner of the life brings to
the world in order to see the world in such a way as to discover meaning. This

cannot happen unless world and person are to some great extent reciprocally suited. And unluckily, all claims of human adaptability notwithstanding, those things are often not well suited to one another.

14. To get beyond here, something now needs to be said about the connection of meaning and happiness. In most moral philosophy, the requirement to treat meaning is commuted into the requirement to specify the end; and the end is usually identified with happiness. One thing that has seemed to make this identi- ✕ fication plausible is the apparent correctness of the claim that happiness is the state of one's life having a point or meaning. But on any natural account of the relation of point and end, this claim is actually inconsistent with the equation 'Happiness = The End'. (Unless happiness can consist in simply having happiness as one's end.) It is also worth observing that, in the very special cases where it is straightforward to say what the point of someone's life is, we may say what he stands for, or may describe his life's work. (I choose these cases not because I think they are specially central but because they are specially clear.) The remark- able thing is that these specifications are not even categorially of a piece with happiness. That does not prove that happiness is *never* the point. The works of practical moralists are replete, however, with warnings of the difficulty or futility of making happiness the aim. If they are right then, by the same token, it would be futile to make it the point.

The misidentification—if misidentification it is—of happiness and end has had a long history. The first fully systematic equation of the end, the good for man, and happiness is Aristotle's. The lamentable and occasionally comical effects of this are much palliated by the close observation and good sense that Aristotle carried to the *specification* of happiness. And it may be said in Aristotle's defence that the charge of misidentification of happiness and the good for man is captious, because his detailed specification of *eudaimonia* can perfectly well stand in—if this be what is required—as a description of the point of human existence: also that Aristotle meant by *eudaimonia* not exactly happiness but a certain kind of success. But that is too quick. Unless we want to walk the primrose path to the solemn conclusion that a meaningful life is just a sum (cf. *Nicomachean Ethics*, 1097b17) of activities worth while in themselves, or self-complete (in the sense of *Metaphysics*, 1048b17), the question is worth taking some trouble over.

Out of good nature a man helps his neighbour dig a drainage ditch. The soil is hard but not impossibly intractable, and together the two of them succeed in digging the ditch. The man who offers to help sees what he is doing in helping dig the ditch as worth while. In so far as meaning is an issue for him, he may see the episode as all of a piece with a life that has meaning. He would not see it so, and he would not have taken on the task, if it were impossible. In the case as we imagine it, the progress of the project is integral to his pleasure in it. But so equally is the fact that he likes his neighbour and enjoys working with him (provided it be on projects that it is within their joint powers to complete).

Shall we say here that the man's helping dig the ditch is instrumental and has the meaning of importance it has for the helper only derivatively? Derivatively from what, on the non-cognitivist view? Or shall we say that the ditch-digging is worth while in itself? But it isn't. It is end-directed. If we cannot say either of these things, can we cut the Gordian knot by saying both? In truth, the embracing of the end depends on the man's feeling for the task of helping someone he likes. But his feeling for the project of helping equally depends on the existence and attainability of the end of digging the ditch.

This is not to deny that Aristotle's doctrine can be restored to plausibility if we allow the meaning of the particular life that accommodates the activity to *confer* intrinsic worth upon the activity. But this is to reverse Aristotle's procedure (which is the only one available to a simple cognitivism). And I doubt we have to choose (*cf.* §6). At its modest and most plausible best the doctrine of cognitive underdetermination can say that we need to be able to think in both directions, down from point to the human activities that answer to it, and up from activities whose intrinsic worth can be demonstrated by Aristotle's consensual method to forms of life in which we are capable by nature of finding point.[54]

15. It might be interesting and fruitful to pick over the wreckage of defunct and discredited ethical theories and see what their negligence of the problem of life's having a meaning contributed to their ruin. I have little to report under this head. But it does seem plain that the failure of naturalistic theories, theories reductively identifying the Good or the End with some natural reality, has been bound up with the question of meaning. Surely the failure of all the reductive naturalisms of the nineteenth century—Pleasure and Pain Utilitarianism, Marxism, Evolutionary Ethics—was precisely the failure to discover in brute nature itself (either in the totality of future pleasures or in the supposedly inevitable advance of various social or biological cum evolutionary processes), anything that the generality of untheoretical men could find reason to invest with overwhelming *importance*. These theories offered nothing that could engage in the right way with human

54. Maybe neither the consensual method nor the argued discussion of such forms would be possible in the absence of the shared neurophysiology that makes possible such community of concepts and such agreement as exists in evaluative and deliberative judgments. Nor would there be such faint prospects as there are of attaining reflective equilibrium or finding a shared mode of criticism. But nature plays only a causal and enabling role here, not the unconvincing speaking part assigned to it by Ethical Naturalism and by Aristotelian Eudaemonism. Aristotle qualified by the addition 'in a complete life' (1098a 16) the equation *eudaimonia = activity of soul in accordance with virtue*. And, tempering somewhat the *sum of goods* conception, he could agree with my strictures on the idea that the philosopher describes a meaning for life by building upwards from the special condition of its meaninglessness. But, as J. L. Austin used to complain, 'If *life* comes in at all, it should not come into Aristotle's argument as an afterthought'. And no help is to be had here from Aristotle's idea that, just as an eye has a function f such that the eye's goodness in respect of f = the good *for* the eye, so a man has his function. Eye:body::man:what? *Cf.* 1194b12. What is it for a man to find some function f that he can *embrace as his,* as giving his life meaning? Nature does not declare.

[162]

You can't reduce the notion of what is valuable to something else

concerns or give point or focus to anyone's life. (This is the cognitivist version of a point that ought to be attributed to David Hume.)

Naturalistic theories have been replaced in our own time by Prescriptivism, Emotivism, Existentialism, and Neutral (satisfaction-based) Utilitarianism. It is misleading to speak of them together. The second and third have had important affinities with moral Pyrrhonism. The first and fourth are very careful and, in the promotion of formal or second-order goods such as equality, tolerance, or consistency, rather earnest. But it is also misleading not to see these positions together.

Suppose that, when pleasure and absence of pain give place in an ethical theory to unspecified merely determinable satisfaction (and when the last drop of mentality is squeezed from the revealed preference theory which is the economic parallel of philosophical Utilitarianism) someone looks to modern Utilitarianism for meaning or happiness. The theory points him towards the greatest satisfaction of human beings' desires. He might embrace that end, if he could understand what that satisfaction consisted in. He might if he could see from his own case what satisfaction consisted in. But that is very likely where he started—unless, more wisely, he started closer to the real issue and was asking himself where he should look to find a point for his life. But, so far as either question is concerned, the theory has crossed out the infantile proposal 'pleasure and lack of pain',[55] and distorted and degraded (in description if not in fact) the complexity of the structure within which a man might have improved upon the childish answer for himself. For all questions of ends, all problems about what constitutes the attainment of given human ends, and all perplexities of meaning, have been studiously but fallaciously transposed by this theory into questions of instrumental means. But means to what? The theory is appreciably further than the nineteenth-century theory was from a conceptual appreciation of the structure of values and focused unfrustrated concerns presupposed to someone's finding a point in his life; and of the need to locate correctly happiness, pleasure, and someone's conception of his own unfolding life within that structure.

If we look to existentialism, we find something curiously similar. Going back to the formation of some of these ideas, I found André Maurois's description in *Call No Man Happy* (trans. Lindley: Cape; London, 1943, p. 43) of his teacher Alain (Emile-Auguste Chartier):

> what I cannot convey by words is the enthusiasm inspired in us by this search, boldly pursued with such a guide; the excitement of those classes which are entered with the persistent hope of discovering, that very morning, the secret of life, and from which one departed with the joy of having understood that perhaps there was no such secret but that nevertheless it was possible to be a human being and to be so with dignity and

---

55. For the thought that this might be literally infantile, I am indebted indirectly to Bradley and directly to Richard Wollheim, 'The Good Self and the Bad Self', *Proceedings of the British Academy*, 1975.

nobility. When I read in *Kim* the story of the Lama who sought so piously for the River of the Arrow, I thought of *our* search.

What happens here—and remember that Alain was the teacher not only of Maurois but also of Sartre—goes wrong even in the question 'What is *the* meaning of life?' We bewitch ourselves to think that we are looking for some one thing like the Garden of the Hesperides, the Holy Grail . . . Then finding nothing like that in the world, no one thing from which all values can be derived and no one focus by which all other concerns can be organized, we console ourselves by looking inwards, but again for some one substitute thing, one thing in us now instead of the world. Of course if the search is conducted in this way it is more or less inevitable that the one consolation will be *dignity* or *nobility* or *commitment:* or more spectatorially *irony, resignation, scorn* . . . But, warm though its proper place is for each of these—important though each of them is in its own non-substitutive capacity—it would be better to go back to the 'the' in the original question; and to interest ourselves afresh in what everybody knows about—the set of concerns he actually has, their objects, and the focus he has formed or seeks to bring to bear upon these: also the prospects of purifying, redeploying or extending this set.(56)

Having brought the matter back to this place, how can a theorist go on? I think he must continue from the point where I myself ought to have begun if the products of philosophy itself had not obstructed the line of sight. Working within an intuitionism or moral phenomenology as tolerant of low-grade non-behaviouristic evidence as is literature (albeit more obsessively elaborative of the commonplace, and more theoretical, in the interpretive sense, than literature), he has to appreciate and describe the working-day complexity of what is experientially involved in a person's seeing a point in living. It is no use to take a going moral theory—Utilitarianism or whatever it is—and paste on to it such *postscripta* as the Millian insight 'It really is of importance not only what men do, but what manner of men they are that do it': or the insight that to see a point in living someone has to be such that he can like himself: or to try to super-impose upon the theory the structure that we have complained that Utilitarianism degrades. If life's having a point is at all central to moral theory then room must be made for these things right from the very beginning. The phenomenological account I advocate would accommodate all these things in conjunction with (1) ordinary anthropocentric objectivity, (2) the elements of value-focus and discov-

---

56. *Cf.* Williams, 'Persons, Character and Morality' (pp. 208ff.), in Amelie Rorty (ed.), *The Identities of Persons* (University of California Press, 1976): "The categorical desires which propel one forward do not have to be even very evident to consciousness, let alone grand or large; one good testimony to one's existence having a point is that the question of its point does not arise, and the propelling concerns may be of a relatively everyday kind such as certainly provide the ground of many sorts of happiness (*cf.* p. 209)."

ery, and (3) the element of invention that it is the non-cognitivist's conspicuous distinction to have imported into the argument.

Let us not underestimate what would have been done if this work were realized. But ought the theorist to be able to do more? Reluctant as one should be to draw any limits to the potentiality or enterprise of discursive reason, I see no reason why he should. Having tamed non-cognitivism and made of it a doctrine of cognitive underdetermination, which allows the world to impinge upon but not to determine the point possessed by individual lives, and which sees value properties not as created but as *lit up* by the focus that the man who lives the life brings to the world; and, having described what finding meaning is, it will not be for the theorist as such to insist on intruding himself further. As Bradley says in *Appearance and Reality* (450):

> If to show theoretical interest in morality and religion is taken as setting oneself up as a teacher or preacher, I would rather leave these subjects to whoever feels that such a character suits him.

8

Values and
Secondary Qualities

John McDowell

1. J. L. Mackie insists that ordinary evaluative thought presents itself as a matter of sensitivity to aspects of the world.[1] And this phenomenological thesis seems correct. When one or another variety of philosophical non-cognitivism claims to capture the truth about what the experience of value is like, or (in a familiar surrogate for phenomenology[2]), about what we mean by our evaluative language, the claim is never based on careful attention to the lived character of evaluative thought or discourse. The idea is, rather, that the very concept of the cognitive or factual rules out the possibility of an undiluted representation of how things are, enjoying, nevertheless, the internal relation to 'attitudes' or the will that would be needed for it to count as evaluative.[3] On this view the phenomenology of value would involve a mere incoherence, if it were—as Mackie says—a possibility that then tends (naturally enough) not to be so much as entertained.

This paper grew out of my contributions to a seminar on J. L. Mackie's *Ethics: Inventing Right and Wrong* (Penguin, Harmondsworth, 1977: hereafter referred to as *E*) which I had the privilege of sharing with Mackie and R. M. Hare in 1978. It was first published in *Morality and Objectivity*, ed. Ted Honderich (London: Routledge and Kegan Paul, 1985), pp. 110–29. I do not believe John Mackie would have found it strange that I should pay tribute to a sadly missed colleague by continuing a strenuous disagreement with him.

1. See *E*, pp. 31–5 (pp. 105–9 in this volume). I shall also abbreviate references to the following other books by Mackie: *Problems from Locke* (Clarendon Press, Oxford, 1976: hereafter *PFL*); and *Hume's Moral Theory* (Routledge and Kegan Paul, London, 1980); hereafter *HMT*.

2. An inferior surrogate: it leads us to exaggerate the extent to which expressions of our sensitivity to values are signalled by the use of a special vocabulary. See my 'Aesthetic Value, Objectivity, and the Fabric of the World', in Eva Schaper, ed., *Pleasure, Preference, and Value* (Cambridge University Press, Cambridge, 1983), pp. 1–16, at pp. 1–2.

3. I am trying here to soften a sharpness of focus that Mackie introduces by stressing the notion of prescriptivity. Mackie's singleness of vision here has the perhaps unfortunate effect of discouraging a distinction such as David Wiggins has drawn between 'valuations' and 'directives or deliberative (or practical) judgements' (see 'Truth, Invention, and the Meaning of Life', *Proceedings of the British Academy* LXII (1976), 331–78, at pp. 338–9). My topic here is really the former of these. (It may be that the distinction does not matter in the way that Wiggins suggests: see n. 35 below.)

But, as Mackie sees, there is no satisfactory justification for supposing that the factual is, by definition, attitudinatively and motivationally neutral. This clears away the only obstacle to accepting his phenomenological claim; and the upshot is that non-cognitivism must offer to correct the phenomenology of value, rather than to give an account of it.[4]

In Mackie's view the correction is called for. In this paper I want to suggest that he attributes an unmerited plausibility to this thesis, by giving a false picture of what one is committed to if one resists it.

2. Given that Mackie is right about the phenomenology of value, an attempt to accept the appearances makes it virtually irresistible to appeal to a perceptual model. Now Mackie holds that the model must be perceptual awareness of *primary* qualities (see *HMT*, pp. 32, 60–1, 73–4). And this makes it comparatively easy to argue that the appearances are misleading. For it seems impossible—at least on reflection—to take seriously the idea of something that is like a primary quality in being simply *there,* independently of human sensibility, but is nevertheless intrinsically (not conditionally on contingencies about human sensibility) such as to elicit some 'attitude' or state of will from someone who becomes aware of it. Moreover, the primary-quality model turns the epistemology of value into mere mystification. The perceptual model is no more than a model: perception, strictly so called, does not mirror the role of reason in evaluative thinking, which seems to require us to regard the apprehension of value as an intellectual rather than a merely sensory matter. But if we are to take account of this, while preserving the model's picture of values as brutely and absolutely *there,* it seems that we need to postulate a faculty—'intuition'—about which all that can be said is that it makes us aware of objective rational connections: the model itself ensures that there is nothing helpful to say about how such a faculty might work, or why its deliverances might deserve to count as knowledge.

But why is it supposed that the model must be awareness of primary qualities rather than secondary qualities? The answer is that Mackie, following Locke, takes secondary-quality perceptions, as conceived by a prephilosophical consciousness, to involve a projective error: one analogous to the error he finds in ordinary evaluative thought. He holds that we are prone to conceive secondary-quality experience in a way that would be appropriate for experience of primary qualities. So a pre-philosophical secondary-quality model for awareness of value

[handwritten margin note: mcdowell believes that how can primary qualities be independent of human sensibility]

4. I do not believe that the 'quasi-realism' that Simon Blackburn has elaborated is a real alternative to this. (See p. 358 of his 'Truth, Realism, and the Regulation of Theory', in Peter A. French, Theodore E. Uehling, Jr., and Howard Wettstein, eds., *Midwest Studies in Philosophy V: Studies in Epistemology* (University of Minnesota Press, Minneapolis, 1980, pp. 353–71.) In so far as the quasi-realist holds that the values, in his thought and speech about which he imitates the practices supposedly characteristic of realism, are *really* products of projecting 'attitudes' into the world, he must have a conception of genuine reality—that which the values lack and the things on to which they are projected have. And the phenomenological claim ought to be that *that* is what the appearances entice us to attribute to values.

[handwritten margin note: doesn't understand the destinction between primary + secondary qualities]

would in effect be, after all, a primary-quality model. And to accept a philosophically corrected secondary-quality model for the awareness of value would be simply to give up trying to go along with the appearances.

I believe, however, that this conception of secondary-quality experience is seriously mistaken.

3. A secondary quality is a property the ascription of which to an object is not adequately understood except as true, if it is true, in virtue of the object's disposition to present a certain sort of perceptual appearance: specifically, an appearance characterizable by using a word for the property itself to say how the object perceptually appears. Thus an object's being red is understood as obtaining in virtue of the object's being such as (in certain circumstances) to look, precisely, red.

[handwritten margin note: account like — Locke the colour as a secondary quality]

This account of secondary qualities is faithful to one key Lockean doctrine, namely the identification of secondary qualities with 'powers to produce various sensations in us'[5]. (The phrase 'perceptual appearance', with its gloss, goes beyond Locke's unspecific 'sensations', but harmlessly; it serves simply to restrict our attention, as Locke's word may not, to properties that are in a certain obvious sense perceptible.[6])

I have written of what property-ascriptions are understood to be true in virtue of, rather than of what they are true in virtue of. No doubt it is true that a given thing is red in virtue of some microscopic textural property of its surface; but a predication understood only in such terms—not in terms of how the object would look—would not be an ascription of the secondary quality of redness.[7]

Secondary-quality experience presents itself as perceptual awareness of properties genuinely possessed by the objects that confront one. And there is no general obstacle to taking that appearance at face value.[8] An object's being such as to look red is independent of its actually looking red to anyone on any particular occasion; so, notwithstanding the conceptual connection between being red and being experienced as red, an experience of something as red can count as a case of being presented with a property that is there anyway—there independently of the

---

5. *An Essay concerning Human Understanding,* II, viii.10.

6. Being stung by a nettle is an actualization of a power in the nettle that conforms to Locke's description, but it seems wrong to regard it as a perception of that power; the experience lacks an intrinsically representational character which that would require. (It is implausible that looking red is intelligible independently of being red; combined with the account of secondary qualities that I am giving, this sets up a circle. But it is quite unclear that we ought to have the sort of analytic or definitional aspirations that would make the circle problematic. See Colin McGinn, *The Subjective View* (Clarendon Press, Oxford, 1983), pp. 6–8.

7. See McGinn, op. cit., pp. 12–14.

8. Of course there is room for the concept of illusion, not only because the senses can malfunction but also because of the need for a modifier like my '(in certain circumstances)', in an account of what it is for something to have a secondary quality. (The latter has no counterpart with primary qualities.)

[handwritten: Colour is independent of experience therefore it cannot be a secondary quality]

experience itself.[9] And there is no evident ground for accusing the appearance of being misleading. What would one expect it to be like to experience something's being such as to look red, if not to experience the thing in question (in the right circumstances) as looking, precisely, red?

On Mackie's account, by contrast, to take experiencing something as red at face value, as a non-misleading awareness of a property that really confronts one, is to attribute to the object a property which is 'thoroughly objective' (*PFL*, p. 18), in the sense that it does not need to be understood in terms of experiences that the object is disposed to give rise to; but which nevertheless resembles redness as it figures in our experience—this to ensure that the phenomenal character of the experience need not stand accused of misleadingness, as it would if the 'thoroughly objective' property of which it consituted an awareness were conceived as a microscopic textural basis for the object's disposition to look red. This use of the notion of resemblance corresponds to one key element in Locke's exposition of the concept of a primary quality.[10] In these Lockean terms Mackie's view amounts to accusing a naïve perceptual consciousness of taking secondary qualities for primary qualities (see *PFL*, p. 16).

According to Mackie, this conception of primary qualities that resemble colours as we see them is coherent; that nothing is characterized by such qualities is established by merely empirical argument (see *PFL*, pp. 17–20). But is the idea coherent? This would require two things: first, that colours figure in perceptual experience neutrally, so to speak, rather than as essentially phenomenal qualities of objects, qualities that could not be adequately conceived except in terms of how their possessors would look; and, second, that we command a concept of resemblance that would enable us to construct notions of possible primary qualities out of the idea of resemblance to such neutral elements of experience. The first of these requirements is quite dubious. (I shall return to this.) But even if we try to let it pass, the second requirement seems impossible. Starting with, say, redness as it (putatively neutrally) figures in our experience, we are asked to form the notion of a feature of objects which resembles that, but which is adequately conceivable otherwise than in terms of how its possessors would look (since if it were adequately conceivable only in those terms it would simply be secondary). But the second part of these instructions leaves it wholly mysterious what to make of the first: it precludes the required resemblance being in phenomenal respects, but it is quite unclear what other sense we could make of the notion of resemblance to redness as it figures in our experience. (If we find no other, we have failed to let the first requirement pass; redness as it figures in our experience proves stubbornly

9. See the discussion of (one interpretation of the notion of) objectivity at pp. 77–8 of Gareth Evans, 'Things without the Mind', in Zak van Straaten, ed., *Philosophical Subjects: Essays Presented to P. F. Strawson* (Clarendon Press, Oxford, 1980), pp. 76–116. Throughout the present section I am heavily indebted to this most important paper.

10. See *Essay*, II, viii.15.

[169]

phenomenal.)[11] I have indicated how we can make error-free sense of the thought that colours are authentic objects of perceptual awareness; in face of that, it seems a gratuitous slur on perceptual 'common sense' to accuse it of this wildly problematic understanding of itself.

Why is Mackie resolved, nevertheless, to convict 'common sense' of error? Secondary qualities are qualities not adequately conceivable except in terms of certain subjective states, and thus subjective themselves in a sense that that characterization defines. In the natural contrast, a primary quality would be objective in the sense that what it is for something to have it can be adequately understood otherwise than in terms of dispositions to give rise to subjective states. Now this contrast between objective and subjective is not a contrast between veridical and illusory experience. But it is easily confused with a different contrast, in which to call a putative object of awareness 'objective' is to say that it is there to be experienced, as opposed to being a mere figment of the subjective state that purports to be an experience of it. If secondary qualities were subjective in the sense that naturally contrasts with this, naïve consciousness would indeed be wrong about them, and we would need something like Mackie's Lockean picture of the error it commits. What is acceptable, though, is only that secondary qualities are subjective in the first sense, and it would be simply wrong to suppose that this gives any support to the idea that they are subjective in the second.[12]

More specifically, Mackie seems insufficiently whole-hearted in an insight of his about perceptual experiences. In the case of 'realistic' depiction, it makes sense to think of veridicality as a matter of resemblance between aspects of a picture and aspects of what it depicts.[13] Mackie's insight is that the best hope of a philosophically hygienic interpretation for Locke's talk of 'ideas', in a perceptual context, is in terms of 'intentional objects': that is, aspects of representational content—aspects of how things seem to one in the enjoyment of a perceptual experience. (See *PFL*, pp. 47–50.) Now it is an illusion to suppose, as Mackie does, that this warrants thinking of the relation between a quality and an 'idea' of it on the model of the relation between a property of a picture's subject and an aspect of the picture. Explaining 'ideas' as 'intentional objects' should direct our attention to the relation between how things are and how an experience represents them as being—in fact, identity, not resemblance, if the representation is veridi-

---

11. Cf. pp. 56–7 of P. F. Strawson, 'Perception and Its Objects', in G. F. Macdonald, ed., *Perception and Identity: Essays Presented to A. J. Ayer* (Macmillan, London, 1979), pp. 41–60.

12. This is a different way of formulating a point made by McGinn, op. cit., p. 121. Mackie's phrase 'the fabric of the world' belongs with the second sense of 'objective', but I think his arguments really address only the first. *Pace* p. 103 of A. W. Price, 'Varieties of Objectivity and Values', *Proceedings of the Aristotelian Society* lxxxii (1982–3), 103–19, I do not think the phrase can be passed over as unhelpful, in favour of what the arguments do succeed in establishing, without missing something that Mackie wanted to say. (A gloss on 'objective' as 'there to be experienced' does not figure in Price's inventory, p. 104. It seems to be the obvious response to his challenge at pp. 118–9.)

13. I do not say it is correct: scepticism about this is very much in point. (See Nelson Goodman, *Languages of Art* [Oxford University Press, London, 1969], chapter I.)

cal.[14] Mackie's Lockean appeal to resemblance fits something quite different: a relation borne to aspects of how things are by intrinsic aspects of a bearer of representational content—not how things are represented to be, but features of an item that does the representing, with particular aspects of its content carried by particular aspects of what it is intrinsically (nonrepresentationally) like.[15] Perceptual experiences have representational content; but nothing in Mackie's defence of the 'intentional objects' gloss on 'ideas' would force us to suppose that they have it in that sort of way.[16]

The temptation to which Mackie succumbs, to suppose that intrinsic features of experience function as vehicles for particular aspects of representational content, is indifferent to any distinction between primary and secondary qualities in the representational significance that these features supposedly carry. What it is for a colour to figure in experience and what it is for a shape to figure in experience would be alike, on this view, in so far as both are a matter of an experience's having a certain intrinsic feature. If one wants, within this framework, to preserve Locke's intuition that primary-quality experience is distinctive in potentially disclosing the objective properties of things, one will be naturally led to Locke's use of the notion of resemblance. But no notion of resemblance could get us from an essentially experiential state of affairs to the concept of a feature of objects intelligible otherwise than in terms of how its possessors would strike us. (A version of this point told against Mackie's idea of possible primary qualities answering to 'colours as we see them'; it tells equally against the Lockean conception of shapes.)

If one gives up the Lockean use of resemblance, but retains the idea that primary and secondary qualities are experientially on a par, one will be led to suppose that the properties attributed to objects in the 'manifest image' are all equally phenomenal—intelligible, that is, only in terms of how their possessors are disposed to appear. Properties that are objective, in the contrasting sense, can

14. When resemblance is in play, it functions as a palliative to lack of veridicality, not as what veridicality consists in.

15. Intrinsic features of experience, functioning as vehicles for aspects of content, seem to be taken for granted in Mackie's discussion of Molyneux's problem (*PFL*, pp. 28–32). The slide from talk of content to talk that fits only bearers of content seems to happen also in Mackie's discussion of truth, in *Truth, Probability, and Paradox* (Clarendon Press, Oxford, 1973), with the idea that a formulation like 'A true statement is one such that the way things are is the way it represents things as being' makes truth consist in a relation of correspondence (rather than identity) between how things are and how things are represented as being; pp. 56–7 come too late to undo the damage done by the earlier talk of 'comparison', e.g., at pp. 50, 51. (A subject matter for the talk that fits bearers is unproblematically available in this case; but Mackie does not mean to be discussing truth as a property of sentences or utterances.)

16. Indeed, this goes against the spirit of a passage about the word 'content' at *PFL*, p. 43. Mackie's failure to profit by his insight emerges particularly strikingly in his remarkable claim (*PFL*, p. 50) that the 'intentional object' conception of the content of experience yields an account of perception that is within the target area of 'the stock objections against an argument from an effect to a supposed cause of a type which is never directly observed'. (Part of the trouble here is a misconception of direct realism as a surely forlorn attempt to make perceptual knowledge unproblematic: *PFL*, p. 43.)

then figure only in the 'scientific image'.[17] On these lines one altogether loses hold of Locke's intuition that primary qualities are distinctive in being both objective and perceptible.[18]

If we want to preserve the intuition, as I believe we should, then we need to exorcize the idea that what it is for a quality to figure in experience is for an experience to have a certain intrinsic feature; in fact I believe that we need to reject these supposed vehicles of content altogether. Then we can say that colours and shapes figure in experience, not as the representational significance carried by features that are—being intrinsic features of experience—indifferently subjective (which makes it hard to see how a difference in respect of objectivity could show up in their representational significance); but simply as properties that objects are represented as having, distinctively phenomenal in the one case and not so in the other. (Without the supposed intrinsic features, we would be immune to the illusion that experiences cannot represent objects as having properties that are not phenomenal—properties that are adequately conceivable otherwise than in terms of dispositions to produce suitable experiences.[19]) What Locke unfelicitously tried to yoke together, with his picture of real resemblances of our 'ideas', can now divide into two notions that we must insist on keeping separate: first, the possible veridicality of experience (the objectivity of its object, in the second of the two senses I distinguished), in respect of which primary and secondary qualities are on all fours; and second, the not essentially phenomenal character of some properties that experience represents objects as having (their objectivity in the first sense), which marks off the primary perceptible qualities from the secondary ones.

In order to deny that a quality's figuring in experience consists in an experience's having a certain intrinsic feature, we do not need to reject the intrinsic features altogether; it would suffice to insist that a quality's figuring in experience consists in an experience's having a certain intrinsic feature *together with* the quality's being the representational significance carried by that feature. But I do not believe that this yields a position in which acceptance of the supposed vehicles of content coheres with a satisfactory account of perception. This position would have it that the fact that an experience represents things as being one way rather

---

17. The phrases 'manifest image' and 'scientific image' are due to Wilfrid Sellars; see 'Philosophy and the Scientific Image of Man', in *Science, Perception and Reality* (Routledge and Kegan Paul, London, 1963).

18. This is the position of Strawson, op. cit. (and see also his 'Reply to Evans', in van Straaten, ed., op. cit., pp. 273–82). I am suggesting a diagnosis, to back up McGinn's complaint, op. cit., p. 124n.

19. Notice Strawson's sleight of hand with phrases like 'shapes-as-seen', at p. 286 of 'Reply to Evans'. Strawson's understanding of what Evans is trying to say fails altogether to accommodate Evans's remark ('Things without the Mind', p. 96) that 'to deny that . . . primary properties are *sensory* is not at all to deny that they are *sensible* or *observable*'. Shapes as seen are *shapes*—that is, non-sensory properties: it is one thing to deny, as Evans does, that experience can furnish us with the concepts of such properties, but quite another to deny that experience can disclose instantiations of them to us.

than another is strictly additional to the experience's intrinsic nature, and so extrinsic to the experience itself (it seems natural to say 'read into it'). There is a phenomenological falsification here. (This brings out a third role for Locke's resemblance, namely to obviate the threat of such a falsification by constituting a sort of instrinsic representationality: Locke's 'ideas' carry the representational significance they do by virtue of what they are like, and this can be glossed both as 'how they are intrinsically' and as 'what they resemble'.) In any case, given that we cannot project ourselves from features of experience to non-phenomenal properties of objects by means of an appeal to resemblance, it is doubtful that the metaphor of representational significance being 'read into' intrinsic features can be spelled out in such a way as to avoid the second horn of our dilemma. How could representational significance be 'read into' intrinsic features of experience in such a way that what was signified did not need to be understood in terms of them? How could a not intrinsically representational feature of experience become imbued with objective significance in such a way that an experience could count, by virtue of having that feature, as a direct awareness of a not essentially phenomenal property of objects?[20]

How things strike someone as being is, in a clear sense, a subjective matter: there is no conceiving it in abstraction from the subject of the experience. Now a motive for insisting on the supposed vehicles of aspects of content might lie in an aspiration, familiar in philosophy, to bring subjectivity within the compass of a fundamentally objective conception of reality.[21] If aspects of content are not carried by elements in an intrinsic structure, their subjectivity is irreducible. By contrast, one might hope to objectivize any 'essential subjectivity' that needs to be attributed to not intrinsically representational features of experience, by exploiting a picture involving special access on a subject's part to something conceived in a broadly objective way—its presence in the world not conceived as constituted by the subject's special access to it.[22] Given this move, it becomes natural to suppose that the phenomenal character of the 'manifest image' can be explained in terms of a certain familiar picture: one in which a confronted 'external' reality, conceived as having only an objective nature, is processed

---

20. Features of physiologically specified states are not to the point here. Such features are not apparent in experience; whereas the supposed features that I am concerned with would have to be aspects of what experience is like for us, in order to function intelligibly as carriers for aspects of the content that experience presents to us. There may be an inclination to ask why it should be any harder for a feature of experience to acquire an objective significance than it is for a word to do so. But the case of language affords no counterpart to the fact that the objective significance in the case we are concerned with is a matter of how things (e.g.) *look* to be: the special problem is how to stop that 'look' having the effect that a supposed intrinsic feature of experience gets taken up into its own representational significance, thus ensuring that the significance is phenomenal and not primary.

21. See Thomas Nagel, 'Subjective and Objective', in *Mortal Questions* (Cambridge University Press, Cambridge, 1979), pp. 196–213.

22. Cf. Bernard Williams, *Descartes: The Project of Pure Enquiry* (Penguin, Harmondsworth, 1978), p. 295.

through a structured 'subjectivity', conceived in this objectivistic manner. This picture seems to capture the essence of Mackie's approach to the secondary qualities.[23] What I have tried to suggest is that the picture is suspect in threatening to cut us off from the *primary* (not essentially phenomenal) qualities of the objects that we perceive: either (with the appeal to resemblance) making it impossible, after all, to keep an essentially phenomenal character out of our conception of the qualities in question, or else making them merely hypothetical, not accessible to perception. If we are to achieve a satisfactory understanding of experience's openness to objective reality, we must put a more radical construction on experience's essential subjectivity. And this removes an insidious obstacle—one whose foundation is summarily captured in Mackie's idea that it is not simply wrong to count 'colours as we see them' as items in our minds (see the diagram at *PFL*, p. 17)—that stands in the way of understanding how secondary-quality experience can be awareness, with nothing misleading about its phenomenal character, of properties genuinely possessed by elements in a not exclusively phenomenal reality.

4. The empirical ground that Mackie thinks we have for not postulating 'thoroughly objective features which resemble our ideas of secondary qualities' (*PFL*, pp. 18–19) is that attributing such features to objects is surplus to the requirements of explaining our experience of secondary qualities (see *PFL*, pp. 17–18). If it would be incoherent to attribute such features to objects, as I believe, this empirical argument falls away as unnecessary. But it is worth considering how an argument from explanatory superfluity might fare against the less extravagant construal I have suggested for the thought that secondary qualities genuinely characterize objects: not because the question is difficult or contentious, but because of the light it casts on how an explanatory test for reality—which is commonly thought to undermine the claims of values—should be applied.

A '*virtus dormitiva*' objection would tell against the idea that one might mount a satisfying explanation of an object's looking red on its being such as to look red. The weight of the explanation would fall through the disposition to its structural ground.[24] Still, however optimistic we are about the prospects for explaining colour experience on the basis of surface textures,[25] it would be obviously wrong

23. Although McGinn, op. cit., is not taken in by the idea that 'external' reality has only objective characteristics, I am not sure that he sufficiently avoids the picture that underlies that idea: see pp. 106–9. (This connects with a suspicion that at pp. 9–10 he partly succumbs to a temptation to objectivize the subjective properties of objects that he countenances: it is not as clear as he seems to suppose that, say, redness can be, so to speak, abstracted from the way things strike *us* by an appeal to relativity. His worry at pp. 132–6, that secondary-quality experience may after all be phenomenologically misleading, seems to betray the influence of the idea of content-bearing intrinsic features of experience.)

24. See McGinn, op. cit., p. 14.

25. There are difficulties over how complete such explanations could aspire to be: see Price, op. cit., pp. 114–5; and my 'Aesthetic Value, Objectivity, and the Fabric of the World', op. cit., pp. 10–12.

to suppose that someone who gave such an explanation could in consistency deny that the object was such as to look red. The right explanatory test is not whether something pulls its own weight in the favoured explanation (it may fail to do so without thereby being explained away), but whether the explainer can consistently deny its reality.[26]

Given Mackie's view about secondary qualities, the thought that values fail an explanatory test for reality is implicit in a parallel that he commonly draws between them (see, for instance, *HMT*, pp. 51–2; *E*, pp. 19–20; p. 99 in this volume). It is nearer the surface in his 'argument from queerness' (*E*, pp. 38–42; pp. 111–12 above), and explicit in his citing 'patterns of objectification' to explain the distinctive phenomenology of value experience (*E*, pp. 42–6; pp. 114–16 above).[27] Now it is, if anything, even more obvious with values than with essentially phenomenal qualities that they cannot be credited with causal efficacy: values would not pull their weight in any explanation of value experience even remotely analogous to the standard explanations of primary-quality experience. But reflection on the case of secondary qualities has already opened a gap between that admission and any concession that values are not genuine aspects of reality. And the point is reinforced by a crucial disanalogy between values and secondary qualities. To press the analogy is to stress that evaluative 'attitudes', or states of will, are like (say) colour experience in being unintelligible except as modifications of a sensibility like ours. The idea of value experience involves taking admiration, say, to represent its object as having a property which (although there in the object) is essentially subjective in much the same way as the property that an object is represented as having by an experience of redness—that is, understood adequately only in terms of the appropriate modification of human (or similar) sensibility. The disanalogy, now, is that a virtue (say) is conceived to be not merely such as to elicit the appropriate 'attitude' (as a colour is merely such as to cause the appropriate experiences), but rather such as to *merit* it. And this makes it doubtful whether merely causal explanations of value experience are relevant to the explanatory test, even to the extent that the question to ask is whether someone could consistently give such explanations while denying that the values involved are real. It looks as if we should be raising that question about explanations of a different kind.

For simplicity's sake, I shall elaborate this point in connection with something that is not a value, though it shares the crucial feature: namely danger or the fearful. On the face of it, this might seem a promising subject for a projectivist

26. Cf. pp. 206–8, especially p. 208, of David Wiggins, 'What Would Be a Substantial Theory of Truth?' in van Straaten, ed., op. cit., pp. 189–221. The test of whether the explanations in question are consistent with rejecting the item in contention is something that Wiggins once mooted, in the course of a continuing attempt to improve that formulation: I am indebted to discussions with him.

27. See also Simon Blackburn, 'Rule-Following and Moral Realism', in Steven Holtzman and Christopher Leich, eds., *Wittgenstein: To Follow a Rule* (Routledge and Kegan Paul, London, 1981), pp. 163–87; and the first chapter of Gilbert Harman, *The Nature of Morality* (Oxford University Press, New York, 1977) (Essay 6 in this volume).

John McDowell

treatment (a treatment that appeals to what Hume called the mind's 'propensity to spread itself on external objects').[28] At any rate the response that, according to such a treatment, is projected into the world can be characterized, without phenomenological falsification, otherwise than in terms of seeming to find the supposed product of projection already there.[29] And it would be obviously grotesque to fancy that a case of fear might be explained as the upshot of a mechanical (or perhaps para-mechanical) process initiated by an instance of 'objective fearfulness'. But if what we are engaged in is an 'attempt to understand ourselves',[30] then merely causal explanations of responses like fear will not be satisfying anyway.[31] What we want here is a style of explanation that makes sense of what is explained (in so far as sense can be made of it). This means that a technique for giving satisfying explanations of cases of fear—which would perhaps amount to a satisfying explanatory theory of danger, though the label is possibly too grand— must allow for the possibility of criticism; we make sense of fear by seeing it as a response to objects that *merit* such a response, or as the intelligibly defective product of a propensity towards responses that would be intelligible in that way.[32] For an object to merit fear just is for it to be fearful. So explanations of fear that manifest our capacity to understand ourselves in this region of our lives will simply not cohere with the claim that reality contains nothing in the way of fearfulness.[33] Any such claim would undermine the intelligibility that the explanations confer on our responses.

The shared crucial features suggests that this disarming of a supposed explanatory argument for unreality should carry over to the case of values. There is, of course, a striking disanalogy in the contentiousness that is typical of values; but I think it would be a mistake to suppose that this spoils the point. In so far as we succeed in achieving the sort of understanding of our responses that is in question, we do so on the basis of preparedness to attribute, to at least some possible objects of the responses, properties that would validate the responses. What the disanalogy makes especially clear is that the explanations that preclude our denying the reality of the special properties that are putatively discernible from some (broadly)

28. *A Treatise of Human Nature*, I.iii.14. 'Projectivist' is Blackburn's useful label: see 'Rule-Following and Moral Realism', op. cit., and 'Opinions and Chances', in D. H. Mellor, ed., *Prospects for Pragmatism* (Cambridge University Press, Cambridge, 1980), pp. 175–96.

29. At pp. 180–1 of 'Opinions and Chances', Blackburn suggests that a projectivist need not mind whether or not this is so; but I think he trades on a slide between 'can . . . only be understood in terms of' and 'our best vocabulary for identifying' (which allows that there may be an admittedly inferior alternative).

30. The phrase is from p. 165 of Blackburn, 'Rule-Following and Moral Realism'.

31. I do not mean that satisfying explanations will not be causal. But they will not be *merely* causal.

32. I am assuming that we are not in the presence of a theory according to which no responses of the kind in question *could* be well-placed. That would have a quite unintended effect. (See *E*, p. 16; p. 96 in this volume.) Notice that it will not meet my point to suggest that calling a response 'well-placed' is to be understood only quasi-realistically. Explanatory indispensability is supposed to be the test for the *genuine* reality supposedly lacked by what warrants only quasi-realistic treatment.

33. Cf. Blackburn, 'Rule-Following and Moral Realism', op. cit., p. 164.

evaluative point of view are themselves constructed from that point of view. (We already had this in the case of the fearful, but the point is brought home when the validation of the responses is controversial.) However, the critical dimension of the explanations that we want means that there is no question of just any actual response pulling itself up by its own bootstraps into counting as an undistorted perception of the relevant special aspect of reality.[34] Indeed, awareness that values are contentious tells against an unreflective contentment with the current state of one's critical outlook, and in favour of a readiness to suppose that there may be something to be learned from people with whom one's first inclination is to disagree. The aspiration to understand oneself is an aspiration to change one's responses, if that is necessary for them to become intelligible otherwise than as defective. But although a sensible person will never be confident that his evaluative outlook is incapable of improvement, that need not stop him supposing, of some of his evaluative responses, that their objects really do merit them. He will be able to back up this supposition with explanations that show how the responses are well-placed; the explanations will share the contentiousness of the values whose reality they certify, but that should not stop him accepting the explanations any more than (what nobody thinks) it should stop him endorsing the values.[35] There is perhaps an air of bootstrapping about this. But if we restrict ourselves to explanations from a more external standpoint, at which values are not in our field of view, we deprive ourselves of a kind of intelligibility that we aspire to; and projectivists have given no reason whatever to suppose that there would be anything better about whatever different kind of self-understanding the restriction would permit.

5. It will be obvious how these considerations undermine the damaging effect of the primary-quality model. Shifting to a secondary-quality analogy renders irrelevant any worry about how something that is brutely *there* could nevertheless stand in an internal relation to some exercise of human sensibility. Values are not brutely there—not there independently of our sensibility—any more than colours

---

34. This will be so even in areas in which there are no materials for constructing standards of criticism except actual responses: something that is not so with fearfulness, although given a not implausible holism it will be so with values.

35. I can see no reason why we should not regard the contentiousness as ineliminable. The effect of this would be to detach the explanatory test of reality from a requirement of convergence (cf. the passage by Wiggins cited in n. 26 above). As far as I can see, this separation would be a good thing. It would enable resistance to projectivism to free itself, with a good conscience, of some unnecessary worries about relativism. It might also discourage a misconception of the appeal to Wittgenstein that comes naturally to such a position. (Blackburn, 'Rule-Following and Moral Realism', pp. 170–4, reads into my 'Non-Cognitivism and Rule-Following', in Holtzman and Leich, eds., op. cit., pp. 141–62, an interpretation of Wittgenstein as, in effect, making truth a matter of consensus, and has no difficulty in arguing that this will not make room for hard cases; but the interpretation is not mine.) With the requirement of convergence dropped, or at least radically relativized to a point of view, the question of the claim to truth of directives may come closer to the question of the truth status of evaluations than Wiggins suggests, at least in 'Truth, Invention, and the Meaning of Life', op. cit.

[handwritten annotation: —has difficulty relating values to colour]

are; though, as with colours, this does not stop us supposing that they are there independently of any particular apparent experience of them. As for the epistemology of value, the epistemology of danger is a good model. (Fearfulness is not a secondary quality, although the model is available only after the primary-quality model has been dislodged. A secondary-quality analogy for value experience gives out at certain points, no less than the primary-quality analogy that Mackie attacks.) To drop the primary-quality model in this case is to give up the idea that fearfulness itself, were it real, would need to be intelligible from a standpoint independent of the propensity to fear; the same must go for the relations of rational consequentiality in which fearfulness stands to more straightforward properties of things.[36] Explanations of fear of the sort I envisaged would not only establish, from a different standpoint, that some of its objects are really fearful, but also make plain, case by case, what it is about that that makes them so; this should leave it quite unmysterious how a fear response rationally grounded in awareness (unproblematic, at least for present purposes) of these 'fearful-making characteristics' can be counted as being, or yielding, knowledge that one is confronted by an instance of real fearfulness.[37]

Simon Blackburn has written, on behalf of a projectivist sentimentalism in ethics, that 'we profit . . . by realizing that a training of the feelings rather than a cultivation of a mysterious ability to spot the immutable fitnesses of things is the foundation of how to live'.[38] This picture of what an opponent of projectivism must hold is of a piece with Mackie's primary-quality model; it simply fails to fit the position I have described.[39] Perhaps with Aristotle's notion of practical wisdom in mind, one might ask why a training of the feelings (as long as the notion of feeling is comprehensive enough) cannot *be* the cultivation of an ability—utterly unmysterious just because of its connections with feelings—to spot (if you like) the fitnesses of things; even 'immutable' may be all right, so long as it is not understood (as I take it Blackburn intends) to suggest a 'platonistic' conception of the fitnesses of things, which would reimport the characteristic ideas of the primary-quality model.[40]

Mackie's response to this suggestion used to be, in effect, that it simply

36. Mackie's question (*E*, p. 41; p. 113 in this volume) 'Just what *in the world* is signified by this "because"?' involves a tendentious notion of 'the world'.

37. See Price, op. cit., pp. 106–7, 115.

38. 'Rule-Following and Moral Realism', p. 186.

39. Blackburn's realist evades the explanatory burdens that sentimentalism discharges, by making the world rich (cf. p. 181) and then picturing it as simply setting its print on us. Cf. *E*, p. 22 (p. 100 in this volume): 'If there were something in the fabric of the world that validated certain kinds of concern, then it would be possible to acquire these merely by finding something out, by letting one's thinking be controlled by how things were'. This saddles an opponent of projectivism with a picture of awareness of value as an exercise of pure receptivity, preventing him from deriving any profit from an analogy with secondary-quality perception.

40. On 'platonism', see my 'Non-Cognitivism and Rule-Following', op. cit., at pp. 156–7. On Aristotle, see M. F. Burnyeat, 'Aristotle on Learning to be Good', in Amelie O. Rorty, ed., *Essays on Aristotle's Ethics* (University of California Press, Berkeley, Los Angeles, London, 1980), pp. 69–92.

conceded his point.[41] Can a projectivist claim that the position I have outlined is at best a notational variant, perhaps an inferior notational variant, of his own position?

It would be inferior if, in eschewing the projectivist metaphysical framework, it obscured some important truth. But what truth would this be? It will not do at this point to answer 'the truth of projectivism'. I have disarmed the explanatory argument for the projectivist's thin conception of genuine reality. What remains is rhetoric expressing what amounts to a now unargued primary-quality model for genuine reality.[42] The picture that this suggests for value experience—objective (value-free) reality processed through a moulded subjectivity—is no less questionable than the picture of secondary-quality experience on which, in Mackie at any rate, it is explicitly modelled. In fact I should be inclined to argue that it is projectivism that is inferior. Deprived of the specious explanatory argument, projectivism has nothing to sustain its thin conception of reality (that on to which the projections are effected) but a contentiously substantial version of the correspondence theory of truth, with the associated picture of genuinely true judgement as something to which the judger makes no contribution at all.[43]

I do not want to argue this now. The point I want to make is that even if projectivism were not actually worse, metaphysically speaking, than the alternative I have described, it would be wrong to regard the issue between them as nothing but a question of metaphysical preference.[44] In the projectivist picture, having one's ethical or aesthetic responses rationally suited to their objects would be a matter of having the relevant processing mechanism functioning acceptably. Now projectivism can of course perfectly well accommodate the idea of assessing one's processing mechanism. But it pictures the mechanism as something that one can contemplate as an object in itself. It would be appropriate to say 'something one can step back from', were it not for the fact that one needs to use the mechanism itself in assessing it; at any rate one is supposed to be able to step back from any naïvely realistic acceptance of the values that the first-level employment of the mechanism has one attribute to items in the world. How, then, are we to understand this pictured availability of the processing mechanism as an object for contemplation, separated off from the world of value? Is there any alternative to thinking of it as capable of being captured, at least in theory, by a set of principles

41. Price, op. cit. p. 107, cites Mackie's response to one of my contributions to the 1978 seminar.
42. We must not let the confusion between the two notions of objectivity distinguished in §3 above seem to support this conception of reality.
43. Blackburn uses the correspondence theorist's pictures for rhetorical effect, but he is properly sceptical about whether this sort of realism makes sense (see 'Truth, Realism, and the Regulation of Theory', op. cit.). His idea is that the explanatory argument makes a counterpart to its metaphysical favouritism safely available to a projectivist about values in particular. Deprived of the explanatory argument, this projectivism should simply wither away. (See 'Rule-Following and Moral Realism', p. 165.) Of course I am not saying that the thin conception of reality that Blackburn's projectivism needs is unattainable, in the sense of being unformulable. What we lack are reasons of a respectable kind to recognize it as a complete conception *of reality*.
44. Something like this seems to be suggested by Price, pp. 107–8.

for superimposing values onto a value-free reality? The upshot is that the search
for an evaluative outlook that one can endorse as rational becomes, virtually
irresistibly, a search for such a set of principles: a search for a *theory* of beauty or
goodness. One comes to count 'intuitions' respectable only in so far as they can be
validated by an approximation to that ideal.[45] (This is the shape that the attempt to
objectivize subjectivity takes here.) I have a hunch that such efforts are mis-
guided; not that we should rest content with an 'anything goes' irrationalism, but
that we need a conception of rationality in evaluation that will cohere with the
possibility that particular cases may stubbornly resist capture in any general net.
Such a conception is straightforwardly available within the alternative to projectivism that I have described. I allowed that being able to explain cases of fear in
the right way might amount to having a theory of danger, but there is no need to
generalize that feature of the case; the explanatory capacity that certifies the
special objects of an evaluative outlook as real, and certifies its responses to them
as rational, would need to be exactly as creative and case-specific as the capacity
to discern those objects itself. (It would be the same capacity: the picture of
'stepping back' does not fit here.)[46] I take it that my hunch poses a question of
moral and aesthetic taste, which—like other questions of taste—should be capable of being argued about. The trouble with projectivism is that it threatens to
bypass that argument, on the basis of a metaphysical picture whose purported
justification falls well short of making it compulsory. We should not let the
question seem to be settled by what stands revealed, in the absence of compelling
argument, as a prejudice claiming the honour due to metaphysical good taste.

45. It is hard to see how a rational *inventing* of values could take a more piecemeal form.

46. Why do I suggest that a particularistic conception of evaluative rationality is unavailable to a
projectivist? (See Blackburn, 'Rule-Following and Moral Realism', pp. 167–70.) In the terms of that
discussion, the point is that (with no good explanatory argument for his metaphysical favouritism) a
projectivist has no alternative to being 'a *real* realist' about the world on which he thinks values are
superimposed. He cannot stop this from generating a quite un-Wittgensteinian picture of what *really*
going on in the same way would be; which means that *he* cannot appeal to Wittgenstein in order to
avert, as Blackburn puts it, 'the threat which shapelessness poses to a respectable notion of consistency' (p. 169). So, at any rate, I meant to argue in my 'Non-Cognitivism and Rule-Following', to
which Blackburn's paper is a reply. Blackburn thinks his projectivism is untouched by the argument,
because he thinks he can sustain its metaphysical favouritism without appealing to '*real* realism', on
the basis of the explanatory argument. But I have argued that this is an illusion. (At p. 181, Blackburn
writes: 'Of course, it is true that our reactions are "simply felt" and, in a sense, not rationally
explicable'. He thinks he can comfortably say this because our conception of reason will go along with
the quasi-realist truth that his projectivism confers on some evaluations. But how can one restrain the
metaphysical favouritism that a projectivist must show from generating some such thought as 'This is
not *real* reason'? If that is allowed to happen, a remark like the one I have quoted will merely
threaten—like an ordinary nihilism—to dislodge us from our ethical and aesthetic convictions.)

9

How to Be a Moral Realist

Richard N. Boyd

1. Introduction

1.1. *Moral realism*

Scientific realism is the doctrine that scientific theories should be understood as putative descriptions of real phenomena, that ordinary scientific methods constitute a reliable procedure for obtaining and improving (approximate) knowledge of the real phenomena which scientific theories describe, and that the reality described by scientific theories is largely independent of our theorizing. Scientific theories describe reality and reality is "prior to thought" (see Boyd 1982).

An early version of this paper, incorporating the naturalistic treatments of the roles of reflective equilibrium and moral intuitions in moral reasoning and a naturalistic conception of the semantics of moral terms (but not the homeostatic property cluster formulation of consequentialism), was presented to the Philosophy Colloquium at Case-Western Reserve University in 1977. I am grateful to the audience at that colloquium for helpful criticisms which greatly influenced my formulation of later versions.

In approximately the version published here, the paper was presented at the University of North Carolina, the University of Chicago, Cornell University, the Universities of California at Berkeley and at Los Angeles, the University of Washington, Dartmouth College, and Tufts University. Papers defending the general homeostatic property cluster account of natural definitions were presented at Oberlin, Cornell, and Stanford. Extremely valuable criticisms from the audiences at these universities helped me in developing the more elaborate defense of moral realism presented in *Realism and the Moral Sciences* and summarized in Part 5 of the present essay.

My interest in the question of moral realism initially arose from my involvement in the anti–Vietnam War movement of the late 1960s and was sustained in significant measure by my participation in subsequent progressive movements. I have long been interested in whether or not moral relativism played a progressive or a reactionary role in such movements; the present essay begins an effort to defend the latter alternative. I wish to acknowledge the important influence on my views of the Students for a Democratic Society (especially its Worker-Student Alliance Caucus), the International Committee against Racism, and the Progressive Labor party. Their optimism about the possibility of social progress and about the rational capacity of ordinary people have played an important role in the development of my views.

I have benefited from discussions with many people about various of the views presented here. I

Richard N. Boyd

By "moral realism" I intend the analogous doctrine about moral judgments, moral statements, and moral theories. According to moral realism:

1. Moral statements are the sorts of statements which are (or which express propositions which are) true or false (or approximately true, largely false, etc.);

2. The truth or falsity (approximate truth . . .) of moral statements is largely independent of our moral opinions, theories, etc.;

3. Ordinary canons of moral reasoning—together with ordinary canons of scientific and everyday factual reasoning—constitute, under many circumstances at least, a reliable method for obtaining and improving (approximate) moral knowledge.

It follows from moral realism that such moral terms as 'good', 'fair', 'just', 'obligatory' usually correspond to real properties or relations and that our ordinary standards for moral reasoning and moral disputation—together with reliable standards for scientific and everyday reasoning—constitute a fairly reliable way of finding out which events, persons, policies, social arrangements, etc. have these properties and enter into these relations. It is *not* a consequence of moral realism that our ordinary procedures are "best possible" for this purpose—just as it is not a consequence of scientific realism that our existing scientific methods are best possible. In the scientific case, improvements in knowledge can be expected to produce improvements in method (Boyd 1980, 1982, 1983, 1985a, 1985b, 1985c), and there is no reason to exclude this possibility in the moral case.

Scientific realism contrasts with instrumentalism and its variants and with views like that of Kuhn (1970) according to which the reality which scientists study is largely constituted by the theories they adopt. Moral realism contrasts with non-cognitivist metaethical theories like emotivism and with views according to which moral principles are largely a reflection of social constructs or conventions.

What I want to do in this essay is to explore the ways in which recent developments in realist philosophy of science, together with related "naturalistic" developments in epistemology and philosophy of language, can be employed in the articulation and defense of moral realism. It will not be my aim here to establish that moral realism is true. Indeed, if moral realism is to be defended along the lines I will indicate here then a thoroughgoing defense of moral realism would be beyond the scope of a single essay. Fortunately a number of extremely important defenses of moral realism have recently been published (see, e.g.,

want especially to thank David Brink, Norman Daniels, Philip Gasper, Paul Gomberg, Kristin Guyot, Terence Irwin, Barbara Koslowski, David Lyons, Christopher McMahon, Richard Miller, Milton Rosen, Sydney Shoemaker, Robert Stalnaker, Stephen Sullivan, Milton Wachsberg, Thomas Weston, and David Whitehouse. My thinking about homeostatic property cluster definitions owes much to conversations with Philip Gasper, David Whitehouse, and especially Kristin Guyot. I am likewise indebted to Richard Miller for discussions about the foundations of non-utilitarian consequentialism. My greatest debt is to Alan Gilbert and Nicholas Sturgeon. I wish to thank the Society for the Humanities at Cornell University for supporting much of the work reflected in Part 5.

Brink 1984, forthcoming; Gilbert 1981b, 1982, 1984b, 1986b, forthcoming; Miller 1984b; Railton 1986; Sturgeon 1984a, 1984b). What I hope to demonstrate in the present essay is that moral realism can be shown to be a more attractive and plausible philosophical position if recent developments in realist philosophy of science are brought to bear in its defense. I intend the general defense of moral realism offered here as a proposal regarding the metaphysical, epistemological, and semantic framework within which arguments for moral realism are best formulated and best understood.

In addition, I am concerned to make an indirect contribution to an important recent debate among Marxist philosophers and Marx scholars concerning the Marxist analysis of moral discourse (see, e.g., Gilbert 1981a, 1981b, 1982, 1984b, 1986a, 1986b; Miller 1979, 1981, 1982, 1983, 1984a, 1984b; Wood 1972, 1979). Two questions are central in this debate: the question of what metaethical views Marx and other Marxist figures actually held or practiced and the question of what metaethical views are appropriate to a Marxist analysis of history and in particular to a Marxist analysis of the role of class ideology in the determination of the content of moral conceptions. I have nothing to contribute to the efforts to answer the first question, which lies outside my competence. About the second, I am convinced that Marxists should be moral realists and that the admirably motivated decision by many antirevisionist Marxists to adopt a non-realist relativist stance in metaethics represents a sectarian (if nonculpable) error. I intend the defense of moral realism presented here to be fully compatible with the recognition of the operation in the history of moral inquiry of just the sort of ideological forces which Marxist historians (among others) have emphasized. A thoroughgoing defense of this compatibility claim is not attempted in the present essay; I develop it in a forthcoming essay.

1.2. *Scientific knowledge and moral skepticism*

One of the characteristic motivations for anti-realistic metaethical positions—either for non-cognitivist views or for views according to which moral knowledge has a strong constructive or conventional component—lies in a presumed epistemological contrast between ethics, on the one hand, and the sciences, on the other. Scientific methods and theories appear to have properties—objectivity, value-neutrality, empirical testability, for example—which are either absent altogether or, at any rate, much less significant in the case of moral beliefs and the procedures by which we form and criticize them. These differences make the methods of science (and of everyday empirical knowledge) seem apt for the *discovery* of facts while the 'methods' of moral reasoning seem, at best, to be appropriate for the rationalization, articulation, and application of preexisting social conventions or individual preferences.

Many philosophers would like to explore the possibility that scientific beliefs and moral beliefs are not so differently situated as this presumed epistemological contrast suggests. We may think of this task as the search for a conception of 'unified knowledge' which will bring scientific and moral knowledge together within the same analytical framework in much the same way as the positivists' conception of 'unified science' sought to provide an integrated treatment of knowledge within the various special sciences. There are, roughly, two plausible general strategies for unifying scientific and moral knowledge and minimizing the apparent epistemological constrast between scientific and moral inquiry:

1. Show that our scientific beliefs and methods actually possess many of the features (e.g., dependence on nonobjective 'values' or upon social conventions) which form the core of our current picture of moral beliefs and methods of moral reasoning.

2. Show that moral beliefs and methods are much more like our current conception of scientific beliefs and methods (more 'objective', 'external', 'empirical', 'intersubjective', for example) than we now think.

The first of these options has already been explored by philosophers who subscribe to a 'constructivist' or neo-Kantian conception of scientific theorizing (see, e.g., Hanson 1958; Kuhn 1970). The aim of the present essay will be to articulate and defend the second alternative. In recent papers (Boyd 1979, 1982, 1983, 1985a, 1985b, 1985c) I have argued that scientific realism is correct, but that its adequate defense requires the systematic adoption of a distinctly naturalistic and realistic conception of knowledge, of natural kinds, and of reference. What I hope to show here is that once such a distinctly naturalistic and realistic conception is adopted, it is possible to offer a corresponding defense of moral realism which has considerable force and plausibility.

My argumentative strategy will be to offer a list of several challenges to moral realism which will, I hope, be representative of those considerations which make it plausible that there is the sort of epistemological contrast between science and ethics which we have been discussing. Next, I will present a summary of some recent work in realistic philosophy of science and related "naturalistic" theories in epistemology and the philosophy of language. Finally, I will indicate how the results of this recent realistic and naturalistic work can be applied to rebut the arguments against moral realism and to sketch the broad outlines of an alternative realistic conception of moral knowledge and of moral language.

2. Some challenges to moral realism

2.1. *Moral intuitions and empirical observations*

In the sciences, we decide between theories on the basis of observations, which have an important degree of objectivity. It appears that in moral reasoning, moral intuitions play the same role which observations do in science: we test general moral principles and moral theories by seeing how their consequences conform (or fail to conform) to our moral intuitions about particular cases. It appears that it is the foundational role of observations in science which makes scientific objectivity possible. How could moral intuitions possibly play the same sort of foundational role in ethics, especially given the known diversity of moral judgments between people? Even if moral intuitions do provide a 'foundation' for moral inquiry, wouldn't the fact that moral 'knowledge' is grounded in intuitions rather than in observation be exactly the sort of fundamental epistemological contrast which the received view postulates, especially since peoples' moral intuitions typically reflect the particular moral theories or traditions which they already accept, or their culture, or their upbringing? Doesn't the role of moral intuitions in moral reasoning call out for a 'constructivist' metaethics? If moral intuitions don't play a foundational role in ethics and if morality is supposed to be epistemologically like science, *then what plays, in moral reasoning, the role played by observation in science?*

2.2. *The role of "reflective equilibrium" in moral reasoning*

We have already seen that moral intuitions play a role in moral reasoning which appears to threaten any attempt to assimilate moral reasoning to the model of objective empirical scientific methodology. Worse yet, as Rawls (1971) has reminded us, what we do with our moral intuitions, our general moral principles, and our moral theories, in order to achieve a coherent moral position, is to engage in 'trading-off' between these various categories of moral belief in order to achieve a harmonious 'equilibrium'. Moral reasoning *begins* with moral *presuppositions,* general as well as particular, and proceeds by negotiating between conflicting *presuppositions*. It is easy to see how this could be a procedure for rationalization of individual or social norms or, to put it in more elevated terms, a procedure for the 'construction' of moral or ethical systems. But if ethical beliefs and ethical reasoning are supposed to be like scientific beliefs and methods, then this procedure would have to be a procedure for *discovering* moral facts! How could any procedure so presupposition-dependent be a *discovery* procedure rather than a *construction procedure?* (See Dworkin 1973.)

2.3. *Moral progress and cultural variability*

If moral judgments are a species of factual judgment, then one would expect to see moral progress, analogous to progress in science. Moreover, one of the characteristics of factual inquiry in science is its relative independence from cultural distortions: scientists with quite different cultural backgrounds can typically agree in assessing scientific evidence. If moral reasoning is reasoning about objective moral *facts,* then what explains our lack of progress in ethics and the persistence of cultural variability in moral beliefs?

[185]

2.4 *Hard cases*

If goodness, fairness, etc. are real and objective properties, then what should one say about the sorts of hard cases in ethics which we can't seem *ever* to resolve? Our experience in science seems to be that hard scientific questions are only *temporarily* rather than permanently unanswerable. Permanent disagreement seems to be very rare indeed. Hard ethical questions seem often to be permanent rather than temporary.

In such hard ethical cases, is there a fact of the matter inaccessible to moral inquiry? If so, then doesn't the existence of such facts constitute a significant epistemological difference between science and ethics? If not, if there are not facts of the matter, then isn't moral realism simply refuted by such indeterminacy?

2.5. *Naturalism and naturalistic definitions*

If goodness, for example, is a real property, then wouldn't it be a *natural* property? If not, then isn't moral realism committed to some unscientific and superstitious belief in the existence of non-natural properties? If goodness would be a natural property, then isn't moral realism committed to the extremely implausible claim that moral terms like 'good' possess naturalistic definitions?

2.6. *Morality, motivation, and rationality*

Ordinary factual judgments often provide us with reasons for action; they serve as constraints on rational choice. But they do so only because of our antecedent interests or desires. If moral judgments are merely factual judgments, as moral realism requires, then the relation of moral judgments to motivation and rationality must be the same. It would be possible in principle for someone, or some thinking thing, to be enetirely rational while finding moral judgments motivationally neutral and irrelevant to choices of action.

If this consequence follows from moral realism, how can the moral realist account for the particularly close connection between moral judgments and judgments about what to do? What about the truism that moral judgments have commendatory force as a matter of their meaning or the plausible claim that the moral preferability of a course of action always provides a reason (even if not an overriding one) for choosing it?

2.7. *The semantics of moral terms*

Moral realism is an anti-subjectivist position. There is, for example, supposed to be a single objective property which we're all talking about when we use the term 'good' in moral contexts. But people's moral concepts differ profoundly. How can it be maintained that our radically different concepts of 'good' are really concepts of one and the same property? Why not a different property for each significantly different conception of the good? Don't the radical differences in our conceptions of the good suggest either a noncognitivist or a constructivist conception of the semantics of ethical terms?

2.8. *Verificationism and anti-realism in ethics*

Anti-realism in ethics, like the rejection of theoretical realism in science, is a standard positivist position. In the case of science, there is a straightforward verificationist objection to realism about alleged "theoretical entities": they are unobservables; statements about them lie beyond the scope of empirical investigation and are thus unverifiable in principle. (See Boyd 1982 for a discussion of various formulations of this key verificationist argument.)

It is interesting to note that the challenges to moral realism rehearsed in 2.1–2.7 do not take the form of so direct an appeal to verificationism. Only in the case of the concern about non-natural moral properties (2.5) might the issue of verifiability be directly relevant, and then only if the objection to non-natural properties is that they would be unobservable. Instead, the arguments in 2.1–2.7 constitute an *indirect* argument against moral realism: they point to features of moral beliefs or of moral reasoning for which, it is suggested, the best explanation would be one which entailed the rejection of moral realism. Moreover, what is true of the challenges to moral realism rehearsed above is typical: by and large positivists, and philosophers influenced by positivism, did not argue directly for the unverifiability of moral statements; they did not make an appeal to the unobservability of alleged moral properties or deny that moral theories had observational consequences. Instead, they seemed to take a non-cognitivist view of ethics to be established by an "inductive inference to the best explanation" of the sort of facts cited in 2.1–2.7.

In this regard, then, the standard arguments against moral realism are more closely analogous to Kuhnian objections to scientific realism than they are to the standard verificationist arguments against the possibility of knowledge of "theoretical entities" Sections 2.1, 2.2, 2.3, and 2.7 rehearse arguments which are importantly similar to Kuhn's arguments from the paradigm dependence of scientific concepts and methods to a constructivist and antirealistic conception of science. I have argued elsewhere (Boyd 1979, 1982, 1983, 1985a) that a systematic rebuttal to the verificationist epistemology and philosophy of language which form the foundations of logical positivism can in fact be extended to a defense of scientific realism against the more constructivist and neo-Kantian considerations represented by Kuhn's work. If the arguments of the present essay are successful, then this conclusion can be generalized: a realist and anti-empiricist account in the philosophy of science can be extended to a defense of *moral* realism as well, even though the challenges to moral realism are apparently only indirectly verificationist.

Richard N. Boyd

3. Realist philosophy of science

3.1. *The primacy of reality*

By "scientific realism" philosophers mean the doctrine that the methods of science are capable of providing (partial or approximate) knowledge of unobservable ('theoretical') entities, such as atoms or electromagnetic fields, in addition to knowledge about the behavior of observable phenomena (and of course, that the properties of these and other entities studied by scientists are largely theory-independent).

Over the past three decades or so, philosophers of science within the empiricist tradition have been increasingly sympathetic toward scientific realism and increasingly inclined to alter their views of science in a realist direction. The reasons for this realist tendency lie largely in the recognition of the extraordinary role which theoretical considerations play in actual (and patently successful) scientific practice. To take the most striking example, scientists routinely modify or extend operational 'measurement' or 'detection' procedures for 'theoretical' magnitudes or entities on the basis of new theoretical developments. This sort of methodology is perfectly explicable on the realist assumption that the operational procedures in question really are procedures for the measurement or detection of unobservable entities and that the relevant theoretical developments reflect increasingly accurate knowledge of such "theoretical" entities. Accounts of the revisability of operational procedures which are compatible with a non-realist position appear inadequate to explain the way in which theory-dependent revisions of 'measurement' and 'detection' procedures make a positive methodological contribution to the progress of science.

This pattern is quite typical: The methodological contribution made by theoretical considerations in scientific methodology is inexplicable on a non-realist conception but easily explicable on the realist assumption that such considerations are a reflection of the growth of *theoretical* knowledge. (For a discussion of this point see Boyd 1982, 1983, 1985a, 1985b.) Systematic development of this realist theme has produced developments in epistemology, metaphysics, and the philosophy of language which go far beyond the mere rejection of verificationism and which point the way toward a distinctly realist conception of the central issues in the philosophy of science. These developments include the articulation of causal or naturalistic theories of reference (Kripke 1971, 1972; Putnam 1975a; Boyd 1979, 1982), of measurement (Byerly and Lazara 1973) of 'natural kinds' and scientific categories (Quine 1969a; Putnam 1975a; Boyd 1979, 1982, 1983, 1985b), of scientific epistemology generally (Boyd 1972, 1979, 1982, 1983, 1985a, 1985b, 1985c), and of causation (Mackie 1974; Shoemaker 1980; Boyd 1982, 1985b).

Closely related to these developments has been the articulation of causal or naturalistic theories of knowledge (see, e.g., Armstrong 1973; Goldman 1967, 1976; Quine 1969b). Such theories represent generalizations of causal theories of perception and reflect a quite distinctly realist stance with respect to the issue of our knowledge of the external world. What all these developments—both within the philosophy of science and in epistemology generally—have in common is that they portray as a posteriori and contingent various matters (such as the operational 'definitions' of theoretical terms, the 'definitions' of natural kinds, or the reliability of the senses) which philosophers in the modern tradition have typically sought to portray as a priori. In an important sense, these developments represent the fuller working out of the philosophical implications of the realist doctrine that reality is prior to thought. (For a further development of this theme see Boyd 1982, 1983, 1985a, 1985b.) It is just this a posteriority and contingency in philosophical matters, I shall argue, which will make possible a plausible defense of moral realism against the challenges outlined in Part 2.

In the remaining sections of Part 3 I will describe some of the relevant features of these naturalistic and realistic developments. These 'results' in recent realistic philosophy are not, of course, uncontroversial, and it is beyond the scope of this essay to defend them. But however much controversy they may occasion, unlike moral realism, they do not occasion incredulity: they represent a plausible and defensible philosophical position. The aim of this essay is to indicate that, if we understand the relevance of these recent developments to issues in moral philosophy, then moral realism should, though controversial, be equally credible.

3.2. *Objective knowledge from theory-dependent methods*

I suggested in the preceding section that the explanation for the movement toward realism in the philosophy of science during the past two or three decades lies in the recognition of the extraordinarily theory-dependent character of scientific methodology and in the inability of any but a realist conception of science to explain why so theory-dependent a methodology should be reliable. The theoretical revisability of measurement and detection procedures, I claimed, played a crucial role in establishing the plausibility of a realist philosophy of science.

If we look more closely at this example, we can recognize two features of scientific methodology which are, in fact, quite general. In the first place, the realist's account of the theoretical revisability of measurement and detection procedures rests upon a conception of scientific research as *cumulative by successive approximations to the truth.*

Second, this cumulative development is possible because *there is a dialectical relationship between current theory and the methodology for its improvement.* The approximate truth of current theories explains why our existing measurement procedures are (approximately) reliable. That reliability, in turn, helps to explain why our experimental or observational investigations are successful in uncovering new theoretical knowledge, which, in turn, may produce improvements in experimental techniques, etc.

These features of scientific methodology are *entirely* general. Not only mea-

[189]

surement and detection procedures but all aspects of scientific methodology—principles of experimental design, choices of research problems, standards for the assessment of experimental evidence, principles governing theory choice, and rules for the use of theoretical language—are highly dependent upon current theoretical commitments (Boyd 1972, 1973, 1979, 1980, 1982, 1983, 1985a, 1985b; Kuhn 1970; van Fraassen 1980). No aspect of scientific method involves the 'presupposition-free' testing of individual laws or theories. Moreover, the theory dependence of scientific methodology *contributes* to its reliability rather than detracting from it.

The only scientifically plausible explanation for the reliability of a scientific methodology which is so theory-dependent is a thorough-goingly realistic explanation: Scientific methodology, dictated by currently accepted theories, is reliable at producing further knowledge precisely *because, and to the extent that, currently accepted theories are relevantly approximately true*. For example, it is because our current theories are approximately true that the canons of experimental design which they dictate are appropriate for the rigorous testing of new (and potentially more accurate) theories. What the scientific method provides is a paradigm-dependent paradigm-modification strategy: a strategy for modifying or amending our existing theories in the light of further research, which is such that its methodological principles at any given time will themselves depend upon the theoretical picture provided by the currently accepted theories. If the body of accepted theories is itself relevantly sufficiently approximately true, then this methodology operates to produce a subsequent dialectical improvement both in our knowledge of the world and in our methodology itself. Both our new theories and the methodology by which we develop and test them depend upon previously acquired theoretical knowledge. It is not possible to explain even the instrumental reliability of actual scientific practice without invoking this explanation and without adopting a realistic conception of scientific knowledge (Boyd 1972, 1973, 1979, 1982, 1983, 1985a, 1985b, 1985c).

The way in which scientific methodology is theory-dependent dictates that we have a strong methodological preference for new theories which are plausible in the light of our existing theoretical commitments; this means that we prefer new theories which relevantly resemble our existing theories (where the determination of the relevant respects of resemblance is itself a theoretical issue). The reliability of such a methodology is explained by the approximate truth of existing theories, and one consequence of this explanation is that *judgments of theoretical plausibility are evidential*. The fact that a proposed theory is itself plausible in the light of previously confirmed theories is evidence for its (approximate) truth (Boyd 1972, 1973, 1979, 1982, 1983, 1985a, 1985b, 1985c). A purely conventionalistic account of the methodological role of considerations of theoretical plausibility cannot be adequate because it cannot explain the contribution which such considerations make to the instrumental reliability of scientific methodology (Boyd 1979, 1982, 1983).

The upshot is this: The theory-dependent conservatism of scientific methodology is *essential* to the rigorous and reliable testing and development of new scientific theories; on balance, theoretical 'presuppositions' play neither a destructive nor a conventionalistic role in scientific methodology. They are essential to its reliability. If by the 'objectivity' of scientific methodology we mean its capacity to lead to the discovery of *theory-independent reality*, then scientific methodology is objective precisely because it *is theory-dependent* (Boyd 1979, 1982, 1983, 1985a, 1985b, 1985c).

3.3. *Naturalism and radical contingency in epistemology*

Modern epistemology has been largely dominated by positions which can be characterized as 'foundationalist': all knowledge is seen as ultimately grounded in certain foundational beliefs which have an epistemically privileged position— they are a priori or self-warranting, incorrigible, or something of the sort. Other true beliefs are instances of knowledge only if they can be justified by appeals to foundational knowledge. Whatever the nature of the foundational beliefs, or whatever their epistemic privilege is suppose to consist in, it is an a priori question which beliefs fall in the privileged class. Similarly, the basic inferential principles which are legitimate for justifying non-foundational knowledge claims, given foundational premises, are such that they can be identified a priori and it can be shown a priori that they are rational principles of inference. We may fruitfully think of foundationalism as consisting of two parts, *premise foundationalism*, which holds that all knowledge is justifiable from an a priori specifiable core of foundational beliefs, and *inference foundationalism*, which holds that the principles of justifiable inference are ultimately reducible to inferential principles which can be shown a priori to be rational.

Recent work in ''naturalistic epistemology'' or ''causal theories of knowing'' (see, e.g., Armstrong 1973; Goldman 1967, 1976; Quine 1969b) strongly suggest that the foundationalist conception of knowledge is fundamentally mistaken. For the crucial case of perceptual knowledge, there seem to be (in typical cases at least) neither premises (foundational or otherwise) nor inferences; instead, perceptual knowledge obtains when perceptual beliefs are produced by epistemically reliable mechanisms. For a variety of other cases, even where premises and inferences occur, it seems to be the reliable production of belief that distinguishes cases of knowledge from other cases of true belief. A variety of naturalistic considerations suggests that there are no beliefs which are epistemically privileged in the way foundationalism seems to require.

I have argued (see Boyd 1982, 1983, 1985a, 1985b, 1985c) that the defense of scientific realism requires an even more thoroughgoing naturalism in epistemology and, consequently, an even more thoroughgoing rejection of foundationalism. In the first place, the fact that scientific knowledge grows cumulatively by successive approximation and the fact that the evaluation of theories is an ongoing social phenomenon require that we take the crucial causal notion in epistemology to be reliable *regulation* of belief rather than reliable belief *production*. The relevant conception of belief regulation must reflect the approximate social and dialectical character of the growth of scientific knowledge. It will thus be true that the causal mechanisms relevant to knowledge will include mechanisms, social and technical as well as psychological, for the criticism, testing, acceptance, modification, and transmission of scientific theories and doctrines. For that reason, an understanding of the role of social factors in science may be relevant not only for the sociology and history of science but for the epistemology of sciences as well. The epistemology of science is in this respect dependent upon empirical knowledge.

There is an even more dramatic respect in which the epistemology of science rests upon empirical foundations. All the significant methodological principles of scientific inquiry (except, perhaps, the rules of deductive logic, but see Boyd 1985c) are profoundly theory-dependent. They are a reliable guide to the truth *only* because, and to the extent that, the body of background theories which determines their application is relevantly approximately true. The rules of rational scientific inference are not reducible to some more basic rules whose reliability as a guide to the truth is independent of the truth of background theories. Since it is a contingent empirical matter which background theories are approximately true, the rationaity of scientific principles of inference ultimately rests on a contingent matter of empirical fact, just as the epistemic role of the senses rests upon the contingent empirical fact that the senses are reliable detectors of external phenomena. Thus inference foundationalism is radically false; there are no a priori justifiable rules of nondeductive inference. The epistemology of empirical science is an empirical science. (Boyd 1982, 1983, 1985a, 1985b, 1985c).

One consequence of this radical contingency of scientific methods is that the emergence of scientific rationality as we know it depended upon the logically, epistemically, and historically contingent emergence of a relevantly approximately true theoretical tradition. It is not possible to understand the initial emergence of such a tradition as the consequence of some more abstractly conceived scientific or rational methodology which itself is theory-independent. There is no such methodology. We must think of the establishment of the corpuscular theory of matter in the seventeenth century as the beginning of rational methodology in chemistry, not as a consequence of it (for a further discussion see Boyd 1982).

3.4. *Scientific intuitions and trained judgment*

Both noninferential perceptual judgments and elaborately argued explicit inferential judgments in theoretical science have a purely contingent a posteriori foundation. Once this is recognized, it is easy to see that there are methodologically important features of scientific practice which are intermediate between noninferential perception and explicit inference. One example is provided by what science textbook authors often refer to as 'physical intuition', 'scientific maturity', or the like. One of the intended consequences of professional training in a scientific discipline (and other disciplines as well) is that the student acquire a ''feel'' for the issues and the actual physical materials which the science studies. As Kuhn (1970) points out, part of the role of experimental work in the training of professional scientists is to provide such a feel for the paradigms or 'worked examples' of good scientific practice. There is very good reason to believe that having good physical (or biological or psychological) intuitions is important to epistemically reliable scientific practice. It is also quite clear both that the acquisition of good scientific intuitions depends on learning explicit theory, as well as on other sorts of training and practice, *and* that scientists are almost never able to make fully explicit the considerations which play a role in their intuitive judgments. The legitimate role of such 'tacit' factors in science has often been taken (especially by philosophically inclined scientists) to be an especially puzzling feature of scientific methodology.

From the perspective of the naturalistic epistemology of science, there need be no puzzle. It is, of course, a question of the very greatest psychological interest just how intuitive judgments in science work and how they are related to explicit theory, on the one hand, and to experimental practice, on the other. But it seems overwhelmingly likely that scientific intuitions should be thought of as trained judgments which resemble perceptual judgments in not involving (or at least not being fully accounted for by) explicit inferences, but which resemble explicit inferences in science in depending for their reliability upon the relevant approximate truth of the explicit theories which help to determine them. This dependence upon the approximate truth of the relevant background theories will obtain even in those cases (which may be typical) in which the tacit judgments reflect a deeper understanding than that currently captured in explicit theory. It is an important and exciting fact that some scientific knowledge can be represented tacitly before it can be represented explicitly, but this fact poses no difficulty for a naturalistic treatment of scientific knowledge. Tacit or intuitive judgments in science are reliable because they are grounded in a theoretical tradition (itself partly tacit) which is, as a matter of contingent empirical fact, relevantly approximately true.

3.5. *Non–Humean conceptions of causation and reduction*

The Humean conception of causal relations according to which they are analyzable in terms of regularity, correlation, or deductive subsumability under laws is defensible only from a verificationist position. If verificationist criticisms of talk about unobservables are rejected—as they should be—then there is nothing more problematical about talk of causal powers than there is about talk of electrons or electromagnetic fields. There is no reason to believe that causal terms have definitions (analytic or natural) in noncausal terms. Instead, 'cause' and its cognates refer to natural phenomena whose analysis is a matter for physicists, chemists, psychologists, historians, etc., rather than a matter of conceptual analysis. In particular, it is perfectly legitimate—as a naturalistic conception of epistemology requires—to employ unreduced causal notions in philosophical analysis. (Boyd, 1982, 1985b, Shoemaker 1980).

One crucial example of the philosophical application of such notions lies in the analysis of 'reductionism'. If a materialist perspective is sound, then *in some sense* all natural phenomena are 'reducible' to basic physical phenomena. The (prephilosophically) natural way of expressing the relevant sort of reduction is to say that all substances are composed of purely physical substances, all forces are composed of physical forces, all causal powers or potentialities are realized in physical substances and their causal powers, etc. This sort of analysis freely employs unreduced causal notions. If it is 'rationally reconstructed' according to the Humean analysis of such notions, we get the classic analysis of reduction in terms of the syntactic reducibility of the theories in the special sciences to the laws of physics, which in turn dictates the conclusion that all natural properties must be definable in the vocabulary of physics. Such an analysis is entirely without justification from the realistic and naturalistic perspective we are considering. Unreduced causal notions are philosophically acceptable, and the Humean reduction of them mistaken. The prephilosophically natural analysis of reduction is also the philosophically appropriate one. In particular, purely physical objects, states, properties, etc. need not have definitions in "the vocabulary of physics" or in any other reductive vocabulary (see Boyd 1982).

3.6. *Natural definitions*

Locke speculates at several places in Book IV of the *Essay* (see, e.g., IV, iii, 25) that when kinds of substances are defined by 'nominal essences', as he thinks they must be, it will be impossible to have a general science of, say, chemistry. The reason is this: nominal essences define kinds of substance in terms of sensible properties, but the factors which govern the behavior (even the observable behavior) of substances are insensible corpuscular real essences. Since there is no reason to suppose that our nominal essences will correspond to categories which reflect uniformities in microstructure, there is no reason to believe that kinds defined by nominal essences provide a basis for obtaining general knowledge of substances. Only if we could sort substances according to their hidden real essences would systematic general knowledge of substances be possible.

Locke was right. Only when kinds are defined by natural rather than conventional definitions is it possible to obtain sound scientific explanations (Putnam 1975a; Boyd 1985b) or sound solutions to the problem of 'projectibility' in inductive inference in science (Quine 1969a; Boyd 1979, 1982, 1983, 1985a, 1985b, 1985c). Indeed this is true not only for the definitions of natural kinds but also for the definitions of the properties, relations, magnitudes, etc. to which we must refer in sound scientific reasoning. In particular, a wide variety of terms do not possess analytic or stipulative definitions and are instead defined in terms of properties, relations, etc. which render them appropriate to particular sorts of scientific or practical reasoning. In the case of such terms, proposed definitions are always in principle revisable in the light of new evidence or new theoretical developments. Similarly, the fact that two people or two linguistic communities apply different definitions in using a term is not, by itself, sufficient to show that they are using the term to refer to different kinds, properties, etc.

3.7. *Reference and epistemic access*

If the traditional empiricist account of definition by nominal essences (or 'operational definitions' or 'criterial attributes') is to be abandoned in favor of a naturalistic account of definitions (at least for some terms) then a naturalistic conception of reference is required for those cases in which the traditional empiricist semantics has been abandoned. Such a naturalistic account is provided by recent causal theories of reference (see, e.g., Feigl 1956; Kripke 1972; Putnam 1975a). The reference of a term is established by causal connections of the right sort between the use of the term and (instances of) its referent.

The connection between causal theories of reference and naturalistic theories of knowledge and of definitions is quite intimate: reference is itself an epistemic notion and the sorts of causal connections which are relevant to reference are just those which are involved in the reliable regulation of belief (Boyd 1979, 1982). *Roughly,* and for nondegenerate cases, a term *t* refers to a kind (property, relation, etc.) *k* just in case there exist causal mechanisms whose tendency is to bring it about, over time, that what is predicated of the term *t* will be approximately true of *k* (excuse the blurring of the use-mention distinction). Such mechanisms will typically include the existence of procedures which are approximately accurate for recognizing members or instances of *k* (at least for easy cases) and which relevantly govern the use of *t,* the social transmission of certain relevantly approximately true beliefs regarding *k,* formulated as claims about *t* (again excuse the slight to the use-mention distinction), a pattern of deference to experts on *k* with respect to the use of *t,* etc. (for a fuller discussion see Boyd 1979, 1982). When relations of this sort obtain, we may think of the properties of *k* as regulating the use of *t* (via such causal relations), and we may think of what is said using *t* as providing us with socially coordinated *epistemic access* to *k; t* refers to *k* (in nondegenerate cases) just in case the socially coordinated use of *t* provides significant epistemic access to *k,* and not to other kinds (properties, etc.) (Boyd 1979, 1982).

Richard N. Boyd

3.8. Homeostatic property-cluster definitions

The sort of natural definition[1] in terms of corpuscular real essences anticipated by Locke is reflected in the natural definitions of chemical kinds by molecular formulas; 'water = H_2O' is by now the standard example (Putnam 1975a). Natural definitions of this sort specify necessary and sufficient conditions for membership in the kind in question. Recent *non*-naturalistic semantic theories in the ordinary language tradition have examined the possibility of definitions which do not provide necessary and sufficient conditions in this way. According to various property-cluster or criterial attribute theories, some terms have definitions which are provided by a collection of properties such that the possession of an adequate number of these properties is sufficient for falling within the extension of the term. It is supposed to be a conceptual (and thus an a priori) matter what properties belong in the cluster and which combinations of them are sufficient for falling under the term. Insofar as different properties in the cluster are differently "weighted" in such judgments, the weighting is determined by our concept of the kind or property being defined. It is characteristically insisted, however, that our concepts of such kinds are 'open textured' so that there is some indeterminacy in extension *legitimately* associated with property-cluster or criterial attribute definitions. The 'imprecision' or 'vagueness' of such definitions is seen as a perfectly appropriate feature of ordinary linguistic usage, in contrast to the artificial precision suggested by rigidly formalistic positivist conceptions of proper language use.

I shall argue (briefly) that—despite the philistine antiscientism often associated with 'ordinary language' philosophy—the property-cluster conception of definitions provides an extremely deep insight into the possible form of *natural* definitions. I shall argue that there are a number of scientifically important kinds, properties, etc. whose natural definitions are very much like the property-cluster definitions postulated by ordinary-language philosophers (for the record, I doubt that there are any terms whose definitions actually fit the ordinary-language model, because I doubt that there are any significant 'conceptual truths' at all). There are natural kinds, properties, etc. whose natural definitions involve a kind of property cluster *together with* an associated indeterminacy in extension. Both the property-cluster form of such definitions and the associated indeterminacy are dictated by the scientific task of employing categories which correspond to inductively and explanatorily relevant causal structures. In particular, the indeterminacy in extension of such natural definitions could not be remedied without rendering the definitions *un*natural in the sense of being scientifically misleading. What I believe is that the following sort of situation is commonplace in the special

---

1. This is the only section of Part 3 which advances naturalistic and realistic positions not already presented in the published literature. It represents a summary of work in progress. For some further developments see Section 5.2

sciences which study complex structurally or functionally characterized phenomena:

1. There is a family F of properties which are 'contingently clustered' in nature in the sense that they co-occur in an important number of cases.

2. Their co-occurrence is not, at least typically, a statistical artifact, but rather the result of what may be metaphorically (sometimes literally) described as a sort of *homeostasis*. Either the presence of some of the properties in F tends (under appropriate conditions) to favor the presence of the others, or there are underlying mechanisms or processes which tend to maintain the presence of the properties in F, or both.

3. The homeostatic clustering of the properties in F is causally important: there are (theoretically or practically) important effects which are produced by a conjoint occurrence of (many of) the properties in F together with (some or all of) the underlying mechanisms in question.

4. There is a kind term t which is applied to things in which the homeostatic clustering of most of the properties in F occurs.

5. This t has no analytic definition; rather all or part of the homeostatic cluster F together with some or all of the mechanisms which underlie it provides the natural definition of t. The question of just which properties and mechanisms belong in the definition of t is an a posteriori question—often a difficult theoretical one.

6. Imperfect homeostasis is nomologically possible or actual: some thing may display some but not all of the properties in $F;$ some but not all of the relevant underlying homeostatic mechanisms may be present.

7. In such cases, the relative importance of the various properties in F and of the various mechanisms in determining whether the thing falls under t—if it can be determined at all—is a theoretical rather than an conceptual issue.

8. In cases in which such a determination is possible, the outcome will typically depend upon quite particular facts about the actual operation of the relevant homeostatic mechanisms, about the relevant background conditions and about the causal efficacy of the partial cluster of properties from F. For this reason the outcome, if any, will typically be different in different possible worlds, even when the partial property cluster is the same and even when it is unproblematical that the kind referred to by t in the actual world exists.

9. Moreover, there will be many cases of extensional vagueness which are such that they are not resolvable, even given all the relevant facts and all the true theories. There will be things which display some but not all of the properties in F (and/or in which some but not all of the relevant homeostatic mechanisms operate) such that no rational considerations dictate whether or not they are to be classed under t, assuming that a dichotomous choice is to be made.

10. The causal importance of the homeostatic property cluster F together with the relevant underlying homeostatic mechanisms is such that the kind or property denoted by t is a natural kind in the sense discussed earlier.

11. No refinement of usage which replaces *t* by a significantly less extensionally vague term will preserve the naturalness of the kind referred to. Any such refinement would either require that we treat as important distinctions which are irrelevant to causal explanation or to induction or that we ignore similarities which are important in just these ways.

The reader is invited to assure herself that 1–11 hold, for example, for the terms 'healthy' and 'is healthier than.' Whether these are taken to be full-blown cases of natural property (relation) terms is not crucial here. They do illustrate almost perfectly the notion of a homeostatic property cluster and the correlative notion of a homeostatic cluster term. It is especially important to see *both* that a posteriori theoretical considerations in medicine can sometimes decide problematical cases of healthiness or of relative healthiness, often in initially counterintuitive ways *and* that nevertheless only highly artificial modifications of the notions of health and relative health could eliminate most or all of the extensional vagueness which they possess. One way to see the latter point is to consider what we would do if, for some statistical study of various medical practices, we were obliged to eliminate most of the vagueness in the notion of relative healthiness even where medical theory was silent. What we would strive to do would be to resolve the vagueness in such a way as not to bias the results of the study—not to favor one finding about the efficacy of medical practices over another. The role of natural kinds is, by contrast, precisely *to bias* (in the pejoratively neutral sense of the term) inductive generalization (Quine 1969a; Boyd 1979, 1981, 1983, 1985a, 1985b). Our concern not to bias the findings reflects our recognition that the resolution of vagueness in question would be *un*natural in the sense relevant to this inquiry.

The paradigm cases of natural kinds—biological species—are examples of homeostatic cluster kinds in this sense. The appropriateness of any particular biological species for induction and explanation in biology depends upon the *imperfectly* shared and homeostatically related morphological, physiological, and behavioral features which characterize its members. The definitional role of mechanisms of homeostasis is reflected in the role of interbreeding in the modern species concept; for sexually reproducing species, the exchange of genetic material between populations is thought by some evolutionary biologists to be essential to the homeostatic unity of the other properties characteristic of the species and it is thus reflected in the species definition which they propose (see Mayr 1970). The *necessary* indeterminacy in extension of species terms is a consequence of evolutionary theory, as Darwin observed: speciation depends on the existence of populations which are intermediate between the parent species and the emerging one. Any 'refinement' of classification which artifically eliminated the resulting indeterminacy in classification would obscure the central fact about heritable variations in phenotype upon which biological evolution depends. More determinate species categories would be scientifically inappropriate and misleading.

It follows that a consistently developed scientific realism *predicts* indeter-

minacy for those natural kind or property terms which refer to complex phenomena; such indeterminacy is a necessary consequence of "cutting the world at
its (largely theory-independent) joints." Thus consistently developed scientific
realism *predicts* that there will be some failures of bivalence for statements which
refer to complex homeostatic phenomena (contrast, e.g., Putnam 1983 on 'metaphysical realism' and vagueness). Precision in describing indeterminate or 'borderline' cases of homeostatic cluster kinds (properties, etc.) consists not in the
introduction of artificial precision in the definitions of such kinds but rather in a
detailed description of the ways in which the indeterminate cases are like and
unlike typical members of the kind (see Boyd 1982 on borderline cases of
knowledge, which are themselves homeostatic cluster phenomena).

4. How to be a moral realist

4.1. *Moral semantics, intuitions, reflective equilibrium, and hard cases*

Some philosophical opportunities are too good to pass up. For many of the
more abstract challenges to moral realism, recent realistic and naturalistic work in
the philosophy of science is suggestive of possible responses in its defense. Thus
for example, it has occurred to many philosophers (see, e.g., Putnam 1975b) that
naturalistic theories of reference and of definitions might be extended to the
analysis of moral language. *If* this could be done successfully *and if* the results
were favorable to a realist conception of morals, then it would be possible to reply
to several anti-realist arguments. For example, against the objection that wide
divergence of moral concepts or opinions between traditions or cultures indicates
that, at best, a constructivist analysis of morals is possible, the moral realist might
reply that differences in conception or in working definitions need not indicate the
absence of shared causally fixed referents for moral terms.

Similarly, consider the objection that a moral realist must hold that goodness is
a natural property, and thus commit the "naturalistic fallacy" of maintaining that
moral terms possess analytic definitions in, say, physical terms. The moral realist
may choose to agree that goodness is probably a physical property but deny that it
has any analytic definition whatsoever. If the realist's critique of the syntactic
analysis of reductionism in science is also accepted, then the moral realist can
deny that it follows from the premise that goodness is a physical property or that
goodness has any physical definition, analytic or otherwise.

If the moral realist takes advantage of naturalistic and realistic conceptions in
epistemology as well as in semantic theory, other rebuttals to anti-realist challenges are suggested. The extent of the potential for rebuttals of this sort can best
be recognized if we consider the objection that the role of reflective equilibrium in
moral reasoning dictates a constructivist rather than a realist conception of morals.
The moral realist might reply that the dialectical interplay of observations, theory,
and methodology which, according to the realist, constitutes the *discovery* pro-

cedure for scientific inquiry *just is* the method of reflective equilibrium, so that the prevalence of that method in moral reasoning cannot *by itself* dictate a non-realist conception of morals.

If the response just envisioned to the concern over reflective equilibrium is successful, then the defender of moral realism will have established that—in moral reasoning as in scientific reasoning—the role of culturally transmitted presuppositions in reasoning does not necessitate a constructivist (or non-cognitivist) rather than a realist analysis of the subject matter. *If* that is established, then the moral realist might defend the epistemic role of culturally determined intuitions in ethics by treating ethical intuitions on the model of theory-determined intuitions in science, which the scientific realist takes to be examples of epistemically reliable trained judgments.

Finally, if the moral realist is inclined to accept the antirealist's claim that the existence of hard cases in ethics provides a reason to doubt that there is a moral fact of the matter which determines the answer in such cases (more on this later), then the scientific realist's conclusion that bivalence fails for some statements involving homeostatic cluster kind terms *might* permit the moral realist to reason that similar failures of bivalence for some ethical statements need not be fatal to moral realism.

In fact, I propose to employ just these rebuttals to the various challenges to moral realism I have been discussing. They represent the application of a coherent naturalistic conception of semantics and of knowledge against the challenges raised by the critic of moral realism. But they do not stand any chance of rebutting moral anti-realism unless they are incorporated into a broader conception of morals and of moral knowledge which meets certain very strong constraints. These constraints are the subject of the next section.

4.2. *Constraints on a realist conception of moral knowledge*

Suppose that a defense of moral realism is to be undertaken along the lines just indicated. What constraints does that particular defensive strategy place on a moral realist's conception of morals and of moral knowledge? Several important constraints are suggested by a careful examination of the realist doctrines in the philosophy of science whose extension to moral philosophy is contemplated.

In the first place, the scientific realist is able to argue that 'reflective equilibrium' in science and a reliance on theory-dependent scientific intuitions are epistemically reliable *only* on the assumption that the theoretical tradition which governs these methodological practices contains theories which are relevantly approximately true. Indeed, the most striking feature of the consistently realistic epistemology of science is the insistence that the epistemic reliability of scientific methodology is contingent upon the establishment of such a theoretical tradition. Moreover, the possibility of offering a realist rather than a constructivist interpretation of reflective equilibrium and of intuition in science rests upon the realist's claim that observations and theory-mediated measurement and detection of 'unobservables' in science represent epistemically relevant causal interactions between scientists and a theory-independent reality. Were the realist unable to treat observation and measurement as providing 'epistemic access' to reality in this way, a constructivist treatment of scientific knowledge would be almost unavoidable.

Similarly, the scientific realist is able to employ a naturalistic conception of definitions and of reference only because (1) it is arguable that the nature of the subject matter of science dictates that kinds, properties, etc. be defined by nonconventional definitions and (2) it is arguable that actual scientific practices result in the establishment of 'epistemic access' to the various 'theoretical entities' which, the realist maintains, are (part of) the subject matter of scientific inquiry.

Finally, the realist can insist that realism not only can tolerate but implies certain failures of bivalence only because it can be argued that homeostatic cluster kinds (properties, etc.) must have indeterminacy in extension in order for reference to them to be scientifically fruitful. These considerations suggest that the following constraints must be satisfied by an account of moral knowledge if it is to be the basis for the proposed defense of moral realism:

1. It must be possible to explain how our moral reasoning *started out* with a stock of relevantly approximately true moral beliefs so that reflective equilibrium in moral reasoning can be treated in a fashion analogous to the scientific realist's treatment of reflective equilibrium in scientific reasoning. Note that this constraint does not require that it be possible to argue that we started out with close approximations to the truth (seventeenth-century corpuscular theory was quite far from the truth). What is required is that the respects of approximation be such that it is possible to see how continued approximations would be forthcoming as a result of subsequent moral and nonmoral reasoning.

2. There must be an answer to the question "What plays, in moral reasoning, the role played by observation in science?" which can form the basis for a realist rather than a constructivist conception of the foundations of reflective equilibrium in moral reasoning.

3. It must be possible to explain why moral properties, say goodness, would require natural rather than conventional definitions.

4. It must be possible to show that our ordinary use of moral terms provides us with epistemic access to moral properties. Moral goodness must, to some extent, regulate the use of the word 'good' in moral reasoning. Here again examination of the corresponding constraint in the philosophy of science indicates that the regulation need not be nearly perfect, but it must be possible to show that sufficient epistemic access is provided to form the basis for the growth of moral knowledge.

5. It must be possible to portray occasional indeterminacy in the extension of moral terms as rationally dictated by the nature of the subject matter in a way analogous to the scientific realist's treatment of such indeterminacy in the case of homeostatic cluster terms.

Richard N. Boyd

In the work of scientific realists, the case that the analogous constraints are satisfied has depended upon examination of the substantive findings of various of the sciences (such as, e.g., the atomic theory of matter or the Darwinian conception of speciation). It is very unlikely that an argument could be mounted in favor of the view that moral knowledge meets the constraints we are considering which does not rely in a similar way on substantive doctrines about the foundations of morals. What I propose to do instead is to *describe* one account of the nature of morals which almost ideally satisfies the constraints in question and to indicate how a defense of moral realism would proceed on the basis of this account.

It will not be my aim here to defend this account of morals against morally plausible rivals. In fact, I am inclined to think—*partly* because of the way in which it allows the constraints we are considering to be satisfied—that *if* there is a truth of the matter about morals (that is, if moral realism is true), then the account I will be offering is close to the truth. But my aim in this paper is merely to establish that moral realism is plausible and defensible. The substantive moral position I will consider is a plausible version of nonutilitarian consequentialism, one which—I believe—captures many of the features which make consequentialism *one* of the standard and plausible positions in moral philosophy. If moral realism is defensible on the basis of a plausible version of consequentialism, then it is a philosophically defensible position which must be taken seriously in metaethics; and that's all I'm trying to establish here.

It is, moreover, pretty clear that a variety of plausible alternative conceptions of the foundations of morals satisfy the constraints we are discussing. If I am successful here in mounting a plausible defense of moral realism, given the substantive conception I will propose, then it is quite likely that the very powerful semantic and epistemic resources of recent realist philosophy of science could be effectively employed to defend moral realism on the basis of many of the alternative conceptions. I leave it to the defenders of alternative conceptions to explore these possibilities. The defense of moral realism offered here is to be thought of as (the outline of) a 'worked example' of the application of the general strategy proposed in 4.1.

One more thing should be said about the substantive conception of morals offered here. Like any naturalistic account, it rests upon potentially controversial empirical claims about human psychology and about social theory. It is a commonplace, I think, that moral realism is an optimistic position (or, perhaps, that it is typically an optimist's position). One nice feature of the substantive analysis of morals upon which my defense of moral realism will be based is that it quite obviously rests upon optimistic claims about human potential. Perhaps in that respect it is well suited to serve as a representative example of the variety of substantive moral views which would satisfy the constraints in question. (For a further discussion of the methodological implications of the moral realist's reliance on particular substantive moral theories see Section 5.3.)

[202]

4.3. *Homeostatic consequentialism*

In broad outline, the conception of morals upon which the sample defense of moral realism will rest goes like this:

1. There are a number of important human goods, things which satisfy important human needs. Some of these needs are physical or medical. Others are psychological or social; these (probably) include the need for love and friendship, the need to engage in cooperative efforts, the need to exercise control over one's own life, the need for intellectual and artistic appreciation and expression, the need for physical recreation, etc. The question of just which important human needs there are is a potentially difficult and complex empirical question.

2. Under a wide variety of (actual and possible) circumstances these human goods (or rather instances of the satisfaction of them) are homeostatically clustered. In part they are clustered because these goods themselves are—when present in balance or moderation—mutually supporting. There are in addition psychological and social mechanisms which when, and to the extent to which, they are present contribute to the homeostasis. They probably include cultivated attitudes of mutual respect, political democracy, egalitarian social relations, various rituals, customs, and rules of courtesy, ready access to education and information, etc. It is a complex and difficult question in psychology and social theory just what these mechanisms are and how they work.

3. Moral goodness is defined by this cluster of goods and the homeostatic mechanisms which unify them. Actions, policies, character traits, etc. are morally good to the extent to which they tend to foster the realization of these goods or to develop and sustain the homeostatic mechanisms upon which their unity depends.

4. In actual practice, a concern for moral goodness can be a guide to action for the morally concerned because the homeostatic unity of moral goodness tends to mitigate possible conflicts between various individual goods. In part, the possible conflicts are mitigated just because various of the important human goods are mutually reinforcing. Moreover, since the existence of effective homeostatic unity among important human goods is part of the moral good, morally concerned choice is constrained by the imperative to balance potentially competing goods in such a way that homeostasis is maintained or strengthened. Finally, the improvement of the psychological and social mechanisms of homeostasis themselves is a moral good whose successful pursuit tends to further mitigate conflicts of the sort in question. In this regard, moral practice resembles good engineering practice in product design. In designing, say, automobiles there are a number of different desiderata (economy, performance, handling, comfort, durability, . . .) which are potentially conflicting but which enjoy a kind of homeostatic unity if developed in moderation. One feature of good automotive design is that it promotes these desiderata within the limits of homeostasis. The other feature of good

[203]

automotive design (or, perhaps, of good automotive engineering) is that it produces technological advances which permit that homeostatic unity to be preserved at higher levels of the various individual desiderata. So it is with good moral practice as well.[2]

I should say something about how the claim that the nature of the constituents of moral goodness is an empirical matter should be understood. I mean the analogy between moral inquiry and scientific inquiry to be taken *very* seriously. It is a commonplace in the history of science that major advances often depend on appropriate social conditions, technological advances, and prior scientific discoveries. Thus, for example, much of eighteenth-century physics and chemistry was possible only because there had developed (a) the social conditions in which work in the physical sciences was economically supported, (b) a technology sufficiently advanced to make the relevant instrumentation possible, and (c) the theoretical legacy of seventeenth-century Newtonian physics and corpuscular chemistry.

Via somewhat different mechanisms the same sort of dependence obtains in the

---

2. Two points of clarification about the proposed homeostatic consequentialist definition of the good are in order. In the first place, I understand the homeostatic cluster which defines moral goodness to be social rather than individual. The properties in homeostasis are to be thought of as instances of the satisfaction of particular human needs among people generally, rather than within the life of a single individual. Thus, the homeostatic consequentialist holds not (or at any rate not merely) that the satisfaction of each of the various human needs within the life of an individual contributes (given relevant homeostatic mechanisms) to the satisfaction of the others in the life of that same individual. Instead, she claims that, given the relevant homeostatic mechanisms, the satisfaction of those needs for one individual tends to be conducive to their satisfaction for others, and it is to the homeostatic unity of human need satisfaction in the society generally that she or he appeals in proposing a definition of the good.

Homeostatic consequentialism as I present it here is, thus, not a version of ethical egoism. I am inclined to think that individual well-being has a homeostatic property cluster definition and thus that a homeostatic property cluster conception of the definition of the good would be appropriate to the formulation of the most plausible versions of egoism, but I do not find even those versions very plausible and it is certainly not a version of egoism to which I mean to appeal in illustrating the proposed strategy for defending moral realism.

Second, I owe to Judith Jarvis Thomson the observation that, strictly speaking, the homeostatic consequentialist conception of the good does not conform to the more abstract account of homeostatic property cluster definitions presented in Section 3.8. According to that account, the homeostatically united properties and the definitionally relevant properties associated with the relevant mechanisms of homeostasis are all properties of the same kind of thing: organisms, let us say, in the case of the homeostatic property cluster definition of a particular biological species.

By contrast, some of the properties which characterize human well-being and the mechanisms upon which its homeostatic unity depends are (on the homeostatic consequentialist conception) in the first instance properties of individuals, whereas others are properties of personal relations between individuals and still others are properties of large-scale social arrangements. Homeostatic unity is postulated between instances of the realization of the relevant properties in objects of different logical type.

It should be obvious that the additional logical complexity of the proposed homeostatic property cluster definition of the good does not vitiate the rebuttals offered here to anti-realist arguments. For the record, it seems to me that Professor Thomson's observation in fact applies to the actual case of species definitions as well: some of the homeostatically united properties and homeostatic mechanisms which define a species are in the first instance properties of individual organisms, some properties of small groups of organisms, some of larger populations (in the standard sense of that term), and some of the relations between such populations.

growth of our knowledge of the good. Knowledge of fundamental human goods and their homeostasis represents basic knowledge about human psychological and social potential. Much of this knowledge is genuinely *experimental* knowledge and the relevant experiments are ('naturally' occurring) political and social experiments whose occurrence and whose interpretation depends both on 'external' factors and upon the current state of our moral understanding. Thus, for example, we would not have been able to explore the dimensions of our needs for artistic expression and appreciation had not social and technological developments made possible cultures in which, for some classes at least, there was the leisure to produce and consume art. We would not have understood the role of political democracy in the homeostasis of the good had the conditions not arisen in which the first limited democracies developed. Only after the moral insights gained from the first democratic experiments were in hand, were we equipped to see the depth of the moral peculiarity of slavery. Only since the establishment of the first socialist societies are we even beginning to obtain the data necessary to assess the role of egalitarian social practices in fostering the good.

It is also true of moral knowledge, as it is in case of knowledge in other 'special sciences', that the improvement of knowledge may depend upon theoretical advances in related disciplines. It is hard, for example, to see how deeper understanding in history or economic theory could fail to add to our understanding of human potential and of the mechanisms underlying the homeostatic unity of the good.

Let us now consider the application of the particular theory of the good presented here as a part of the strategy for the defense of moral realism indicated in the preceding section. I shall be primarily concerned to defend the realist position that moral goodness is a real property of actions, policies, states of affairs, etc. and that our moral judgments are, often enough, reflections of truths about the good. A complete realist treatment of the semantics of moral terms would of course require examining notions like obligation and justice as well. I will not attempt this examination here, in part because the aim of this essay is merely to indicate briefly how a plausible defense of moral realism might be carried out rather than to carry out the defense in detail. Moreover, on a consequentialist conception of morals such notions as obligation and justice are derivative ones, and it is doubtful if the details of the derivations are relevant to the defense of moral realism in the way that the defense of a realist conception of the good is.

In the remaining sections of the essay I shall offer a defense of homeostatic consequentialist moral realism against the representative anti-realist challenges discussed in Part 2. The claim that the term 'good' in its moral uses refers to the homeostatic cluster property just described (or even the claim that there is such a property) represents a complex and controversial philosophical and empirical hypothesis. For each of the responses to anti-realist challenges which I will present, there are a variety of possible anti-realist rebuttals, both empirical and philosophical. It is beyond the scope of this essay to explore these rebuttals and possible moral realist responses to them in any detail. Instead, I shall merely indicate how plausible realist rebuttals to the relevant challenges can be defended. Once again, the aim of the present paper is not to establish moral realism but merely to establish its plausibility and to offer a general framework within which further defenses of moral realism might be understood.

4.4. *Observations, intuitions, and reflective equilibrium*

Of the challenges to moral realism we are considering, two are straightforwardly epistemological. They suggest that the role of moral intuitions and of reflective equilibrium in moral reasoning dictate (at best) a constructivist interpretation of morals. As we saw in Section 4.2, it would be possible for the moral realist to respond by assimilating the role of moral intuitions and reflective equilibrium to the role of scientific intuitions and theory-dependent methodological factors in the realist account of scientific knowledge, but this response is viable only if it is possible to portray many of our background moral beliefs and judgments as relevantly approximately true and only if there is a satisfactory answer to the question: "What plays, in moral reasoning, the role played in science by observation?" Let us turn first to the latter question.

I propose the answer: 'Observation'.

According to the homeostatic consequentialist conception of morals (indeed, according to any naturalistic conception) goodness is an ordinary natural property, and it would be odd indeed if observations didn't play the same role in the study of this property that they play in the study of all the others. According to the homeostatic consequentialist conception, goodness is a property quite similar to the other properties studied by psychologists, historians, and social scientists, and observations will play the same role in moral inquiry that they play in the other kinds of empirical inquiry about people.

It is worth remarking that in the case of any of the human sciences *some* of what must count as observation is observation of oneself, and *some* is the sort of self-observation involved in introspection. Moreover, *some* of our observations of other people will involve trained judgment and the operation of sympathy. No reasonable naturalistic account of the foundations of psychological or social knowledge *or* of our technical knowledge in psychology or the social sciences will fail to treat such sources of belief—when they are generally reliable—as cases of observation in the relevant sense.

It is true, of course, that both the content and the evidential assessment of observations of this sort will be influenced by theoretical considerations, but this does not distinguish observations in the human sciences from those in other branches of empirical inquiry. The theory dependence of observations and their interpretation is simply one aspect of the pervasive theory dependence of methodology in science which the scientific realist cheerfully acknowledges (since it plays a crucial role in arguments for scientific realism). It is possible to defend a

[206]

realist interpretation of the human sciences because it is possible to argue that actual features in the world constrain the findings in those sciences sufficiently that the relevant background theories will be approximately true enough for theory-dependent observations to play a reliable epistemic role.

In the case of moral reasoning, observations and their interpretation will be subject to just the same sort of theory-dependent influences. This theory dependence is one aspect of the general phenomenon of theory dependence of methodology in moral reasoning which we, following Rawls, have been describing as reflective equilibrium. We will be able to follow the example of scientific realists and to treat the observations which play a role in moral reasoning as sufficiently reliable for the defense of moral realism just in case we are able to portray the theories upon which they and their interpretation depend as relevantly approximately true—that is, just in case we are able to carry out the other part of the moral realist's response to epistemic challenges and to argue that our background moral beliefs are sufficiently near the truth to form the foundations for a reliable empirical investigation of moral matters. Let us turn now to that issue.

What we need to know is whether it is reasonable to suppose that, for quite some time, we have had background moral beliefs sufficiently near the truth that they could form the basis for subsequent improvement of moral knowledge in the light of further experience and further historical developments. Assuming, as we shall, a homeostatic consequentialist conception of morals, this amounts to the question whether our background beliefs about human goods and the psychological and social mechanisms which unite them have been good enough to guide the gradual process of expansion of moral knowledge envisioned in that conception. Have our beliefs about our own needs and capacities been good enough—since, say the emergence of moral and political philosophy in ancient Greece—that we have been able to respond to new evidence and to the results of new social developments by expanding and improving our understanding of those needs and capacities even when doing so required rejecting some of our earlier views in favor of new ones? It is hard to escape the conclusion that this is simply the question ''Has the rational empirical study of human kind proven to be possible?'' Pretty plainly the answer is that such study has proven to be possible, though difficult. In particular we have improved our understanding of our own needs and our individual and social capacities by just the sort of historically complex process envisioned in the homeostatic consequentialist conception. I conclude therefore that there is no reason to think that reflective equilibrium—which is just the standard methodology of any empirical inquiry, social or otherwise—raises any epistemological problems for the defense of moral realism.

Similarly, we may now treat moral intuitions exactly on a par with scientific intuitions, as a species of trained judgment. Such intuitions are *not* assigned a foundational role in moral inquiry; in particular they do not substitute for observations. Moral intuitions are simply one cognitive manifestation of our moral understanding, just as physical intuitions, say, are a cognitive manifestation of

[207]

physicists' understanding of their subject matter. Moral intuitions, like physical intuitions, play a limited but legitimate role in empirical inquiry *precisely because* they are linked to theory *and* to observations in a generally reliable process of reflective equilibrium.

It may be useful by way of explaining the epistemic points made here to consider very briefly how the moral realist might respond to one of the many possible anti-realist rebuttals to what has just been said. Consider the following objection: The realist treatment of reflective equilibrium requires that our background moral beliefs have been for some time relevantly approximately true. As a matter of fact, the overwhelming majority of people have probably always believed in some sort of theistic foundation of morals: moral laws are God's laws; the psychological capacities which underlie moral practice are a reflection of God's design; etc. According to the homeostatic consequentialism which we are supposed to accept for the sake of argument, moral facts are mere natural facts. Therefore, according to homeostatic consequentialism, most people have always had profoundly mistaken moral beliefs. How then can it be claimed that our background beliefs have been relevantly approximately true?

I reply that—assuming that people have typically held theistic beliefs of the sort in question—it does follow from homeostatic consequentialism that they have been *in that respect* very wrong indeed. But being wrong in that respect does not preclude their moral judgments having been relatively reliable reflections of facts about the homeostatic cluster of fundamental human goods, according to the model of the development of moral knowledge discussed earlier. Until Darwin, essentially all biologists attributed the organization and the adaptive features of the physiology, anatomy, and behavior of plants and animals to God's direct planning. That attribution did not prevent biologists from accumulating the truly astonishing body of knowledge about anatomy, physiology, and animal behavior upon which Darwin's discovery of evolution by natural selection depended; nor did it prevent their recognizing the profound biological insights of Darwin's theory. Similarly, seventeenth-century corpuscular chemistry did provide the basis for the development of modern chemistry in a way that earlier quasi-animistic 'renaissance naturalism' in chemistry could not. Early corpuscular theory was right that the chemical properties of substances are determined by the fundamental properties of stable 'corpuscles'; it was wrong about almost everything else, but what it got right was enough to point chemistry in a fruitful direction. I understand the analogy between the development of scientific knowledge and the development of moral knowledge to be very nearly exact.

There may indeed be one important respect in which the analogy between the development of scientific knowledge and the development of moral knowledge is *in*exact, but oddly, this respect of disanalogy makes the case for moral realism stronger. One of the striking consequences of a full-blown naturalistic and realistic conception of knowledge is that our knowledge, even our most basic knowledge, rests upon logically contingent 'foundations'. Our perceptual knowledge, for example, rests upon the logically contingent a posteriori fact that our senses are reliable detectors of certain sorts of external objects. In the case of perceptual knowledge, however, there is a sense in which it is nonaccidental, noncontingent, that our senses are reliable detectors. The approximate reliability of our senses (with respect to some applications) is explained by evolutionary theory in a quite fundamental way (Quine 1969a). By contrast, the reliability of our methodology in chemistry is much more dramatically contingent. As a matter of fact, early thinkers tried to explain features of the natural world by analogy to sorts of order they already partly understood: mathematical, psychological, and mechanical. The atomic theory of matter represents one such attempt to assimilate chemical order to the better-understood mechanical order. In several important senses it was highly contingent that the microstructure of matter turned out to be particulate and mechanical enough that the atomic (or 'corpuscular') *guess* could provide the foundation for epistemically reliable research in chemistry. The accuracy of our guess in this regard is not, for example, explained by either evolutionary necessity or by deep facts about our psychology. In an important sense, the seventeenth-century belief in the corpuscular theory of matter was not reliably produced. It was not produced by an antecedent generally reliable methodology: reasoning by analogy is *not* generally reliable except in contexts where a rich and approximately accurate body of theory *already* exists to guide us in finding the right respects of analogy (see Boyd 1982).

By contrast, the emergence of relevantly approximately true beliefs about the homeostatic cluster of fundamental human goods—although logically contingent—was much less strikingly 'accidental'. From the point of view either of evolutionary theory or of basic human psychology it is hardly accidental that we are able to recognize many of our own and others' fundamental needs. Moreover, it is probably not accidental from an evolutionary point of view that we were able to recognize some features of the homeostasis of these needs. Our initial relevantly approximately accurate beliefs about the good may well have been produced by generally reliable psychological and perceptual mechanisms and thus may have been clear instances of knowledge in a way in which our initial corpuscular beliefs were not (for a discussion of the latter point see Boyd 1982). It is *easier*, not *harder*, to explain how moral knowledge is possible than it is to explain how scientific knowledge is possible. Locke was right that we are fitted by nature for moral knowledge (in both the seventeenth- and the twentieth-century senses of the term) in a way that we are not so fitted for scientific knowledge of other sorts.

4.5. *Moral semantics*

We have earlier considered two objections to the moral realist's account of the semantics of moral terms. According to the first, the observed diversity of moral concepts—between cultures as well as between individuals and groups within a culture—suggests that it will not be possible to assign a single objective subject matter to their moral disputes. The divergence of concepts suggests divergence of reference of a sort which constructivist relativism is best suited to explain. According to the second objection, moral realism is commited to the absurd position that moral terms possess definitions in the vocabulary of the natural sciences. We have seen that a moral realist rebuttal to these challenges is possible which assimilates moral terms to naturalistically and nonreductively definable terms in the sciences. Such a response can be successful only if (1) there are good reasons to think that moral terms must possess natural rather than stipulative definitions and (2) there are good reasons to think that ordinary uses of moral terms provides us with epistemic access to moral properties, so that, for example, moral goodness to some extent regulates our use of the word 'good' in moral contexts.

The homeostatic consequentialist conception of morals provides a justification for the first of these claims. If the good is defined by a homeostatic phenomenon the details of which we still do not entirely know, then it is a paradigm case of a property whose 'essence' is given by a natural rather than a stipulative definition.

Is it plausible that the homeostatic cluster of fundamental human goods has, to a significant extent, regulated the use of the term 'good' so that there is a general tendency, of the sort indicated by the homeostatic consequentialist conception of the growth of moral knowledge, for what we say about the good to be true of that cluster? If what I have already said about the possibility of defending a realist conception of reflective equilibrium in moral reasoning is right, the answer must be "yes." Such a tendency is guaranteed by basic evolutionary and psychological facts, and it is just such a tendency which we can observe in the ways in which our conception of the good has changed in the light of new evidence concerning human needs and potential. Indeed, the way we ('preanalytically') recognize moral uses of the term 'good' and the way we identify moral terms in other languages are precisely by recourse to the idea that moral terms are those involved in discussions of human goods and harms. We tacitly assume *something like* the proposed natural definition of 'good' in the practice of translation of moral discourse. I think it will help to clarify this realist response if we consider two possible objections to it. The first objection reflects the same concern about the relation between moral and theological reasoning that we examined in the preceding section. It goes like this: How is it possible for the moral realist who adopts homeostatic consequentialism to hold that there is a general tendency for our beliefs about the good to get truer? After all, the error of thinking of the good as being defined by God's will persists unabated and is—according to the homeostatic consequentialist's conception—a very important falsehood.

I reply, first, that the sort of tendency to the truth required by the epistemic access account of reference is not such that it must preclude serious errors. Newtonians were talking about mass, energy, momentum, etc. all along, even though they were massively wrong about the structure of space-time. We might be irretrievably wrong about some other issue in physics and still use the terms of physical theory to refer to real entities, magnitudes, etc. All that is required is a significant epistemically relevant causal connection between the use of a term and its referent.

Moreover, as I suggested earlier, it is characteristic of what we recognize as moral discourse (whether in English or in some other language) that considerations of human well-being play a significant role in determining what is said to be 'good'. The moral realist need not deny that other considerations—perhaps profoundly false ones—also influence what we say is good. After all, the historian of biology need not deny that the term 'species' has relatively constant reference throughout the nineteenth century, even though, prior to Darwin, religious considerations injected profound errors into biologists' conception of species. Remember that we do not ordinarily treat a theological theory as a theory *of* moral goodness at all unless it says something about what we independently recognize as human well-being. The role of religious considerations in moral reasoning provides a challenge for moral realists, but exactly the same challenge faces a realist interpretation of biological or psychological theorizing before the twentieth century, and it can surely be met.

The second objection I want to consider represents a criticism of moral realism often attributed to Marx (see, e.g., Wood 1972; for the record I believe that Marx's position on this matter was confused and that he vacillated between an explicit commitment to the relativist position, which Wood discusses, and a tacit commitment to a position whose reconstruction would look something like the position defended here). The objection goes like this: The moral realist—in the guise of the homeostatic consequentialist, say—holds that what regulate the use of moral terms are facts about human well-being. But this is simply not so. Consider, for example, sixteenth-century discussions of rights. One widely acknowledged 'right' was the divine right of kings. Something surely regulated the use of the language of rights in the sixteenth century, but it clearly wasn't human well-being construed in the way the moral realist intends. Instead, it was the well-being of kings and of the aristocratic class of which they were a part.

I agree with the analysis of the origin of the doctrine of the divine right of kings; indeed, I believe that such class determination of moral beliefs is a commonplace phenomenon. But I do not believe that this analysis undermines the claim that moral terms refer to aspects of human well-being. Consider, for example, the psychology of thinking and intelligence. It is extremely well documented (see, e.g., Gould 1981; Kamin 1974) that the content of much of the literature in this area is determined by class interests rather than by the facts. Nevertheless, the psychological terms occurring in the most egregiously prejudiced papers refer to real features of human psychology; this is so because, in other contexts, their use is relevantly regulated by such features. Indeed—and this is the important point— if there were not such an epistemic (and thus referential) connection to real psychological phenomena, the ideological rationalization of class structures represented by the class-distorted literature would be ineffective. It's only when people come to believe, for example, that Blacks lack a trait, *familiar in other contexts as 'intelligence'*, that racist theories can serve to rationalize the socioeconomic role to which Blacks are largely confined.

Similarly, I argue, in order for the doctrine of the divine right of kings to serve a class function, it had to be the case that moral language was often enough connected to issues regarding the satisfaction of real human needs. Otherwise, an appeal to such a supposed right would be ideologically ineffective. Only when rights-talk has *some* real connection to the satisfaction of the needs of non-aristocrats could this instance of rights-talk be useful to kings and their allies.

Once again, when the analogy between moral inquiry and scientific inquiry is fully exploited, it becomes possible to defend the doctrines upon which moral realism rests.

4.6. *Hard cases and divergent views*

Two of the challenges to moral realism we are considering are grounded in the recognition that some moral issues seem very hard to resolve. On the one hand, there seem to be moral dilemmas which resist resolution even for people who share a common moral culture. Especially with respect to the sort of possible cases often considered by moral philosophers, there often seems to be no rational way of deciding between morally quite distinct courses of action. Our difficulty in resolving moral issues appears even greater when we consider the divergence in moral views that exists between people from different backgrounds or cultures. The anti-realist proposes to explain the difficulties involved by denying that there is a common objective subject matter which determines answers to moral questions.

We have seen that—to the extent that she chooses to take the difficulties in resolving moral issues as evidence for the existence of moral statements for which bivalence fails—the moral realist can try to assimilate such failures to the failures of bivalence which realist philosophy *predicts* in the case, for example, of some statements involving homeostatic cluster terms. Such a response will work only to the extent that moral terms can be shown to possess natural definitions relevantly like homeostatic cluster definitions. Of course, according to homeostatic consequentialism, moral terms (or 'good' at any rate) just are homeostatic cluster terms, so this constraint is satisfied. What I want to emphasize is that a moral realist *need not* invoke failures of bivalence in every case in which difficulties arise in resolving moral disputes.

Recall that on the conception we are considering moral inquiry is about a complex and difficult subject matter, proceeds often by the analysis of complex and "messy" naturally occurring social experiments, and is subject to a very high level of social distortion by the influence of class interests and other cultural factors. In this regard moral inquiry resembles inquiry in any of the complex and politically controversial social sciences. In such cases, even where there is no reason to expect failures of bivalence, one would predict that the resolution of some issues will prove difficult or, in some particular social setting, impossible. Thus the moral realist can point to the fact that moral inquiry is a species of social inquiry to explain much of the observed divergence in moral views and the apparent intractability of many moral issues.

Similarly, the complexity and controversiality of moral issues can be invoked to explain the especially sharp divergence of moral views often taken to obtain between different cultures. For the homeostatic consequentialist version of moral realism to be true it must be the case that in each culture in which moral inquiry takes place the homeostatically clustered human goods epistemically regulate moral discourse to an appreciable extent. On the realistic and naturalistic conception of the growth of knowledge, this will in turn require that the moral tradition of the culture in question embody some significant approximations to the truth about moral matters. It is, however, by no means required that two such cultural traditions have started with initial views which approximated the truth to the same extent or along the same dimensions, nor is it required that they have been subjected to the same sorts of social distortion, nor that they have embodied the same sorts of naturally occurring social experimentation. It would thus be entirely unsurprising if two such traditions of moral inquiry should have, about some important moral questions, reached conclusions so divergent that no resolution of their disagreement will be possible within the theoretical and methodological framework which the two traditions *currently* have in common, even though these issues may possess objective answers eventually discoverable from within either tradition or from within a broader tradition which incorporates insights from both.

In this regard it is useful to remember the plausibility with which it can be argued that, if there were agreement on all the nonmoral issues (including theological ones), then there would be no moral disagreements. I'm not sure that this is exactly right. For one thing, the sort of moral agreement which philosophers typically have in mind when they say this sort of thing probably does not include agreement that some question has an indeterminate answer, which is something predicted by homeostatic consequentialism. Nevertheless, careful philosophical examination will reveal, I believe, that agreement on nonmoral issues would eliminate *almost all* disagreement about the sorts of moral issues which arise in ordinary moral practice. Moral realism of the homeostatic consequentialist variety provides a quite plausible explanation for this phenomenon.

It is nevertheless true that, for some few real-world cases and for *lots* of the contrived cases so prevalent in the philosophical literature, there does appear to be serious difficulty in finding rational resolutions—assuming as we typically do that an appeal to indeterminacy of the extension of 'good' doesn't count as a resolution. In such cases the strategy available to the moral realist *is* to insist that failures of bivalence do occur just as a homeostatic consequentialist moral realist predicts.

Philosophers often suggest that the major normative ethical theories will yield the same evaluations in almost all actual cases. Often it is suggested that this fact supports the claim that there is some sort of objectivity in ethics, but it is very difficult to see just why this should be so. Homeostatic consequentialist moral realism provides the basis for a satisfactory treatment of this question. Major theories in normative ethics have almost always sought to provide definitions for moral terms with almost completely definite extensions. This is, of course, in fact a mistake; moral terms possess homeostatic cluster definitions instead. The appearance of sharp divergence between major normative theories, with respect to the variety of possible cases considered by philosophers, arises from the fact that they offer different putative resolutions to issues which lack any resolution *at all* of the sort anticipated in those theories. The general agreement of major normative theories on almost all actual cases is explained both by the fact that the actual features of the good regulate the use of the term 'good' in philosophical discourse *and* by the homeostatic character of the good: when different normative theories put different weight on different components of the good, the fact that such components are—in actual cases—linked by reliable homeostatic mechanisms tends to mitigate, in real-world cases, the effects of the differences in the weights assigned. Homeostatic consequentialism represents the common grain of truth in other normative theories. (For further discussion of the resulting case for moral realism see Section 5.4.)

4.7. *Morality, motivation, and rationality*

There remains but one of the challenges to moral realism which we are here considering. It has often been objected against moral realism that there is some sort of logical connection between moral judgments and reasons for action which a moral realist cannot account for. It might be held, for example, that the recognition that one course of action is morally preferable to another *necessarily* provides a reason (even if not a decisive one) to prefer the morally better course of action. Mere facts (especially mere *natural* facts) cannot have this sort of logical connection to rational choice or reasons for action. Therefore, so the objection goes, there cannot be moral facts; moral realism (or at least naturalistic moral realism) is impossible.

It is of course true that the naturalistic moral realist must deny that moral judgments necessarily provide reasons for action; surely, for example, there could be nonhuman cognizing systems which could understand the natural facts about moral goodness but be entirely indifferent to them in choosing how to act. Moral judgments might provide for them no reasons for action whatsoever. Moreover, it is hard to see how the naturalistic moral realist can escape the conclusion that it would be *logically possible* for there to be a human being for whom moral judgments provided no reasons for action. The moral realist must therefore deny that the connection between morality and reasons for action is so strong as the objection we are considering maintains. The appearance of an especially intimate connection must be explained in some other way.

[214]

The standard naturalist response is to explain the apparent intimacy of the connection by arguing that the natural property moral goodness is one such that for psychologically normal humans, the fact that one of two choices is morally preferable will in fact provide some reason for preferring it. The homeostatic consequentialist conception of the good is especially well suited to this response since it defines the good in terms of the homeostatic unity of fundamental human needs. It seems to me that this explanation of the close connection between moral judgments and reasons for action is basically right, but it ignores—it seems to me—one important source of the anti-realist's intuition that the connection between moral judgments and rational choice must be a necessary one. What I have in mind is the very strong intuition which many philosophers share that the person for whom moral judgments are motivationally indifferent would not only be psychologically atypical but would have some sort of *cognitive* deficit with respect to moral reasoning as well. The anti-realist diagnoses this deficit as a failure to recognize a definitional or otherwise necessary connection between moral goodness and reasons for action.

I think that there is a deep insight in the view that people for whom questions of moral goodness are irrelevant to how they would choose to act suffer a cognitive deficit. I propose that the deficit is not—as the anti-realist would have it—a failure to recognize a necessary connection between moral judgments and reasons for action. Instead, I suggest, if we adopt a naturalistic conception of moral knowledge we can diagnose in such people a deficit in the capacity to make moral judgments somewhat akin to a perceptual deficit. What I have in mind is the application of a causal theory of moral knowledge to the examination of a feature of moral reasoning which has been well understood in the empiricist tradition since Hume, that is, the role of sympathy in moral understanding.

It is extremely plausible that for normal human beings the capacity to access human goods and harms—the capacity to *recognize* the extent to which others are well or poorly off with respect to the homeostatic cluster of moral goods and the capacity to *anticipate correctly* the probable effect on others' well-being of various counterfactual circumstances—depends upon their capacity for sympathy, their capacity to imagine themselves in the situation of others or even to find themselves involuntarily doing so in cases in which others are especially well or badly off. The idea that sympathy plays this sort of cognitive role is a truism of nineteenth-century faculty psychology, and it is very probably right.

It is also very probably right, as Hume insists, that the operation of sympathy is *motivationally* important: as a matter of contingent psychological fact, when we put ourselves in the place of others in imagination, the effects of our doing so include our taking pleasure in others' pleasures and our feeling distress at their misfortune, and we are thus motivated to care for the well-being of others. The psychological mechanisms by which all this takes place may be more complicated than Hume imagined, but the fact remains that one and the same psychological mechanism—sympathy—plays *both* a cognitive *and* a motivational role in normal human beings. We are now in a position to see why the morally unconcerned

[215]

person, the person for whom moral facts are motivationally irrelevant, probably suffers a *cognitive* deficit with respect to moral reasoning. Such a person would have to be deficient in sympathy, because the motivational role of sympathy is precisely to make moral facts motivationally relevant. In consequence, she or he would be deficient with respect to a cognitive capacity (sympathy) which is ordinarily important for the correct assessment of moral facts. The motivational deficiency would, as a matter of contingent fact about human psychology, be a cognitive deficiency as well.

Of course it does not follow that there could not be cognizing systems which are quite capable of assessing moral facts without recourse to anything like sympathy; they might, for example, rely on the application of a powerful tacit or explicit theory of human psychology instead. Indeed it does not follow that there are not actual people—some sociopaths and con artists, for example—who rely on such theories instead of sympathy. But it is true, just as the critic of moral realism insists, that there is generally a cognitive deficit associated with moral indifference. The full resources of naturalistic epistemology permit the moral realist to acknowledge and explain this important insight of moral antirealists.

4.8. *Conclusion*

I have argued that if the full resources of naturalistic and realistic conceptions of scientific knowledge and scientific language are deployed and if the right sort of positive theory of the good is advanced, then it is possible to make a plausible case for moral realism in response to typical anti-realist challenges. Two methodological remarks about the arguments I have offered may be useful. In the first place, the rebuttals I have offered to challenges to moral realism really do depend strongly upon the naturalistic and nonfoundational aspects of current (scientific) realist philosophy of science. They depend, roughly, upon the aspects of the scientific realist's program which make it plausible for the scientific realist to claim that philosophy is an empirical inquiry continuous with the sciences and with, e.g., history and empirical social theory. I have argued elsewhere (Boyd 1981, 1982, 1983, 1985a, 1985b, 1985c) that these aspects of scientific realism are essential to the defense of scientific realism against powerful empiricist and constructivist arguments.

If we now ask how one should decide between scientific realism and its rivals, I am inclined to think that the answer is that the details of particular technical arguments will not be sufficient to decide the question rationally; instead, one must assess the overall conceptions of knowledge, language, and understanding which go with the rival conceptions of science (I argue for this claim in Boyd 1983). *One* important constraint on an acceptable philosophical conception in these areas is that it permit us to understand the obvious fact that moral reasoning is not nearly so different from scientific or other factual reasoning as logical positivists have led us to believe. It is initially plausible, I think, that a constructivist conception of science is favored over both empiricist and realist conceptions insofar as we confine our attention to this constraint. If what I have said here is correct, this may well not be so. Thus the successful development of the arguments presented here may be relevant not only to our assessment of moral realism but to our assessment of scientific realism as well. Here is a kind of methodological unity of philosophy analogous to (whatever it was which positivists called) 'unity of science'.

My second methodological point is that the arguments for moral realism presented here depend upon optimistic empirical claims both about the organic unity of human goods and about the possibility of reliable knowledge in the 'human sciences' generally. Although I have not argued for this claim here, I believe strongly that any plausible defense of naturalistic moral realism would require similarly optimistic empirical assumptions. I am also inclined to believe that insofar as moral anti-realism is plausible its plausibility rests not only upon technical philosophical arguments but also upon relatively pessimistic empirical beliefs about the same issues. I suggest, therefore, that our philosophical examination of the issues of moral realism should include, in addition to the examination of technical arguments on both sides, the careful examination of empirical claims about the unity and diversity of human goods and about our capacity for knowledge of ourselves. That much of philosophy ought surely to be at least partly empirical.

5. Addendum

5.1. *History*

This paper, in the form in which it appears here, was written in 1982. Since that time it has undergone a transformation into a work, *Realism and the Moral Sciences* (Boyd, forthcoming, henceforth *RMS*) much too long to publish or excerpt for the present volume. I do however want to indicate briefly the direction in which the line of argument presented here has been developed in that later work. I shall briefly summarize three ways in which *RMS* goes beyond the argumentative strategy of this essay: a further characterization of homeostatic property cluster definitions, a response to an apparent circularity resulting from the employment of a sample substantive moral theory, and an indication of the most general evidence favoring moral realism.

5.2. *Homeostatic property clusters again*

In *RMS* I add an additional clause to the account of homeostatic property cluster definitions as follows:

12. The homeostatic property cluster which serves to define *t* is not individuated extensionally. Instead, property clusters are individuated like (type or token)

historical objects or precesses: certain changes over time (or in space) in the property cluster or in the underlying homeostatic mechanisms preserve the identity of the defining cluster. In consequence, the properties which determine the conditions for falling under *t* may vary over time (or space), *while t continues to have the same definition*. To fall under *t* is to participate in the (current temporal and spatial stage of) the relevant property clustering. The historicity of the individuation criteria for definitional property clusters of this sort reflects the explanatory or inductive significance (for the relevant branches of theoretical or practical inquiry) of the historical development of the property cluster and of the causal factors which produce it, and considerations of explanatory and inductive significance determine the appropriate standards of individuation for the property cluster itself. The historicity of the individuation conditions for the property cluster is thus essential for the naturalness of the kind to which *t* refers.

This modification is suggested by the example of biological species definitions. The property cluster and homeostatic mechanisms which define a species must in general be individuated nonextensionally as a process-like historical entity. This is so because the mechanisms of reproductive isolation which are fundamentially definitional for many sexually reproducing species may vary significantly over the life of a species. Indeed, it is universally recognized that selection for characters which enhance reproductive isolation from related species is a significant factor in phyletic evolution, and it is one which necessarily alters over time a species' defining property cluster and homeostatic mechanisms (Mayr 1970).

I propose in *RMS* that the homeostatic property cluster definition of moral goodness exhibits this sort of historicality. This additional factor increases the complexity of that definition considerably. Moreover there are failures of bivalence in the individuation of homeostatic property cluster definitions, especially across possible worlds, just as there are for other sorts of historically individuated entities. These bivalence failures with respect to the individuation of the definition of moral goodness increases the range of counterfactual cases for which there will be failures of bivalence in the application of the term 'good'. The resources available to the moral realist for explaining divergent moral opinions, especially with respect to counterfactual cases, are thus enhanced.

5.3. *Hard cases, cultural variability, and an apparent circularity of argumentation*

In Part 4 of the present essay, I defend moral realism from the perspective of a sample substantive moral theory, homeostatic consequentialism. I argue that since this substantive theory is defensible and since it affords the basis for a reasonable defense of a moral realism, moral realism is itself philosophically defensible. In *RMS* I consider a possible objection to this argumentative strategy. According to the objection, the defense of moral realism offered here requires a realistic understanding of homeostatic consequentialism, since otherwise, for example, the moral properties to which epistemic access is demonstrated might be purely socially constructed or conventional, as constructivist anti-realists in ethics maintain. The realist understanding of homeostatic consequentialism, so the objection goes, begs the question against the anti-realist so that the defense of moral realism is not even prima facie successful.

I examine this objection in the light of the corresponding objection to arguments for scientific realism. I argue that a defense of scientific realism requires that the realist articulate and defend a *theory of epistemic contact* and a *theory of error* for those traditions of inquiry for which she offers a realist account. Each of these theories must necessarily rest upon the best available theories of the relevant subject matter, realistically understood. In order to see why the question is not necessarily begged against the anti-realist we need to distinguish two sorts of anti-realist arguments from the diversity of opinions or intractability of issues within the relevant area of inquiry.

The first sort of argument, which I call the *external argument from theoretical diversity,* challenges the realist to explain the diversity of theoretical conceptions and the difficulty of their resolution within the relevant tradition of inquiry. An adequate realist response to this challenge will consist of an account of the epistemically significant causal relations between inquirers and the supposed theory-independent subject matter of that tradition, together with a theory of the sources of error within the tradition which account for the observed diversity of theoretical conceptions and for whatever difficulty exists in resolving the resulting theoretical disputes. These theories of epistemic contact and of error will necessarily and properly reflect the best available current theories of the subject matter in question realistically understood. No question is begged against the anti-realist because the realistically understood theories of epistemic contact and of error do not *by themselves* constitute the argument for realism. Instead, the philosophical contest is between larger-scale *philosophical packages:* a realist package which incorporates the realistically understood theories of epistemic contact and of error into a larger account of the metaphysics, epistemology, semantics, methodology, and historical development of the relevant areas of inquiry and various competing anti-realist packages of comparable scope.

Indeed, I argue, an understanding of scientific realism according to which it is grounded in realistically understood theories of epistemic contact and of error is essential not only for a fair presentation of the case *for* realism but also for a fair presentation of the various cases *against* it. I conclude, by analogy, that no questions are begged with respect to the external argument from theoretical diversity in the moral realist's reliance upon realistically understood theories of epistemic contact and of error.

I also identify a fundamentally different anti-realist argument from theoretical diversity, which I call the *internal argument from theoretical diversity* and which represents, I suggest, an important but largely inexplicit consideration in arguments against realism, whether scientific or moral. The internal argument is directed against the evidential acceptability of the realist's philosophical package even by by its own standards. The argument procedes by identifying a widely accepted and (so far as I can see) unobjectionable methodological principle according to which the prevalence of a variety of competing theoretical conceptions within a subject area should reduce one's confidence in the truth (or approximate truth) of any one of them. This principle is then applied against the realist's theories of epistemic contact and of error. It is argued that the existence of a diversity of competing theories in the relevant area(s) of inquiry renders epistemically illegitimate the particular theoretical commitments which underlie whichever theories of epistemic contact and of error the realist chooses to adopt.

The problem raised by this criticism of the realist's philosophical package is not that it does a poorer job than that of some anti-realist competitors in explaining theoretical diversity; rather the objection is that the realist must, by her own standards of evidence, hold that there is little evidence favoring the theoretical conceptions which underlie her own philosophical position.

It is important to recognize that because it appeals to standards of evidence which are prephilosophically generally accepted and are presumed to be internal to the relevant discipline(s), the internal argument from theoretical diversity is cogent only with respect to the theoretical diversity represented by those competing theories which are plausible candidates given the best current methodological standards in the relevant disciplines. It thus contrasts sharply with the external argument. To respond appropriately to the latter, the realist must be in a position to adequately explain *all* of the diverse opinions (however implausible by current standards) within the history of the tradition of inquiry regarding which she defends a realist conception. In responding to the internal argument, by contrast, she need only respond to the challenge to her theoretical commitments which is raised by the diversity within currently plausible theoretical conceptions, where plausibility is assessed by the best available contemporary standards.

I suggest in *RMS* that the way in which scientific realists have tacitly met this (itself inexplicit) objection can be reconstructed as follows: Instead of offering, for the discipline(s) in question, a single theory of epistemic contact and a single theory of error derived from one of the plausible alternative theoretical conceptions, the realist should be thought of as offering a family of such pairs of theories, one pair grounded in each of the alternative conceptions. She then should be thought of as arguing that these alternative theories of epistemic contact and of error participate sufficiently in a relationship of (*partial*) *mutual ratification* sufficiently deep that an adequate realist philosophical package can be grounded in their disjunction. By their partial mutual ratification I intend the relationship which obtains if the rival theories of epistemic contact and of error agree about a large number of particular cases of epistemic success and of error and if they give similar accounts about the nature of evidential relationships between data and doctrines within the relevant field(s) without, of course, agreeing in all of the theoretical details of their accounts of those relationships.

The levels of mutual ratification between competing theories of epistemic contact and of error which are required for the defense of realism will depend on the broader dialectical interactions of the competing philosophical packages, realist and anti-realist. It is nevertheless possible to identify an extremely strong pattern of mutual ratification which seems to characterize the methodological situation of those mature sciences regarding which realism is agreed to be an especially plausible position.

For any particular body of inquiry we may construct a conditionalized theory of epistemic contact and of error as follows. First, we form each of the propositions of the form "If *T* then (*C* and *E*)", where *T* is one of the currently plausible theoretical conceptions in the relevant field and *C* and *E* are the theories of epistemic contact and of error for the history of the relevant body of inquiry which are best suited to a defense of a realist conception of that body of inquiry, on the assumption that *T* is the (largely) correct choice from among the competing plausible alternatives. We then form the conditionalized theory of epistemic contact and of error by taking the conjunction of each of these propositions. Let us say that a situation of mutual conditional ratification obtains if (a) the individual theories of epistemic contact and of error obtained from the various plausible theories agree on many actual cases of evidential judgments and (b) the conditionalized theory of epistemic contact is rationally acceptable given the standards of evidence common to *all* the competing theoretical conceptions. It is this strong pattern of *conditional mutual ratification* which seems to characterize those areas of inquiry about which realism seems especially plausible.

Having characterized mutual conditional ratification in *RMS*, I then develop the claim of the present essay that a defense of moral realism along roughly the lines developed in Part 4 is possible on the basis of any of the genuinely plausible general moral theories. I consider the theories of epistemic contact and of error which would be appropriate to such theories, and I conclude that to a *very* good first approximation conditional mutual ratification obtains with respect to the spectrum of general moral theories which are genuinely plausible by the best current standards. Indeed, I argue, an especially strong form of conditional mutual ratification obtains which is characterized by three additional features:

1. To an extremely good first approximation, moral judgments regarding actual cases of actions, policies, character traits, etc. are—given prevailing standards of moral argument—dictated by judgments regarding nonmoral factual questions (including, for example, questions about human nature, about the nature of social, political, and economic processes, about whether or not there are any gods, and about their natures if there are any, . . .). In consequence, moral disagreements regarding such actual cases can be seen (on a philosophically appropriate rational reconstruction) as stemming from disagreements over nonmoral factual matters. (I call this relationship the *rational supervenience* of the relevant moral judgments on nonmoral factual judgments.)

2. Rational supervenience appears to fail for a few actual cases and for many

counterfactual ones. For almost all of these it is plausible to argue that the cases in question are ones in which there is unrecognized failure of bivalence. For the few remaining cases of apparent failures of rational supervenience realist explanations in terms of nonculpable inadequacies in methodology or theoretical understanding are readily available. This conception of the sources of failures of rational supervenience is itself ratified by all of the genuinely plausible competing general moral theories.

3. The conditionalized theory of epistemic contact and of error upon which the plausible competing general moral theories agree is such that it attributes differences in judgments regarding *general* moral theories to differences over nonmoral factual matters. Thus, rational supervenience upon nonmoral factual judgments obtains for general moral theories as well as for particular moral judgments.

It is upon this quite striking form of conditional mutual ratification that, in my view, the moral realist's response to the internal argument from theoretical diversity properly rests.

5.4. *The evidence for moral realism*

In the present essay I argue that moral realism can be defended on the basis of a particular substantive moral theory (homeostatic consequentialism) which is itself defensible. I conclude that moral realism is itself a defensible position worthy of further development, and of criticisms appropriate to the epistemological, semantic, and metaphysical arguments in its favor which the analogy with scientific realism suggests. I suggest here (and argue in *RMS*) that the same sort of defense can be formulated on the basis of any of the other plausible competing moral theories. Thus, if the arguments I offer are correct, there is reason to believe that the defender of any of the currently plausible general moral theories should defend her theory on a realist understanding of its content and should herself be a moral realist. The question remains what the attitude toward moral realism should be of the philosopher who is, as yet, not committed to any particular general moral theory.

I address this question in *RMS*. I maintain that the best argument for moral realism in the present philosophical context probably would consist of a more thoroughgoing defense of a particular naturalistic and realistic substantive moral theory much like homeostatic consequentialism. I also conclude, however, that there is powerful evidence favoring moral realism whose persuasive force does not depend upon establishing the case for any particular moral theory. Indeed, I suggest that the strongest such evidence is provided by the phenomenon of conditional mutual ratification just discussed and especially by the apparent rational supervenience of moral opinion upon nonmoral factual opinion which it reveals.

Three considerations suggest that the phenomenon in question provides especially good prima facie evidence for moral realism. In the first placed, the current philosophical setting is one in which answers are seen to be readily available to the more abstract epistemological and semantic objections to moral realism (those raised by the issues of the nature of the analog in moral inquiry of observations in science, of the epistemic roles of moral intuitions and of reflective equilibrium, of the nature of the definitional and referential semantics of moral terms). In such a setting the arguments from the diversity of moral theories and from the corresponding intractability of moral disputes—just the arguments addressed by the articulation of a family of conditionally mutually ratifying theories of epistemic contact and of error—emerge as the strongest arguments against moral realism.

Second, the anti-realist arguments from diversity and intractability are especially persuasive because they appear to establish that even the philosopher with substantial initial moral commitments will be forced to the conclusion that the non-reality (or the purely socially constructed nature) of her subject matter provides the only plausible explanation for the diversity of moral opinions and the intractability of moral disputes. What the finding of conditional mutual ratification of theories of epistemic contact and of error and the associated rational supervenience of moral opinions upon nonmoral factual opinions indicates is that, by contrast, there is an alternative realist explanation for divergence and intractability which is ratified by all the currently plausible moral theories.

Finally, the most convincing evidence against moral realism stemming from divergence and intractability seems (at least for many professional philosophers) to come from an examination of the many counterfactual cases regarding which ''moral intuitions'' sharply diverge. The foundational role which many philosophers assign (if only tacitly) to philosophical intuitions and especially to moral intuitions makes this evidence against moral realism seem especially strong. It is precisely with respect to such cases that the treatment of the epistemic role of moral intuitions and the identification of sources of bivalence failures for counterfactual cases which are incorporated in the various conditionally mutually ratifying theories of epistemic contact and of error are most effective. Thus, the realist resources for explaining divergence and intractability reflected in those theories seem especially well suited to rebut the most convincing of the anti-realist arguments in question.

I should add that in *RMS* I examine in detail a related objection to moral realism: that the moral realist is (in contrast to the constructivist moral irrealist) compelled to adopt an implausible and objectionable chauvinist attitude toward moral communities (especially prescientific communities) whose moral views depart sharply from her own. A tendency toward such chauvinism was certainly a feature of logical positivist treatments of scientific objectivity and it is initially plausible to conclude that it will mark the moral realist's conception as well.

By way of examining the question of chauvinism, I define three relations of commensurability which might obtain within a tradition of inquiry. *Semantic commensurability* obtains just in case there is a common subject matter for all the temporal stages of the tradition and its various subtraditions. *Global methodologi-*

cal commensurability obtains just in case the differences between the prevailing theoretical conceptions between any two tradition (or subtradition) stages are always resolvable by the appropriate application of research methods endorsed by each. *Local methodological commensurability* obtains just in case this sort of resolution is always possible for the differences between consecutive tradition stages or between contemporaneous stages of different subtraditions within the tradition of inquiry in question.

I argue that the tendency toward chauvinism within positivist philosophy of science—insofar as its origins were internal to technical philosophy rather than more broadly social—stemmed from a tendency for positivists to hold that semantic commensurability entails (or at any rate strongly suggests) global methodological commensurability and from a consequent tendency to apply contemporary standards of scientific methodology when assessing the rationality of members of different earlier communities of inquirers. By contrast, I argue, scientific realism predicts wholesale failures of global methodological commensurability and makes only highly qualified predictions of local methodological commensurability, even where global semantic commensurability obtains. Thus, the chauvinist tendencies internal to the positivist tradition are not only absent from the realist tradition but corrected within it.

I conclude, by analogy, that contemporary moral realism likewise embodies an appropriate antichauvinist conception of methodology, which is not to say that it is proof against chauvinism deriving from external social influences. Finally, I argue that the alternative constructivist relativist approach is in important respects chauvinist and uncritical. It holds the current stages in the relevant research traditions just as much immune from criticism as it does earlier and prescientific stages and it precludes the diagnosis of culpable methodological errors (culturally chauvinist errors among them) when these do occur, whether in the current stages of the relevant tradition or in its earlier stages. If it is otherwise defensible, realism then represents the preferred antidote to cultural chauvinism.

Finally, I further develop the theme suggested in the present essay that moral realism is an optimistic position. I argue that, given available evidence, the most plausible way in which the doctrine here identified as moral realism could prove to be wrong would be for the broad family of basic human goods to be incapable of a suitably strong homeostatic unity. The non-realist alternative I envision as most plausible would have ''relativist'' features and would entail the dependence of (some) moral truths upon the moral beliefs actually held in the relevant moral communities. What I have in mind is a situation in which the following are both true:

1. (The relativist component) The sorts of fundamental human goods typically recognized as relevant in moral reasoning lack the sort of homeostatic unity tacitly presupposed in moral discourse: there is no psychologically and socially stable way of ameliorating the conflicts between them and adjudicating those which remain which are satisfactory by reasonable prevailing moral standards. Instead,

[224]

there are two (or more) stable ways of achieving homeostasis between those goods, each capable of sustaining a morality (and moral progress) of sorts, but in each (all) of them certain human goods are necessarily slighted with respect to the others in a way certainly unacceptable by contemporary moral standards. This plurality of morally compromised forms of moral homeostasis is not remediable by future moral, economic, or political developments: it reflects nonmaleable features of human nature. Most difficult disagreements in substantive moral philosophy reflect the tacit adherence of the disputants to one or the other of these stable 'moralities' or their unsuccessful attempts to formulate viable alternatives comprising the best features of both (several), or both. Resolution of those disagreements requires that we recognize the conflation of moral standards that caused them and that we (relativistically) disambiguate our uses of moral terms.

2. (The belief dependence component) Actually practiced stable moral arrangements will necessarily approximate one rather than the other(s) of the available stable forms of moral homeostasis. Insofar as we think of participants in such an arrangement as reasoning about the features of their own particular form of moral homeostasis when they engage in moral reasoning (as the first component suggests that we should), we will find that the truth of some of their important moral beliefs (so construed) will depend quite strongly on their having generally adopted the moral beliefs peculiarly appropriate to the tradition of moral practice in which they function. This will be so for two reasons. First, it will be generally true on their moral conception that the goodness (justice, permissibility, . . .) of actions, practices, policies, character traits, etc. will depend upon the ways in which they contribute (or fail to contribute) to the satisfaction of fundamental human needs. Second, the nature of fundamental human needs (at least within the relevant moral communities) will be significantly determined by the moral beliefs held within the community: needs accorded a prominent role in the community's moral scheme will (in consequence of the effects of moral and social teaching on individual development) be more strongly felt than those needs assigned a less prominent role, even when, for those raised in (one of) the alternative sort(s) of moral community, the psychological importance of the needs might be reversed. Morally important human needs (and their relative importance) are thus significantly created by one's participation on one or the (an) other sort of moral community: such communities make among their members the moral psychology appropriate to their moral practices. Because of the limitation of homeostasis between human goods specified in (1), no more encompassing moral psychology is possible.

It is, I think, evident why the conception of our moral situation envisioned in (1) and (2) is properly described as pessimistic. What I argue in *RMS* is that it is nevertheless only in a relatively uninteresting sense *non-realistic*. The dependence of the truth of moral propositions upon moral beliefs envisioned in (2) would be, I argue, an ordinary case of causal dependence and not the sort of logical dependence required by a constructivist conception of morals analogous to a Kuhnian neo-Kantian conception of the dependence of scientific truth on the

Richard N. Boyd

adoption of theories or paradigms. The subject matter of moral inquiry in each of the relevant communities would be theory-and-belief-independent in the sense relevant to the dispute between realist and social constructivists.

The relativism envisioned in (1) would then, I argue, properly be seen as an ordinary realist case of partial denotation (in the sense of Field [1973]). Thus, although the situation envisioned in (1) and (2) would refute moral realism as that doctrine is ordinarily construed (and as it is construed in the present essay), it would not undermine a generally realistic conception of moral language in favor of a constructivist one. The case for the former conception, I suggest, is quite strong indeed.

Bibliography

Armstrong, D. M. 1973. *Belief, Truth and Knowledge.* Cambridge: Cambridge University Press.

Boyd, R. 1972. "Determinism, Laws and Predictability in Principle." *Philosophy of Science* (39): 431–50.

——. 1973. "Realism, Underdetermination and a Causal Theory of Evidence." *Noûs* (7): 1–12.

——. 1979. "Metaphor and Theory Change." In A. Ortony, ed., *Metaphor and Thought.* Cambridge: Cambridge University Press.

——. 1980. "Materialism without Reductionism: What Physicalism Does Not Entail." In N. Block, ed., *Readings in Philosophy of Psychology,* vol. 1. Cambridge: Harvard University Press.

——. 1982. "Scientific Realism and Naturalistic Epistemology." In P. D. Asquith and R. N. Giere, eds., *PSA 1980,* vol. 2. East Lansing: Philosophy of Science Association.

——. 1983. "On the Current Status of the Issue of Scientific Realism." *Erkenntnis* (19): 45–90.

——. 1985a. "Lex Orendi Est Lex Credendi." In Paul Churchland and Clifford Hooker, eds., *Images of Science: Scientific Realism Versus Constructive Empiricism.* Chicago: University of Chicago Press.

——. 1985b. "Observations, Explanatory Power, and Simplicity." In P. Achinstein and O. Hannaway, eds., *Observation, Experiment, and Hypothesis in Modern Physical Science.* Cambridge: MIT Press.

——. 1985c. "The Logician's Dilemma: Deductive Logic, Inductive Inference and Logical Empiricism." *Erkenntnis* (22): 197–252.

——. forthcoming. *Realism and the Moral Sciences* (unpublished manuscript).

Brink, D. 1984. "Moral Realism and the Skeptical Arguments from Disagreement and Queerness." *Australasian Journal of Philosophy* (62.2): 111–25.

——. forthcoming. *Moral Realism and the Foundation of Ethics.* Cambridge: Cambridge University Press.

Byerly, H., and V. Lazara. 1973. "Realist Foundations of Measurement." *Philosophy of Science* (40): 10–28.

Carnap, R. 1934. *The Unity of Science.* Trans. M. Black. London: Kegan Paul.

Dworkin, R. 1973. "The Original Position." *University of Chicago Law Review* (40): 500–33.

Feigl, H. 1956. "Some Major Issues and Developments in the Philosophy of Science of Logical Empiricism." In H. Feigl and M. Scriven, eds., *Minnesota Studies in the Philosophy of Science,* vol. 1. Minneapolis: University of Minnesota Press.

[226]

Field, H. 1973. "Theory Change and the Indeterminacy of Reference." *Journal of Philosophy* (70): 462–81.

Gilbert, A. 1981a. *Marx's Politics: Communists and Citizens*. New Brunswick, N.J.: Rutgers University Press.

——. 1981b. "Historical Theory and the Structure of Moral Argument in Marx," *Political Theory* (9): 173–205.

——. 1982. "An Ambiguity in Marx's and Engel's Account of Justice and Equality," *Americal Political Science Review* (76): 328–46.

——. 1984a. "The Storming of Heaven: Capital and Marx's Politics." In J. R. Pennock, ed., *Marxism Today*, Nomos (26). New York: New York University Press.

——. 1984b. "Marx's Moral Realism: Eudaimonism and Moral Progress." In J. Farr and T. Ball, eds., *After Marx*, Cambridge: Cambridge University Press.

——. 1986a. "Moral Realism, Individuality and Justice in War," *Political Theory* (14): 105–35.

——. 1986b. "Democracy and Individuality," *Social Philosophy and Policy* (3): 19–58.

——. forthcoming. *Equality and Objectivity*.

Goldman, A. 1967. "A Causal Theory of Knowing." *Journal of Philosophy* (64): 357–72.

——. 1976. "Discrimination and Perceptual Knowledge." *Journal of Philosophy* (73): 771–91.

Goodman, N. 1973. *Fact, Fiction, and Forecast*. 3d edition. Indianapolis: Bobbs-Merrill.

Gould, S. J. 1981. *The Mismeasure of Man*. New York: W. W. Norton.

Hanson, N. R. 1958. *Patterns of Discovery*. Cambridge: Cambridge University Press.

Kamin, L. J. 1974. *The Science and Politics of I.Q.* Potomac, Md.: Lawrence Erlbaum Associates.

Kripke, S. A. 1971. "Identity and Necessity." In M. K. Munitz, ed., *Identity and Individuation*. New York: New York University Press.

——. 1972. "Naming and Necessity." In D. Davidson and G. Harman, eds., *The Semantics of Natural Language*. Dordrecht: D. Reidel.

Kuhn, T. 1970. *The Structure of Scientific Revolutions*. 2d edition. Chicago: University of Chicago Press.

Mackie, J. L. 1974. *The Cement of the Universe*. Oxford: Oxford University Press.

Mayr, E. 1970. *Populations, Species and Evolution*. Cambridge: Harvard University Press.

Miller, R. 1978. "Methodological Individualism and Social Explanation." *Philosophy of Science* (45): 387–414.

——. 1979. "Reason and Committment in the Social Sciences." *Philosophy and Public Affairs* (8): 241–66.

——. 1981. "Rights and Reality." *Philosophical Review* (90): 383–407.

——. 1982. "Rights and Consequences." *Midwest Studies in Philosophy* (7): 151–74.

——. 1983. "Marx and Morality." *Nomos* (26): 3–32.

——. 1984a. *Analyzing Marx*. Princeton: Princeton University Press.

——. 1984b. "Ways of Moral Learning." *Philosophical Review* (94): 507–56.

Putnam, H. 1975a. "The Meaning of 'Meaning'." In H. Putnam, *Mind, Language and Reality*. Cambridge: Cambridge University Press.

——. 1975b. "Language and Reality." In H. Putnam, *Mind, Language and Reality*. Cambridge: Cambridge University Press.

——. 1983. "Vagueness and Alternative Logic." In H. Putnam, *Realism and Reason*. Cambridge: Cambridge University Press.

Quine, W. V. O. 1969a. "Natural Kinds." In W. V. O. Quine, *Ontological Relativity and Other Essays*. New York: Columbia University Press.

——. 1969b. "Epistemology Naturalized." In W. V. O. Quine, *Ontological Relativity and Other Essays*. New York: Columbia University Press.

Railton, P. 1986. "Moral Realism." *Philosophical Review* (95): 163–207.

Rawls, J. 1971. *A Theory of Justice*. Cambridge: Harvard University Press.

Shoemaker, S. 1980. "Causality and Properties." In P. van Inwagen, ed., *Time and Cause*. Dordecht: D. Reidel.

Sturgeon, N. 1984a. "Moral Explanations." In D. Copp and D. Zimmerman, eds., *Morality, Reason and Truth*. Totowa, N.J.: Rowman and Allanheld. Reprinted in this volume.

———. 1984b. "Review of P. Foot, *Moral Relativism* and *Virtues and Vices*." *Journal of Philosophy* (81): 326–33.

van Fraassen, B. 1980. *The Scientific Image*. Oxford: Oxford University Press.

Wood, A. 1972. "The Marxian Critique of Justice." *Philosophy and Public Affairs* (1): 244–82.

———. 1979. "Marx on Right and Justice: A reply to Husami." *Philosophy and Public Affairs* (8): 267–95.

———. 1984. "A Marxian Approach to 'The Problem of Justice'." *Philosophica* (33): 9–32.

10

Moral Explanations

Nicholas L. Sturgeon

There is one argument for moral skepticism that I respect even though I remain unconvinced. It has sometimes been called the argument from moral diversity or relativity, but that is somewhat misleading, for the problem arises not from the diversity of moral views, but from the apparent difficulty of *settling* moral disagreements, or even of knowing what would be required to settle them, a difficulty thought to be noticeably greater than any found in settling disagreements that arise in, for example, the sciences. This provides an argument for moral skepticism because one obviously possible explanation for our difficulty in set- tling moral disagreements is that they are really unsettleable, that there is no way of justifying one rather than another competing view on these issues; and a possible further explanation for the unsettleability of moral disagreements, in turn, is moral nihilism, the view that on these issues there just is no fact of the matter, that the impossibility of discovering and establishing moral truths is due to there not being any.

I am, as I say, unconvinced: partly because I think this argument exaggerates the difficulty we actually find in settling moral disagreements, partly because there are alternative explanations to be considered for the difficulty we do find. Under the latter heading, for example, it certainly matters to what extent moral disagreements depend on disagreements about other questions which, however disputed they may be, are nevertheless regarded as having objective answers: questions such as which, if any, religion is true, which account of human psychol- ogy, which theory of human society. And it also matters to what extent consider- ation of moral questions is in practice skewed by distorting factors such as

This essay first appeared in *Morality, Reason and Truth,* ed. David Copp and David Zimmerman (Totowa, N.J.: Rowman and Allanheld, 1985), pp. 49–78. Copyright © 1984 by Rowman and Allanheld Publishers.

personal interest and social ideology. These are large issues. Although it is possible to say some useful things to put them in perspective,[1] it appears impossible to settle them quickly or in any a priori way. Consideration of them is likely to have to be piecemeal, and, in the short run at least, frustratingly indecisive.

These large issues are not my topic here. But I mention them, and the difficulty of settling them, to show why it is natural that moral skeptics have hoped to find some quicker way of establishing their thesis. I doubt that any exist, but some have of course been proposed. Verificationist attacks on ethics should no doubt be seen in this light, and J. L. Mackie's "argument from queerness" is a clear instance (Mackie, *Ethics: Inventing Right and Wrong* [Harmondsworth, England, 1977], pp. 38–42 [pp. 111–14 of this volume]). The quicker argument on which I shall concentrate, however, is neither of these, but instead an argument by Gilbert Harman designed to bring out the "basic problem" about morality, which in his view is "its apparent immunity from observational testing" and "the seeming irrelevance of observational evidence" (Harman, *The Nature of Morality: An Introduction to Ethics* [New York, 1977], pp. vii, viii. Parenthetical page references are to this work). The argument is that reference to moral facts appears unnecessary for the *explanation* of our moral observations and beliefs.

Harman's view, I should say at once, is not in the end a skeptical one, and he does not view the argument I shall discuss as a decisive defense of moral skepticism or moral nihilism. Someone else might easily so regard it, however. For Harman himself regards it as creating a strong *prima facie* case for skepticism and nihilism, strong enough to justify calling it "the problem with ethics."[2] And he believes it shows that the only recourse for someone who wishes to avoid moral skepticism is to find defensible reductive definitions for ethical terms; so skepticism would be the obvious conclusion for anyone to draw who doubted the possibility of such definitions. I believe, however, that Harman is mistaken on both counts. I shall show that his argument for skepticism either rests on claims that most people would find quite implausible (and so cannot be what constitutes, for *them*, the problem with ethics); or else it becomes just the application to ethics of a familiar *general* skeptical strategy, one which, if it works for ethics, will work equally well for unobservable theoretical entities, or for other minds, or for an external world (and so, again, can hardly be what constitutes the distinctive problem with *ethics*). In the course of my argument, moreover, I shall suggest that one can in any case be a moral realist, and indeed an ethical naturalist, without believing that we are now or ever will be in possession of reductive naturalistic definitions for ethical terms.

---

1. As, for example, in Alan Gerwirth "Positive 'Ethics' and Normative 'Science'," *Philosophical Review* 69 (1960), 311–30, in which there are some useful remarks about the first of them.
2. Harman's title for the entire first section of his book.

I. The problem with ethics

Moral theories are often tested in thought experiments, against imagined examples; and, as Harman notes, trained researchers often test scientific theories in the same way. The problem, though, is that scientific theories can also be tested against the world, by observations or real experiments; and, Harman asks, "can moral principles be tested in the same way, out in the world?" (p. 4 [p. 120 of this volume]).

This would not be a very interesting or impressive challenge, of course, if it were merely a resurrection of standard verificationist worries about whether moral assertions and theories have any testable empirical implications, implications statable in some relatively austere "observational" vocabulary. One problem with that form of the challenge, as Harman points out, is that there are no "pure" observations, and in consequence no purely observational vocabulary either. But there is also a deeper problem that Harman does not mention, one that remains even if we shelve worries about "pure" observations and, at least for the sake of argument, grant the verificationist his observational language, pretty much as it was usually conceived: that is, as lacking at the very least any obviously theoretical terminology from any recognized science, and of course as lacking any moral terminology. For then the difficulty is that moral principles fare just as well (or just as badly) against the verificationist challenge as do typical scientific principles. For it is by now a familiar point about scientific principles—principles such as Newton's law of universal gravitation or Darwin's theory of evolution—that they are entirely devoid of empirical implications when considered in isolation.[3] We do of course base observational predictions on such theories and so test them against experience, but that is because we do *not* consider them in isolation. For we can derive these predictions only by relying at the same time on a large background of additional assumptions, many of which are equally theoretical and equally incapable of being tested in isolation. A less familiar point, because less often spelled out, is that the relation of moral principles to observation is similar in *both* these respects. Candidate moral principles—for example, that an action is wrong just in case there is something else the agent could have done that would have produced a greater net balance of pleasure over pain—lack empirical implications when considered in isolation. But it is easy to derive empirical consequences from them, and thus to test them against experience, if we allow ourselves, as we do in the scientific case, to rely on a background of other assumptions of comparable status.

---

3. This point is generally credited to Pierre Duhem, *The Aim and Structure of Physical Theory,* trans. Philip P. Wiener (Princeton, 1954). It is a prominent theme in the influential writings of W. V. O. Quine. For an especially clear application of it, see Hilary Putnam, "The 'Corroboration' of Theories," in *Mathematics, Matter, and Method. Philosophical Papers,* vol. I, 2d ed. (Cambridge, 1977).

Thus, if we conjoin the act-utilitarian principle I just cited with the further view, also untestable in isolation, that it is always wrong deliberately to kill a human being, we can deduce from these two premises together the consequence that deliberately killing a human being always produces a lesser balance of pleasure over pain than some available alternative act; and this claim is one any positivist would have conceded we know, in principle at least, how to test. If we found it to be false, moreover, then we would be forced by this empirical test to abandon at least one of the moral claims from which we derived it.

It might be thought a worrisome feature of this example, however, and a further opening for skepticism, that there could be controversy about which moral premise to abandon, and that we have not explained how our empirical test can provide an answer to *this* question. And this may be a problem. It should be a familiar problem, however, because the Duhemian commentary includes a precisely corresponding point about the scientific case: that if we are at all cautious in characterizing what we observe, then the requirement that our theories merely be *consistent* with observation is an astoundingly weak one. There are always many, perhaps indefinitely many, different mutually inconsistent ways to adjust our views to meet this constraint. Of course, in practice we are often confident of how to do it: If you are a freshman chemistry student, you do not conclude from your failure to obtain the predicted value in an experiment that it is all over for the atomic theory of gases. And the decision can be equally easy, one should note, in a moral case. Consider two examples. From the surprising moral thesis that Adolf Hitler was a morally admirable person, together with a modest piece of moral theory to the effect that no morally admirable person would, for example, instigate and oversee the degradation and death of millions of persons, one can derive the testable consequence that Hitler did not do this. But he did, so we must give up one of our premises; and the choice of which to abandon is neither difficult nor controversial.

Or, to take a less monumental example, contrived around one of Harman's own, suppose you have been thinking yourself lucky enough to live in a neighborhood in which no one would do anything wrong, at least not in public; and that the modest piece of theory you accept, this time, is that malicious cruelty, just for the hell of it, is wrong. Then, as in Harman's example, "you round a corner and see a group of young hoodlums pour gasoline on a cat and ignite it." At this point, either your confidence in the neighborhood or your principle about cruelty has got to give way. But the choice is easy, if dispiriting, so easy as hardly to require thought. As Harman says, "You do not need to *conclude* that what they are doing is wrong; you do not need to figure anything out; you can *see* that it is wrong" (p. 4 [p. 120 of this volume]). But a skeptic can still wonder whether this practical confidence, or this "seeing," rests in either sort of case on anything more than deeply ingrained conventions of thought—respect for scientific experts, say, and for certain moral traditions—as opposed to anything answerable to the facts of the matter, any reliable strategy for getting it right about the world.

Now, Harman's challenge is interesting partly because it does not rest on these verificationist doubts about whether moral beliefs have observational implications, but even more because what it does rest on is a partial answer to the kind of general skepticism to which, as we have seen, reflection on the verificationist picture can lead. Many of our beliefs are justified, in Harman's view, by their providing or helping to provide a reasonable *explanation* of our observing what we do. It would be consistent with your failure, as a beginning student, to obtain the experimental result predicted by the gas laws, that the laws are mistaken. That would even be one explanation of your failure. But a better explanation, in light of your inexperience and the general success experts have had in confirming and applying these laws, is that you made some mistake in running the experiment. So our scientific beliefs can be justified by their explanatory role; and so too, in Harman's view, can mathematical beliefs and many commonsense beliefs about the world.

Not so, however moral beliefs: They appear to have no such explanatory role. That is "the problem with ethics." Harman spells out his version of this contrast:

> You need to make assumptions about certain physical facts to explain the occurrence of the observations that support a scientific theory, but you do not seem to need to make assumptions about any moral facts to explain the occurrence of the so-called moral observations I have been talking about. In the moral case, it would seem that you need only make assumptions about the psychology or moral sensibility of the person making the moral observation. (p. 6 [p. 121 of this volume])

More precisely, and applied to his own example, it might be reasonable, in order to explain your judging that the hoodlums are wrong to set the cat on fire, to assume "that the children really are pouring gasoline on a cat and you are seeing them do it." But there is no

> obvious reason to assume anything about "moral facts," such as that it is really wrong to set the cat on fire. . . . Indeed, an assumption about moral facts would seem to be totally irrelevant to the explanation of your making the judgment you make. It would seem that all we need assume is that you have certain more or less well articulated moral principles that are reflected in the judgments you make, based on your moral sensibility. (p. 7 [p. 122 of this volume])

And Harman thinks that if we accept this conclusion, suitably generalized, then, subject to a possible qualification I shall come to shortly, we must conclude that moral theories cannot be tested against the world as scientific theories can, and that we have no reason to believe that moral facts are part of the order of nature or that there is any moral knowledge (pp. 23, 35).

My own view is that Harman is quite wrong, not in thinking that the explanatory role of our beliefs is important to their justification, but in thinking that moral beliefs play no such role.[4] I shall have to say something about the initial plausibility of Harman's thesis as applied to his own example, but part of my reason for dissenting should be apparent from the other example I just gave. We find it easy (and so does Harman [p. 108]) to conclude from the evidence not just that Hitler was not morally admirable, but that he was morally depraved. But isn't it plausible that Hitler's moral depravity—the fact of his really having been morally depraved—forms part of a reasonable explanation of why we believe he was depraved? I think so, and I shall argue concerning this and other examples that moral beliefs commonly play the explanatory role Harman denies them. Before I can press my case, however, I need to clear up several preliminary points about just what Harman is claiming and just how his argument is intended to work.

II. Observation, explanation, and reduction

(1) For there are several ways in which Harman's argument invites misunderstanding. One results from his focusing at the start on the question of whether there can be moral *observations*.[5] But this question turns out to be a side issue, in no way central to his argument that moral principles cannot be tested against the world. There are a couple of reasons for this, of which the more important[6] by far is that Harman does not really require of moral facts, if belief in them is to be justified, that they figure in the explanation of moral observations. It would be enough, on the one hand, if they were needed for the explanation of moral beliefs that are not in any interesting sense observations. For example, Harman thinks belief in moral facts would be vindicated if they were needed to explain our drawing the moral conclusions we do when we reflect on hypothetical cases, but I

---

4. Harman is careful always to say only that moral beliefs *appear* to play no such role; and since he eventually concludes that there *are* moral facts (p. 132), this caution may be more than stylistic. I shall argue that this more cautious claim, too, is mistaken (indeed, that is my central thesis). But to avoid issues about Harman's intent, I shall simply mean by "Harman's argument" the skeptical argument of his first two chapters, whether or not he means to endorse all of it. This argument surely deserves discussion in its own right in either case, especially since Harman himself never explains what is wrong with it.

5. He asks: "Can moral principles be tested in the same way [as scientific hypotheses can], out in the world? You can observe someone do something, but can you ever perceive the rightness or wrongness of what he does?" (p. 4 [p. 120 of this volume]).

6. The other is that Harman appears to use "observe" and ("perceive" and "see") in a surprising way. One would normally take observing (or perceiving, or seeing) something to involve knowing it was the case. But Harman apparently takes an observation to be *any* opinion arrived at as "a direct result of perception" (p. 5) or, at any rate (see next footnote), "immediately and without conscious reasoning" (p. 7). This means that observations need not even be true, much less known to be true. A consequence is that the existence of moral observations, in Harman's sense, would not be sufficient to show that moral theories can be tested against the world, or to show that there is moral knowledge, although this *would* be sufficient if "observe" were being used in a more standard sense. What I argue in the text is that the existence of moral observations (in either Harman's or the standard sense) is not *necessary* for showing this about moral theories either. (See pages 121, and 122 of this volume.)

think there is no illumination in calling these conclusions observations.[7] It would also be enough, on the other hand, if moral facts were needed for the explanation of what were clearly observations, but not moral observations. Harman thinks mathematical beliefs are justified, but he does not suggest that there are mathematical observations; it is rather that appeal to mathematical truths helps to explain why we make the physical observations we do (p. 10 [p. 124 of this volume]). Moral beliefs would surely be justified, too, if they played such a role, whether or not there are any moral observations.

So the claim is that moral facts are not needed to explain our having any of the moral beliefs we do, whether or not those beliefs are observations, and are equally unneeded to explain any of the observations we make, whether or not those observations are moral. In fact, Harman's view appears to be that moral facts aren't needed to explain anything at all: although it would perhaps be question-begging for him to begin with this strong a claim, since he grants that if there were any moral facts, then appeal to other moral facts, more general ones, for example, might be needed to explain *them* (p. 8 [p. 123 of this volume]). But he is certainly claiming, at the very least, that moral facts aren't needed to explain any nonmoral facts we have any reason to believe in.

This claim has seemed plausible even to some philosophers who wish to defend the existence of moral facts and the possibility of moral knowledge. Thus, Thomas Nagel has recently retreated to the reply that

> it begs the question to assume that *explanatory* necessity is the test of reality in this area. . . . To assume that only what has to be included in the best explanatory picture of the world is real, is to assume that there are no irreducibly normative truths.[8]

But this retreat will certainly make it more difficult to fit moral knowledge into anything like a causal theory of knowledge, which seems plausible for many other cases, or to follow Hilary Putnam's suggestion that we "apply a generally causal account of reference . . . to moral terms" (Putnam, "Language and Reality," in

---

7. This sort of case does not meet Harman's characterization of an observation as an opinion that is "a direct result of perception" (p. 5), but he is surely right that moral facts would be as well vindicated if they were needed to explain our drawing conclusions about hypothetical cases as they would be if they were needed to explain observations in the narrower sense. To be sure, Harman is still confining his attention to cases in which we draw the moral conclusion from our thought experiment "immediately and without conscious reasoning" (p. 7), and it is no doubt the existence of such cases that gives purchase to talk of a "moral sense." But this feature, again, can hardly matter to the argument: Would belief in moral facts be less justified if they were needed only to explain the instances in which we draw the moral conclusion *slowly?* Nor can it make any difference for that matter whether the case we are reflecting on is hypothetical. So my example in which we, quickly or slowly, draw a moral conclusion about Hitler from what we know of him, is surely relevant.

8. Thomas Nagel, "The Limits of Objectivity," in Sterling M. McMurrin, ed., *The Tanner Lectures on Human Values* (Salt Lake City and Cambridge, 1980), p. 114n. Nagel actually directs this reply to J. L. Mackie.

Mind, Language, and Reality. Philosophical Papers, vol. 2 [Cambridge, 1975], p. 290). In addition, the concession is premature in any case, for I shall argue that moral facts do fit into our explanatory view of the world, and in particular into explanations of many moral observations and beliefs.

(2) Other possible misunderstandings concern what is meant in asking whether reference to moral facts is *needed* to explain moral beliefs. One warning about this question I save for my comments on reduction below; but another, about what Harman is clearly *not* asking, and about what sort of answer I can attempt to defend to the question he is asking, can be spelled out first. For, to begin with, Harman's question is clearly not just whether there is *an* explanation of our moral beliefs that does not mention moral facts. Almost surely there is. Equally surely, however, there is *an* explanation of our commonsense nonmoral beliefs that does not mention an external world: one which cites only our sensory experience, for example, together with whatever needs to be said about our psychology to explain why with that history of experience we would form just the beliefs we do. Harman means to be asking a question that will lead to skepticism about moral facts, but not to skepticism about the existence of material bodies or about well-established scientific theories of the world.

Harman illustrates the kind of question he is asking, and the kind of answer he is seeking, with an example from physics which it will be useful to keep in mind. A physicist sees a vapor trail in a cloud chamber and thinks, ''There goes a proton.'' What explains his thinking this? Partly, of course, his psychological set, which largely depends on his beliefs about the apparatus and all the theory he has learned; but partly also, perhaps, the hypothesis that ''there really was a proton going through the cloud chamber, causing the vapor trail, which he saw as a proton.'' We will *not* need this latter assumption, however, ''if his having made that observation could have been equally well explained by his psychological set alone, without the need for any assumption about a proton'' (p. 6 [p. 122 of this volume]).[9] So for reference to moral facts to be *needed* in the explanation of our beliefs and observations, is for this reference to be required for an explanation that is somehow *better* than competing explanations. Correspondingly, reference to moral facts will be unnecessary to an explanation, in Harman's view, not just because we can find some explanation that does not appeal to them, but because *no* explanation that appeals to them is any better than some competing explanation that does not.

Now, fine discriminations among competing explanations of almost anything are likely to be difficult, controversial, and provisional. Fortunately, however, my discussion of Harman's argument will not require any fine discriminations.

---

9. It is surprising that Harman does not mention the obvious intermediate possibility, which would occur to any instrumentalist: to cite the physicist's psychological set *and* the vapor trail, but say nothing about protons or other unobservables. It is *this* explanation that is most closely parallel to an explanation of beliefs about an external world in terms of sensory experience and psychological makeup, or of moral beliefs in terms of nonmoral facts together with our ''moral sensibility.''

This is because Harman's thesis, as we have seen, is *not* that moral explanations lose out by a small margin; nor is it that moral explanations, although sometimes initially promising, always turn out on further examination to be inferior to nonmoral ones. It is, rather, that reference to moral facts always looks, right from the start, to be "completely irrelevant" to the explanation of any of our observations and beliefs. And my argument will be that this is mistaken: that many moral explanations appear to be good explanations, or components in good explanations, that are not obviously undermined by anything else that we know. My suspicion, in fact, is that moral facts are needed in the sense explained, that they will turn out to belong in our best overall explanatory picture of the world, even in the long run, but I shall not attempt to establish that here. Indeed, it should be clear why I could not pretend to do so. For I have explicitly put to one side the issue (which I regard as incapable in any case of quick resolution) of whether and to what extent actual moral disagreements can be settled satisfactorily. But I assume it would count as a defect in any sort of explanation to rely on claims about which rational agreement proved unattainable. So I concede that it *could* turn out, for anything I say here, that moral explanations are all defective and should be discarded. What I shall try to show is merely that many moral explanations look reasonable enough to be in the running; and, more specifically, that nothing Harman says provides any reason for thinking they are not. This claim is surely strong enough (and controversial enough) to be worth defending.

(3) It is implicit in this statement of my project, but worth noting separately, that I take Harman to be proposing an *independent* skeptical argument—independent not merely of the argument from the difficulty of settling disputed moral questions, but also of other standard arguments for moral skepticism. Otherwise his argument is not worth independent discussion. For *any* of these more familiar skeptical arguments will of course imply that moral explanations are defective, on the reasonable assumption that it would be a defect in any explanation to rely on claims as doubtful as these arguments attempt to show all moral claims to be. But if *that* is why there is a problem with moral explanations, one should surely just cite the relevant skeptical argument, rather than this derivative difficulty about moral explanations, as the basic "problem with ethics," and it is that argument we should discuss. So I take Harman's interesting suggestion to be that there is a *different* difficulty that remains even if we put other arguments for moral skepticism aside and *assume,* for the sake of argument, that there are moral facts (for example, that what the children in his example are doing is really wrong): namely, that these assumed facts *still* seem to play no explanatory role.

This understanding of Harman's thesis crucially affects my argumentative strategy in a way to which I should alert the reader in advance. For it should be clear that assessment of this thesis not merely permits, but *requires,* that we provisionally assume the existence of moral facts. I can see no way of evaluating the claim that *even if* we assumed the existence of moral facts they would still appear explanatorily irrelevant, without assuming the existence of some, to see how they would look. So I do freely assume this in each of the examples I discuss in the next section. (I have tried to choose plausible examples, moreover, moral facts most of us would be inclined to believe in if we did believe in moral facts, since those are the easiest to think about; but the precise examples don't matter, and anyone who would prefer others should feel free to substitute his own.) I grant, furthermore, that if Harman were right about the outcome of this thought experiment—that even after we assumed these facts they still looked irrelevant to the explanation of our moral beliefs and of other nonmoral facts—then we might conclude with him that there were, after all, no such facts. But I claim he is wrong: Once we have provisionally assumed the existence of moral facts, they *do* appear relevant, by perfectly ordinary standards, to the explanation of moral beliefs and of a good deal else besides. Does this prove that there *are* such facts? Well of course it helps support that view, but here I carefully make no claim to have shown so much. What I *show* is that any remaining reservations about the existence of moral facts must be based on those *other* skeptical arguments, of which Harman's argument is independent. In short, there may still be a "problem with ethics," but it has *nothing* special to do with moral explanations.

(4) A final preliminary point concerns a qualification Harman adds himself. As I have explained his argument so far, it assumes that we could have reason to believe in moral facts only if this helped us "explain why we observe what we observe" (p. 13); but, he says, this assumption is too strong, for we can have evidence for the truth of some beliefs that play no such explanatory role. We might, for example, come to be able to explain color perception without saying that objects have colors, by citing certain physical and psychological facts. But this would not show that there are no colors; it would show only that facts about color are "somehow reducible" to these physical and psychological facts. And this leaves the possibility that moral facts, too, even if they ultimately play no explanatory role themselves, might be "reducible to certain other facts that can help explain our observations" (p. 14). So a crucial question is: What would justify a belief in reducibility? What makes us think color facts might be reducible to physical (or physical and psychological) facts, and what would justify us in thinking moral facts reducible to explanatory natural facts of some kind?

Harman's answer is that it is still the *apparent* explanatory role of color facts, or of moral facts, that matters; and hence that this qualification to his argument is not so great as it might seem. We know of no precise reduction for facts of either sort. We believe even so that reduction is possible for color facts because even when we are able to explain color perception without saying that objects are colored,

> we will still *sometimes* refer to the actual colors of objects in explaining color perception, if only for the sake of simplicity. . . . We will continue to believe that objects have colors because we will continue to refer to the actual colors of objects in the explanations that we will in practice give.

[238]

But Harman thinks that no comparable point holds for moral facts. "There does not ever seem to be, even in practice, any point to explaining someone's moral observations by appeal to what is actually right or wrong, just or unjust, good or bad" (p. 22).

Now I shall argue shortly that this is just wrong: that sober people frequently offer such explanations of moral observations and beliefs, and that many of these explanations look plausible enough on the evidence to be worth taking seriously. So a quick reply to Harman, strictly adequate for my purpose, would be simply to accept his concession that this by itself should lead us to regard moral facts as (at worst) reducible to explanatory facts.[10] Concern about the need for, and the role of, reductive definitions has been so central to meta-ethical discussion in this century, however, and has also proved enough of a sticking point in discussions I have had of the topic of this essay, that I should say a bit more.

As a philosophical naturalist, I take natural facts to be the only facts there are.[11] If I am prepared to recognize moral facts, therefore, I must take them, too, to be natural facts: But which natural facts? It is widely thought that an ethical naturalist must answer this question by providing reductive naturalistic definitions[12] for moral terms and, indeed, that until one has supplied such definitions one's credentials as a *naturalist* about any supposed moral facts must be in doubt. Once such definitions are in hand, however, it seems that moral explanations should be dispensable, since any such explanations can then be paraphrased in nonmoral terms; so it is hard to see why an ethical naturalist should attach any importance to them. Now, there are several problems with this reasoning, but the main one is that the widely held view on which it is based is mistaken: mistaken about where a scheme of reductive naturalistic definitions would be found, if there were to be one, but also about whether, on a naturalistic view of ethics, one should expect there to be such a thing at all. I shall take up these points in reverse order, arguing first (a) that it is a mistake to require of ethical naturalism that it even promise reductive definitions for moral terms, and then (b) that even if such definitions are to be forthcoming it is, at the very least, no special problem for ethical naturalism that we are not *now* in confident possession of them.

(a) Naturalism is in one clear sense a "reductionist" doctrine of course, for it holds that moral facts are nothing but natural facts. What I deny, however, is that

10. And it is hard to see how facts could be reducible to explanatory facts without being themselves explanatory. Opaque objects often look red to normally sighted observers in white light because they *are* red; it amplifies this explanation, but hardly undermines it, if their redness turns out to be an electronic property of the matter composing their surfaces.

11. Some of what I say could no doubt be appropriated by believers in supernatural facts, but I leave the details to them. For an account I could largely accept, if I believed any of the theology, see R. M. Adams, "Divine Command Metaethics as Necessary A Posteriori," in Paul Helm, ed., *Divine Commands and Morality* (Oxford, 1981), pp. 109–18.

12. Or, at any rate, a reductive scheme of translation. It surely needn't provide explicit term-by-term definitions. Since this qualification does not affect my argument, I shall henceforth ignore it.

Nicholas L. Sturgeon

from this metaphysical doctrine about what sort of facts moral facts are, anything follows about the possibility of reduction in another sense (to which I shall henceforth confine the term) more familiar from the philosophical literature: that is, about whether moral expressions can be given reductive definitions in some distinctive nonmoral vocabulary, in which any plausible moral explanations could then be recast. The difficulty with supposing naturalism to require this can be seen by pressing the question of just what this distinctive vocabularly is supposed to be. It is common to say merely that this reducing terminology must be "factual" or "descriptive" or must designate natural properties; but unless ethical naturalism has already been ruled out, this is no help, for what naturalists of course contend is that moral discourse is *itself* factual and descriptive (although it may be other things as well), and that moral terms themselves stand for natural properties. The idea, clearly, is supposed to be that the *test* of whether these naturalistic claims about moral discourse are correct is whether this discourse is reducible to some other; but what other? I consider two possibilities.

(i) Many would agree that it is too restrictive to understand ethical naturalism as requiring that moral terms be definable in the terminology of fundamental physics. One reason it is too restrictive is that philosophical naturalism might be true even if physicalism, the view that everything is physical, is not. Some form of emergent dualism might be correct, for example. A different reason, which I find more interesting (because I think physicalism *is* true), is that physicalism entails nothing in any case about whether even biology or psychology, let alone ethics, is reducible to physics. There are a number of reasons for this, but a cardinality problem noted by Richard Boyd is sufficient to secure the point ("Materialism without Reductionism: Non-Humean Causation and the Evidence for Physicalism," in *The Physical Basis of Mind* [Cambridge, Mass., forthcoming]). If there are (as there appear to be) any continuous physical parameters, then there are continuum many physical states of the world, but there are at most countably many predicates in any language, including that of even ideal physics; so there are more physical properties than there are physical expressions to represent them. Thus, although physicalism certainly entails that biological and psychological properties (and ethical properties, too, if there are any) are physical, nothing follows about whether we have any but biological or psychological or ethical terminology for representing these particular physical properties.

(ii) Of course, not many discussions of ethical naturalism have focused on the possibility of reducing ethics to physics; social theory, psychology, and occasionally biology have appeared more promising possibilities. But that facts might be *physical* whether or not all the disciplines that deal with them are reducible to *physics,* helps give point to my question of why we should think that if all ethical facts are *natural* (or, for that matter, *social* or *psychological* or *biological*), it follows that they can equally well be expressed in some other, nonmoral idiom; and it also returns us to the question of just what this alternative idiom is supposed to be. The answer to this latter question simply assumed in most discussions of ethical naturalism, I think, is that there are a number of disciplines that we pretty well know to deal with a single natural world, for example, physics, biology, psychology, and social theory; that it is a matter of no great concern whether any of *these* disciplines is reducible to some one of the others or to anything else; but that the test of whether ethical naturalism is true *is* whether ethics is reducible to some (nonmoral) combination of *them*.[13]

But what rationale is there for holding ethics alone to this reductive test? Perhaps there would be one if ethics appeared in some salient respect strikingly dissimilar to these other disciplines: if, for example, Harman were right what whereas physics, biology, and the rest offer plausible explanations of many obviously natural facts, including facts about our beliefs and observations, ethics never does. Perhaps ethics could then plausibly be required to earn its place by some alternative route. But I shall of course argue that Harman is wrong about this alleged dissimilarity, and I take my argument to provide part of the defense required for a naturalistic but nonreductive view of ethics.

(b) A naturalist, however, will certainly want (and a critic of naturalism will likely demand) a fuller account than this of just where moral facts are supposed to fit in the natural world. For all I have shown, moreover, this account might even provide a scheme of reduction for moral discourse: My argument has been not that ethical naturalism could not take this form, but only that it need not. So where should one look for such a fuller account or (if it is to be had) such a reduction? The answer is that the account will have to be derived from our best moral theory, together with our best theory of the rest of the natural world—exactly as, for example, any reductive account of colors will have to be based on all we know about colors, including our best optical theory together with other parts of physics and perhaps psychology. If hedonistic act-utilitarianism (and enough of its associated psychology) turns out to be true, for example, then we can define the good as pleasure and the absence of pain, and a right action as one that produces at least as much good as any other, and that will be where the moral facts fit. If, more plausibly, some other moral theory turns out to be correct, we will get a different account and (if the theory takes the right form) different reductive definitions. It would of course be a serious objection to ethical *naturalism* if we discovered that the *only* plausible moral theories had to invoke supernatural facts of some kind, by making right and wrong depend on the will of a deity, for example, or by implying that only persons with immortal souls could have moral obligations. We would then have to choose between a naturalistic world view and a belief in moral facts. But an ethical naturalist can point out that there are familiar moral theories that lack implications of this sort and that appear defensible in the light of all we know

13. *Nonmoral* because ethics (or large parts of it) will be trivially reducible to psychology and social theory if we take otherwise unreduced talk of moral character traits just to be *part* of psychology and take social theory to *include,* for example, a theory of justice. As an ethical naturalist, I see nothing objectionable or unscientific about conceiving of psychology and social theory in this way, but of course this is not usually how they are understood when questions about reduction are raised.

[241]

about the natural world; and any of them, if correct, could provide a naturalistic account of moral facts and even (if one is to be had) a naturalistic reduction of moral discourse.

Many philosophers will balk at this confident talk of our discovering some moral theory to be correct. But their objection is just the familiar one whose importance I acknowledged at the outset, before putting it to one side: For I grant that the difficulty we experience in settling moral issues, including issues in moral theory, is a problem (although perhaps not an insuperable one) for any version of moral realism. All I contend here is that there is not, in addition to this acknowledged difficulty, any special further (or prior) problem of finding reductive definitions for moral terms or of figuring out where moral facts fit in the natural world. Our moral theory, if once we get it, will provide whatever reduction is to be had and will tell us where the moral facts fit. The suspicion that there must be more than this to the search for reductive definitions almost always rests, I believe, on the view that these definitions must be suited to a special epistemic role: for example, that they will have to be analytic or conceptual truths and so provide a privileged basis for the rest of our theory. But I am confident that moral reasoning, like reasoning in the sciences, is inevitably dialectical and lacks a priori foundations of this sort. I am also sure that no ethical naturalist need think otherwise.[14]

The relevance of these points is this: It is true that if we once obtained correct reductive definitions for moral terms, moral explanations would be in principle dispensable; so if ethical naturalism had to promise such definitions, it would also have to promise the eliminability in principle of explanations couched in moral terms. But note three points. First, it should be no surprise, and should be regarded as no special difficulty for naturalism even on a reductionist conception of it, that we are not now in possession of such definitions, and so not *now* in a position to dispense with any moral explanations that seem plausible. To be confident of such definitions we would need to know just which moral theory is correct; but ethics is an area of great controversy, and I am sure we do not yet know this. Second, if some moral explanations do seem plausible, as I shall argue, then one important step toward improving this situation in ethics will be to see what sort of theory emerges if we attempt to refine these explanations in the light both of empirical evidence and theoretical criticism. So it is easy to see,

14. For more on this view of moral reasoning, see Nicholas L. Sturgeon, ''Brandt's Moral Empiricism,'' *Philosophical Review* 91 (1982), 389–422. On scientific reasoning see Richard N. Boyd, *Realism and Scientific Epistemology* (Cambridge, forthcoming).

G. E. Moore, *Principia Ethica*, (Cambridge, 1903) thought that the *metaphysical* thesis that moral facts are natural facts entailed that moral theory would have a priori foundations. For he took the metaphysical thesis to require not merely that there be a reductive scheme of translation for moral terminology, but that this reduction include explicit property-identities (such as ''goodness = pleasure and the absence of pain''); and these he assumed could be true only if analytic. I of course reject the view that naturalism requires any sort of reductive definitions; but even if it required this sort, it is by now widely acknowledged that reductive property-identities (such as ''temperature = mean molecular kinetic energy'') can be true without being analytic. See Hilary Putnam, ''On Properties,'' in *Mathematics, Matter and Method. Philosophical Papers,* vol. I, 2d. ed. (Cambridge, 1977).

again even on a reductionist understanding of naturalism that promises the eliminability of moral explanations in the long run, why any naturalist will think that
for the foreseeable short run such explanations should be taken seriously on their
own terms.

The third and most important point, finally, is that the eliminability of moral
explanations for *this* reason, if actually demonstrated, would of course not represent a triumph of ethical skepticism but would rather derive from its defeat. So we
must add one further caution, as I promised, concerning Harman's thesis that no
reference to moral facts is *needed* in the explanation of moral beliefs. For there
are, as we can now see, two very different reasons one might have for thinking
this. One—Harman's reason, and my target in the remainder of this essay—is
that no moral explanations even seem plausible, that reference to moral facts
always strikes us as ''completely irrelevant'' to the explanation of moral beliefs.
This claim, if true, would tend to support moral skepticism. The other reason—
which I have just been considering, and with which I also disagree—is that any
moral explanations that *do* seem plausible can be paraphrased without explanatory
loss in entirely nonmoral terms. I have argued that it is a mistake to understand
ethical naturalism as promising this kind of reduction even in principle; and I think
it in any case absurd overconfidence to suppose that anyone can spell out an
adequate reduction now. But any reader unconvinced by my arguments should
note also that this *second* reason is no version of moral skepticism: For what
anyone convinced by it must think, is that we either are or will be able to say, in
entirely nonmoral terms, exactly which natural properties moral terms refer to.[15]
So Harman is right to present reductionism as an alternative to skepticism; part of
what I have tried to show is just that it is neither the only nor the most plausible
such alternative, and that no ethical naturalist need be committed to it.

III. Moral explanations

With these preliminary points aside, I turn to my arguments against Harman's
thesis. I shall first add to my example of Hitler's moral character several more in
which it seems plausible to cite moral facts as part of an explanation of nonmoral
facts, and in particular of people's forming the moral opinions they do. I shall then
argue that Harman gives us no plausible reason to reject or ignore these explanations; I shall claim, in fact, that the same is true for his own example of the
children igniting the cat. I shall conclude, finally, by attempting to diagnose the
source of the disagreement between Harman and me on these issues.

My Hitler example suggests a whole range of extremely common cases that
appear not to have occurred to Harman, cases in which we cite someone's moral

15. Nor does this view really promise that we can do without reference to moral facts; it merely says
that we can achieve this reference without using moral terms. For we would surely have as much
reason to think that the facts expressed by *these* nonmoral terms were moral facts as we would for
thinking that our reductive definitions were correct.

character as part of an explanation of his or her deeds, and in which that whole story is then available as a plausible further explanation of someone's arriving at a correct assessment of that moral character. Take just one other example. Bernard DeVoto, in *The Year of Decision: 1846,* describes the efforts of American emigrants already in California to rescue another party of emigrants, the Donner Party, trapped by snows in the High Sierras, once their plight became known. At a meeting in Yerba Buena (now San Francisco), the relief efforts were put under the direction of a recent arrival, Passed Midshipman Selim Woodworth, described by a previous acquaintance as "a great busybody and ambitious of taking a command among the emigrants."[16] But Woodworth not only failed to lead rescue parties into the mountains himself, where other rescuers were counting on him (leaving children to be picked up by him, for example), but had to be "shamed, threatened, and bullied" even into organizing the efforts of others willing to take the risk; he spent time arranging comforts for himself in camp, preening himself on the importance of his position; and as a predictable result of his cowardice and his exercises in vainglory, many died who might have been saved, including four known still to be alive when he turned back for the last time in mid-March. DeVoto concludes: "Passed Midshipman Woodworth was just no damned good" (1942, p. 442). I cite this case partly because it has so clearly the structure of an inference to a reasonable explanation. One can think of competing explanations, but the evidence points against them. It isn't, for example, that Woodworth was a basically decent person who simply proved too weak when thrust into a situation that placed heroic demands on him. He volunteered, he put no serious effort even into tasks that required no heroism, and it seems clear that concern for his own position and reputation played a much larger role in his motivation than did any concern for the people he was expected to save. If DeVoto is right about this evidence, moreover, it seems reasonable that part of the explanation of his believing that Woodworth was no damned good is just that Woodworth *was* no damned good.

DeVoto writes of course with more moral intensity (and with more of a flourish) than academic historians usually permit themselves, but it would be difficult to find a serious work of biography, for example, in which actions are not explained by appeal to moral character: sometimes by appeal to specific virtues and vices, but often enough also by appeal to a more general assessment. A different question, and perhaps a more difficult one, concerns the sort of example on which Harman concentrates, the explanation of judgments of right and wrong. Here again Harman appears just to have overlooked explanations in terms of moral character: A judge's thinking that it would be wrong to sentence a particular offender to the maximum prison term the law allows, for example, may be due in part to her decency and fairmindedness, which I take to be moral facts if any are.

---

16. DeVoto, *The Year of Decision: 1846* (Boston, 1942), p. 426; a quotation from the notebooks of Francis Parkman. The account of the entire rescue effort is on pp. 424–44.

But do moral features of the action or institution being judged ever play an explanatory role? Here is an example in which they appear to. An interesting historical question is why vigorous and reasonably widespread moral opposition to slavery arose for the first time in the eighteenth and nineteenth centuries, even though slavery was a very old institution; and why this opposition arose primarily in Britain, France, and in French- and English-speaking North America, even though slavery existed throughout the New World.[17] There is a standard answer to this question. It is that chattel slavery in British and French America, and then in the United States, was much *worse* than previous forms of slavery, and much worse than slavery in Latin America. This is, I should add, a controversial explanation. But as is often the case with historical explanations, its proponents do not claim it is the whole story, and many of its opponents grant that there may be some truth in these comparisons, and that they may after all form a small part of a larger explanation.[18] This latter concession is all I require for my example. Equally good for my purpose would be the more limited thesis that explains the growth of antislavery sentiment in the United States, between the Revolution and the Civil War, in part by saying that slavery in the United States became a more oppressive institution during that time. The appeal in these standard explanations is straightforwardly to moral facts.

What is supposed to be wrong with all these explanations? Harman says that assumptions about moral facts seem "completely irrelevant" in explaining moral observations and moral beliefs (p. 7 [p. 122 of this volume]), but on its more natural reading that claim seems pretty obviously mistaken about these examples. For it is natural to think that if a particular assumption is completely irrelevant to the explanation of a certain fact, then the fact would have obtained, and we could have explained it just as well, even if the assumption had been false.[19] But I do not believe that Hitler would have done all he did if he had not been morally depraved,

17. What is being explained, of course, is not just why people came to think slavery wrong, but why people who were not themselves slaves or in danger of being enslaved came to think it so seriously wrong as to be intolerable. There is a much larger and longer history of people who thought it wrong but tolerable, and an even longer one of people who appear not to have got past the thought that the world would be a better place without it. See David Brion Davis, *The Problem of Slavery in Western Culture* (Ithaca, 1966).

18. For a version of what I am calling the standard view about slavery in the Americas, see Frank Tannenbaum, *Slave and Citizen* (New York, 1947). For an argument against both halves of the standard view, see Davis, esp. pp. 60–61, 223–25, 262–63.

19. This counterfactual test no doubt requires qualification. When there are concomitant effects that in the circumstances could each only have been brought about by their single cause, it may be true that if the one effect had not occurred, then neither would the other, but the occurrence of the one is not relevant to the explanation of the other. The test will also be unreliable if it employs backtracking or "that-would-have-had-to-be-because" counterfactuals. (I take these to include ones in which what is tracked back to is not so much a cause as a condition that partly constitutes another: as when someone's winning a race is part of what constitutes her winning five events in one day, and it is true that if she hadn't won five events, that would have had to be because she didn't win that particular race.) So it should not be relied on in cases of either of these sorts. But none of my examples falls into either of these categories.

nor, on the assumption that he was not depraved, can I think of any plausible alternative explanation for his doing those things. Nor is it plausible that we would all have believed he was morally depraved even if he hadn't been. Granted, there is a tendency for writers who do not attach much weight to fascism as a social movement to want to blame its evils on a single maniacal leader, so perhaps some of them would have painted Hitler as a moral monster even if he had not been one. But this is only a tendency, and one for which many people know how to discount, so I doubt that our moral belief really is overdetermined in this way. Nor, similarly, do I believe that Woodworth's actions were overdetermined, so that he would have done just as he did even if he had been a more admirable person. I suppose one could have doubts about DeVoto's objectivity and reliability; it is obvious he dislikes Woodworth, so perhaps he would have thought him a moral loss and convinced his readers of this no matter what the man was really like. But is more plausible that the dislike is mostly based on the same evidence that supports DeVoto's moral view of him, and that very different evidence, at any rate, would have produced a different verdict. If so, then Woodworth's moral character is part of the explanation of DeVoto's belief about his moral character.

It is more plausible of course that serious moral opposition to slavery would have emerged in Britain, France, and the United States even if slavery hadn't been worse in the modern period than before, and worse in the United States than in Latin America, and that the American antislavery movement would have grown even if slavery had not become more oppressive as the nineteenth century progressed. But that is because these moral facts are offered as at best a partial explanation of these developments in moral opinion. And if they really *are* part of the explanation, as seems plausible, then it is also plausible that whatever effect they produced was not entirely overdetermined; that, for example, the growth of the antislavery movement in the United States would at least have been somewhat slower if slavery had been and remained less bad an institution. Here again it hardly seems "completely irrelevant" to the explanation whether or not these moral facts obtained.

It is more puzzling, I grant, to consider Harman's own example in which you see the children igniting a cat and react immediately with the thought that this is wrong. Is it true, as Harman claims, that the assumption that the children are really doing something wrong is "totally irrelevant" to any reasonable explanation of your making that judgment? Would you, for example, have reacted in just the same way, with the thought that the action is wrong, even if what they were doing *hadn't* been wrong, and could we explain your reaction equally well on this assumption? Now, there is more than one way to understand this counterfactual question, and I shall return below to a reading of it that might appear favorable to Harman's view. What I wish to point out for now is merely that there is a natural way of taking it, parallel to the way in which I have been understanding similar counterfactual questions about my own examples, on which the answer to it has to be simply: It depends. For to answer the question, I take it, we must consider a situation in which what the children are doing is not wrong, but which is otherwise as much like the actual situation as possible, and then decide what your reaction would be in that situation. But since what makes their action wrong, what its wrongness *consists* in, is presumably something like its being an act of gratuitious cruelty (or, perhaps we should add, of intense cruelty, and to a helpless victim), to imagine them not doing something wrong we are going to have to imagine their action different in this respect. More cautiously and more generally, if what they are actually doing is wrong, and if moral properties are, as many writers have held, supervenient on natural ones,[20] then in order to imagine them not doing something wrong we are going to have to suppose their action different from the actual one in some of its natural features as well. So our question becomes: Even if the children had been doing something else, something just different enough not to be wrong, would you have taken them even so to be doing something wrong?

Surely there is no one answer to this question: It depends on a lot about you, including your moral views and how good you are at seeing at a glance what some children are doing. It probably depends also on a debatable moral issue; namely, just *how* different the children's action would have to be in order not to be wrong. (Is unkindness to animals, for example, also wrong?) I believe we can see how, in a case in which the answer was clearly affirmative, we might be tempted to agree with Harman that the wrongness of the action was no part of the explanation of your reaction. For suppose you are like this. You hate children. What you especially hate, moreover, is the sight of children enjoying themselves; so much so that whenever you see children having fun, you immediately assume they are up to no good. The more they seem to be enjoying themselves, furthermore, the readier you are to fasten on any pretext for thinking them engaged in real wickedness. Then it is true that even if the children had been engaged in some robust but innocent fun, you would have thought they were doing something wrong; and Harman is perhaps right[21] about you that the actual wrongness of the

---

20. What would be generally granted is just that *if* there are moral properties they supervene on natural properties. But, remember, we are assuming for the sake of argument that there are.

From my view that moral properties *are* natural properties, it of course follows trivially that they supervene on natural properties: that, necessarily, nothing could differ in its moral properties without differing in some natural respect. But I also accept the more interesting thesis usually intended by the claim about supervenience—that there are more basic natural features such that, necessarily, once they are fixed, so are the moral properties. (In supervening on more basic facts of some sort, moral facts are like *most* natural facts. Social facts like unemployment, for example, supervene on complex histories of many individuals and their relations; and facts about the existence and properties of macroscopic physical objects—colliding billiard balls, say—clearly supervene on the microphysical constitution of the situations that include them.)

21. Not *certainly* right, because there is still the possibility that your reaction is to some extent overdetermined and is to be explained partly by your sympathy for the cat and your dislike of cruelty, as well as by your hatred for children (although this last alone would have been sufficient to produce it).

We could of course rule out this possibility by making you an even less attractive character, indifferent to the suffering of animals and not offended by cruelty. But it may then be hard to imagine that such a person (whom I shall cease calling "you") could retain enough of a grip on moral thought

action you see is irrelevant to your thinking it wrong. This is because your reaction is due to a feature of the action that coincides only very accidentally with the ones that make it wrong.[22] But, of course, and fortunately, many people aren't like this (nor does Harman argue that they are). It isn't true of them that, in general, if the children had been doing something similar, although different enough not to be wrong, they would still have thought the children were doing something wrong. And it isn't true either, therefore, that the wrongness of the action is irrelevant to the explanation of why they think it wrong.

Now, one might have the sense from my discussion of all these examples—but perhaps especially from my discussion of this last one, Harman's own—that I have perversely been refusing to understand his claim about the explanatory irrelevance of moral facts in the way he intends. And perhaps I have not been understanding it as he wishes. In any case, I agree, I have certainly not been understanding the crucial counterfactual question, of whether we would have drawn the same moral conclusion even if the moral facts had been different, in the way he must intend. But I am not being perverse. I believe, as I said, that my way of taking the question is the more natural one. And more important, although there is, I grant, a reading of that question on which it will always yield the answer Harman wants—namely, that a difference in the moral facts would *not* have made a difference in our judgment—I do not believe this can support his argument. I must now explain why.

It will help if I contrast my general approach with his. I am addressing questions about the justification of belief in the spirit of what Quine has called "epistemology naturalized."[23] I take this to mean that we have in general no a priori way of knowing which strategies for forming and refining our beliefs are likely to take us closer to the truth. The only way we have of proceeding is to assume the approximate truth of what seems to us the best overall theory we already have of what we are like and what the world is like, and to decide in the light of *that* what strategies of research and reasoning are likely to be reliable in producing a more nearly true overall theory. One result of applying these procedures, in turn, is likely to be the refinement or perhaps even the abandonment of parts of the tentative theory with which we began.

I take Harman's approach, too, to be an instance of this one. He says we are

---

for us to be willing to say he thought the action wrong, as opposed to saying that he merely pretended to do so. This difficulty is perhaps not insuperable, but it is revealing. Harman says that the actual wrongness of the action is "completely irrelevant" to the explanation of the observer's reaction. Notice that what is in fact true, however, is that it is *very hard* to imagine someone who reacts in the way Harman describes, but whose reaction is *not* due, at least in part, to the actual wrongness of the action.

22. Perhaps deliberate cruelty is worse the more one enjoys it (a standard counterexample to hedonism). If so, the fact that the children are enjoying themselves makes their action worse, but presumably isn't what makes it wrong to begin with.

23. W. V. O. Quine, "Epistemology Naturalized," in *Ontological Relativity and Other Essays* (New York, 1969), pp. 69–90. See also Quine, "Natural Kinds," in the same volume.

justified in believing in those facts that we need to assume to explain why we observe what we do. But he does not think that our knowledge of this principle about justification is a priori. Furthermore, as he knows, we cannot decide whether one explanation is better than another without relying on beliefs we already have about the world. Is it really a better explanation of the vapor trail the physicist sees in the cloud chamber to suppose that a proton caused it, as Harman suggests in his example, rather than some other charged particle? Would there, for example, have been no vapor trail in the absence of that proton? There is obviously no hope of answering such questions without assuming at least the approximate truth of some quite far-reaching microphysical theory, and our knowledge of such theories is not a priori.

But my approach differs from Harman's in one crucial way. For among the beliefs in which I have enough confidence to rely on in evaluating explanations, at least at the outset, are some moral beliefs. And I have been relying on them in the following way.[24] Harman's thesis implies that the supposed moral fact of Hitler's being morally depraved is irrelevant to the explanation of Hitler's doing what he did. (For we may suppose that if it explains his doing what he did, it also helps explain, at greater remove, Harman's belief and mine in his moral depravity.) To assess this claim, we need to conceive a situation in which Hitler was *not* morally depraved and consider the question whether in that situation he would still have done what he did. My answer is that he would not, and this answer relies on a (not very controversial) moral view: that in any world at all like the actual one, only a morally depraved person could have initiated a world war, ordered the ''final solution,'' and done any number of other things Hitler did. That is why I believe that, if Hitler hadn't been morally depraved, he wouldn't have done those things, and hence that the fact of his moral depravity is relevant to an explanation of what he did.

Harman, however, cannot want us to rely on any such moral views in answering this counterfactual question. This comes out most clearly if we return to his example of the children igniting the cat. He claims that the wrongness of this act is irrelevant to an explanation of your thinking it wrong, that you would have *thought* it wrong even if it wasn't. My reply was that in order for the action not to be wrong it would have had to lack the feature of deliberate, intense, pointless cruelty, and that if it had differed in this way you might very well *not* have thought it wrong. I also suggested a more cautious version of this reply: that since the action is in fact wrong, and since moral properties supervene on more basic natural ones, it would have had to be different in *some* further natural respect in

---

24. Harman of course allows us to assume the moral facts whose explanatory relevance is being assessed: that Hitler was depraved, or that what the children in his example are doing is wrong. But I have been assuming something more—something about what depravity *is,* and about what *makes* the children's action wrong. (At a minimum, in the more cautious version of my argument, I have been assuming that *something* about its more basic features makes it wrong, so that it could not have differed in its moral quality without differing in those other features as well.)

order not to be wrong; and that we do not know whether if it had so differed you would still have thought it wrong. Both of these replies, again, rely on moral views, the latter merely on the view that there is *something* about the natural features of the action in Harman's example that makes it wrong, the former on a more specific view as to which of these features do this.

But Harman, it is fairly clear, intends for us *not* to rely on any such moral views in evaluating his counterfactual claim. His claim is not that if the action had not been one of deliberate cruelty (or had otherwise differed in whatever way would be required to remove its wrongness), you would still have thought it wrong. It is, instead, that if the action were one of deliberate, pointless cruelty, but this *did not make it wrong,* you would still have thought it was wrong. And to return to the example of Hitler's moral character, the counterfactual claim that Harman will need in order to defend a comparable conclusion about that case is not that if Hitler had been, for example, humane and fair-minded, free of nationalistic pride and racial hatred, he would still have done exactly as he did. It is, rather, that if Hitler's psychology, and anything else about his situation that could strike us as morally relevant, had been exactly as it in fact was, but this had *not constituted moral depravity,* he would still have done exactly what he did.

Now the antecedents of these two conditionals are puzzling. For one thing, both are, I believe, necessarily false. I am fairly confident, for example, that Hitler really was morally depraved,[25] and since I also accept the view that moral features supervene on more basic natural properties,[26] I take this to imply that there is no

---

25. And anyway, remember, this is the sort of fact Harman allows us to assume in order to see whether, if we assume it, it will look explanatory.

26. It is about here that I have several times encountered the objection: but surely *supervenient* properties aren't needed to explain anything. It is a little hard, however, to see just what this objection is supposed to come to. If it includes endorsement of the conditional I here attribute to Harman, then I believe the remainder of my discussion is an adequate reply to it. If it is the claim that, because moral properties are supervenient, we can always exploit the insights in any moral explanations, however plausible they may seem, without resort to moral *language,* then I have already dealt with it in my discussion of reduction: The claim is probably false, but even if it is true, it is no support for Harman's view, which is not that moral explanations are plausible but reducible, but that they are totally implausible. And doubts about the causal efficacy of supervenient facts seem misplaced in any case, as attention to my earlier examples (note 20) illustrates. High unemployment causes widespread hardship, and can also bring down the rate of inflation. The masses and velocities of two colliding billiard balls causally influence the subsequent trajectories of the two balls. There is no doubt some sense in which these facts are causally efficacious *in virtue of* the way they supervene on—that is, are constituted out of, or causally realized by—more basic facts, but this hardly shows them *inefficacious.* (Nor does Harman appear to think it does: for his *favored* explanation of your moral belief about the burning cat, recall, appeals to psychological facts (about your moral sensibility), a biological fact (that it's a cat), and macrophysical facts (that it's on fire)—supervenient facts all, on his physicalist view and mine.) If anyone does hold to a general suspicion of causation by supervenient facts and properties, however, as Jaegwon Kim appears to ("Causality, Identity and Supervenience in the Mind Body Problem," in *Midwest Studies* 4 (Morris 1979), pp. 47–48.), it is enough here to note that this suspicion cannot diagnose any special difficulty with *moral* explanations, any distinctive "problem with ethics." The "problem," arguably, will be with every discipline but fundamental physics. On this point, see Richard W. Miller, "Reason and Commitment in the Social Sciences," *Philosophy & Public Affairs,* 8 (1979), esp. 252–55.

possible world in which Hitler has just the personality he in fact did, in just the situation he was in, but is not morally depraved. Any attempt to describe such a situation, moreover, will surely run up against the limits of our moral concepts— what Harman calls our "moral sensibility"—and this is no accident. For what Harman is asking us to do, in general, is to consider cases in which absolutely *everything* about the nonmoral facts that could seem morally relevant to us, in light of whatever moral theory we accept and of the concepts required for our understanding of that theory, is held fixed, but in which the moral judgment that our theory yields about the case is nevertheless mistaken. So it is hardly surprising that, using that theory and those concepts, we should find it difficult to conceive in any detail what such a situation would be like. It is especially not surprising when the cases in question are as paradigmatic in light of the moral outlook we in fact have as is Harman's example or as is, even more so, mine of Hitler's moral character. The only way we could be wrong about this latter case (assuming we have the nonmoral facts right) would be for our whole moral theory to be hopelessly wrong, so radically mistaken that there could be no hope of straightening it out through adjustments from within.

But I do not believe we should conclude, as we might be temped to,[27] that we therefore know a priori that this is not so, or that we cannot understand these conditionals that are crucial to Harman's argument. Rather, now that we have seen how we have to understand them, we should grant that they are true: that if our moral theory were somehow hopelessly mistaken, but all the nonmoral facts remained exactly as they in fact are, then, since we do *accept* that moral theory, we would still draw exactly the moral conclusions we in fact do. But we should deny that any skeptical conclusion follows from this. In particular, we should deny that it follows that moral facts play no role in explaining our moral judgments.

For consider what follows from the parallel claim about microphysics, in particular about Harman's example in which a physicist concludes from his observation of a vapor trail in a cloud chamber, and from the microphysical theory he accepts, that a free proton has passed through the chamber. The parallel claim, notice, is *not* just that if the proton had not been there the physicist would have thought it was. This claim is implausible, for we may assume that the physicist's theory is generally correct, and it follows from that theory that if there hadn't been a proton there, then there wouldn't have been a vapor trail. But in a perfectly similar way it is implausible that if Hitler hadn't been morally depraved we would still have thought he was: for we may assume that our moral theory also is at least roughly correct, and it follows from the most central features of that theory that if Hitler hadn't been morally depraved, he wouldn't have done what he did. The *parallel* claim about the microphysical example is, instead, that if there hadn't been a proton there, but there *had* been a vapor trail, the physicist would still have

27. And as I take it Philippa Foot, in *Moral Relativism,* the Lindley Lectures (Lawrence, Kans., 1978), for example, is still prepared to do, at least about paradigmatic cases.

concluded that a proton was present. More precisely, to maintain a perfect parallel with Harman's claims about the moral cases, the antecedent must specify that although no proton is present, absolutely *all* the nonmicrophysical facts that the physicist, in light of his theory, might take to be relevant to the question of whether or not a proton is present, are exactly as in the actual case. (These macrophysical facts, as I shall for convenience call them, surely include everything one would normally think of as an observable fact.) Of course, we shall be unable to imagine this without imagining that the physicist's theory is pretty badly mistaken,[28] but I believe we should grant that, *if* the physicist's theory were somehow this badly mistaken, but all the macrophysical facts (including all the observable facts) were held fixed, then the physicist, since he does accept that theory, would still draw all the same conclusions that he actually does. That is, this conditional claim, like Harman's parallel claims about the moral cases, is true.

But no skeptical conclusions follow; nor can Harman, since he does not intend to be a skeptic about physics, think that they do. It does not follow, in the first place, that we have any reason to think the physicist's theory *is* generally mistaken. Nor does it follow, furthermore, that the hypothesis that a proton really did pass through the cloud chamber is not part of a good explanation of the vapor trail, and hence of the physicist's thinking this has happened. This looks like a reasonable explanation, of course, only on the assumption that the physicist's theory is at least roughly true, for it is this theory that tells us, for example, what happens when charged particles pass through a supersaturated atmosphere, what other causes (if any) there might be for a similar phenomenon, and so on. But, as I say,

---

28. If we imagine the physicist *regularly* mistaken in this way, moreover, we will have to imagine his theory not just mistaken but hopelessly so. And we can easily reproduce the other notable feature of Harman's claims about the moral cases, that what we are imagining is *necessarily* false, if we suppose that one of the physicist's (or better, chemist's) conclusions is about the microstructure of some common substance, such as water. For I agree with Saul Kripke that whatever microstructure water actually has is essential to it, that it has this structure in every possible world in which it exists. (S. A. Kripke, *Naming and Necessity* [Cambridge, Mass., 1980], pp. 115–44.) If we are right (as we have every reason to suppose) in thinking that water is actually H_2O, therefore, the conditional "If water were not H_2O, but all the observable, macrophysical facts were just as they actually are, chemists would still have come to *think* it was H_2O" has a necessarily false antecedent; just as, if we are right (as we also have good reason to suppose) in thinking that Hitler was actually morally depraved, the conditional "If Hitler were just as he was in all natural respects, but not morally depraved, we would still have *thought* he was depraved" has a necessarily false antecedent. Of course, I am not suggesting that in either case our knowledge that the antecedent is false is a priori.

These counterfactuals, because of their impossible antecedents, will have to be interpreted over worlds that are (at best) only "epistemically" possible; and, as Richard Boyd has pointed out to me, this helps to explain why anyone who accepts a causal theory of knowledge (or any theory according to which the justification of our beliefs depends on what explains our holding them) will find their truth irrelevant to the question of how much we know, either in chemistry or in morals. For although there certainly are counterfactuals that are relevant to questions about what causes what (and, hence, about what explains what), these have to be counterfactuals about real possibilities, not merely epistemic ones.

we have not been provided with any reason for not trusting the theory to this extent.

Similarly, I conclude, we should draw no skeptical conclusions from Harman's claims about the moral cases. It is true, I grant, that if our moral theory were seriously mistaken, but we still believed it, and the nonmoral facts were held fixed, we would still make just the moral judgments we do. But *this* fact by itself provides us with no reason for thinking that our moral theory *is* generally mistaken. Nor, again, does it imply that the fact of Hitler's really having been morally depraved forms no part of a good explanation of his doing what he did and hence, at greater remove, of our thinking him depraved. This explanation will appear reasonable, of course, only on the assumption that our accepted moral theory is at least roughly correct, for it is this theory that assures us that only a depraved person could have thought, felt, and acted as Hitler did. But, as I say, Harman's argument has provided us with no reason for not trusting our moral views to this extent, and hence with no reason for doubting that it is sometimes moral facts that explain our moral judgments.

I conclude with three comments about my argument.

(1) I have tried to show that Harman's claim—that we would have held the particular moral beliefs we do even if those beliefs were untrue—admits of two readings, one of which makes it implausible, and the other of which reduces it to an application of a general skeptical strategy, which could as easily be used to produce doubt about microphysical as about moral facts. The general strategy is this. Consider any conclusion C we arrive at by relying both on some distinguishable "theory" T and on some body of evidence not being challenged, and ask whether we would have believed C even if it had been false. The plausible answer, *if* we are allowed to rely on T, will often be no: for if C had been false, then (according to T) the evidence would have had to be different, and in that case we wouldn't have believed C. (I have illustrated the plausibility of this sort of reply for all my moral examples, as well as for the microphysical one.) But the skeptic intends us *not* to rely on T in this way, and so rephrases the question: Would we have believed C even if it were false *but* all the evidence had been exactly as it in fact was? Now the answer has to be yes, and the skeptic concludes that C is doubtful. (It should be obvious how to extend this strategy to belief in other minds, or in an external world.) I am of course not convinced: I do not think answers to the rephrased question show anything interesting about what we know or justifiably believe. But it is enough for my purposes here that no such *general* skeptical strategy could pretend to reveal any problems peculiar to belief in *moral* facts.

(2) My conclusion about Harman's argument, although it is not exactly the same as, is nevertheless similar to and very much in the spirit of the Duhemian point I invoked earlier against verificationism. There the question was whether typical moral assertions have testable implications, and the answer was that they

do, so long as you include additional moral assumptions of the right sort among the background theories on which you rely in evaluating these assertions. Harman's more important question is whether we should ever regard moral facts as relevant to the explanation of nonmoral facts, and in particular of our having the moral beliefs we do. But the answer, again, is that we should, so long as we are willing to hold the right sorts of *other* moral assumptions fixed in answering counterfactual questions. Neither answer shows morality to be on any shakier ground than, say, physics: for typical microphysical hypotheses, too, have testable implications, and appear relevant to explanations, only if we are willing to assume at least the approximate truth of an elaborate microphysical theory and to hold this assumption fixed in answering counterfactual questions.

(3) Of course, this picture of how explanations depend on background theories, and moral explanations in particular on moral background theories, does show why someone already tempted toward moral skepticism on other grounds (such as those mentioned at the beginning of this essay) might find Harman's claim about moral explanations plausible. To the extent that you already have pervasive doubts about moral theories, you will also find moral facts nonexplanatory. So I grant that Harman may have located a natural symptom of moral skepticism; but I am sure he has neither traced this skepticism to its roots nor provided any independent argument for it. His claim that we do not *in fact* cite moral facts in explanation of moral beliefs and observations cannot provide such an argument, for that claim is false. So, too, is the claim that assumptions about moral facts seem irrelevant to such explanations, for many do not. The claim that we *should* not rely on such assumptions because they *are* irrelevant, on the other hand, unless it is supported by some independent argument for moral skepticism, will just be question-begging: for the principal test of whether they are relevant, in any situation in which it appears they might be, is a counterfactual question about what would have happened if the moral fact had not obtained, and how we answer that question depends precisely upon whether we *do* rely on moral assumptions in answering it.

A different concern, to which Harman only alludes in the passages I have discussed, is that belief in moral facts may be difficult to render consistent with a naturalistic world view. Since I share a naturalistic viewpoint, I agree that it is important to show that belief in moral facts need not be belief in anything supernatural or "nonnatural." I have of course not dealt with every argument from this direction, but I *have* argued for the important point that naturalism in ethics does not require commitment to reductive definitions for moral terms, any more than physicalism about psychology and biology requires a commitment to reductive definitions for the terminology of those sciences.

My own view I stated at the outset: that the only argument for moral skepticism with any independent weight is the argument from the difficulty of settling disputed moral questions. I have shown that anyone who finds Harman's claim about moral explanations plausible must already have been tempted toward skepticism by some other considerations, and I suspect that the other considerations will always just be the ones I sketched. So that is where discussion should focus. I also suggested that those considerations may provide less support for moral skepticism than is sometimes supposed, but I must reserve a thorough defense of that thesis for another occasion.[29]

29. This essay has benefited from helpful discussion of earlier versions read at the University of Virginia, Cornell University, Franklin and Marshall College, Wayne State University, and the University of Michigan. I have been aided by a useful correspondence with Gilbert Harman; and I am grateful also for specific comments from Richard Boyd, David Brink, David Copp, Stephen Darwall, Terence Irwin, Norman Kretzmann, Ronald Nash, Peter Railton, Bruce Russell, Sydney Shoemaker, and Judith Slein.
Only after this essay had appeared in print did I notice that several parallel points about *aesthetic* explanations had been made by Michael Slote in "The Rationality of Aesthetic Value Judgments," *Journal of Philosophy* 68 (1971), 821–39; interested readers should see that paper.

11

Moral Theory and Explanatory Impotence

Geoffrey Sayre-McCord

1. Introduction

Among the most enduring and compelling worries about moral theory is that it is disastrously isolated from confirmation. The exact nature of this isolation has been subject to two interpretations. According to one, moral theory is totally insulated from observational consequences and is therefore in principle untestable. According to the other, moral theory enjoys the privilege of testability but suffers the embarrassment of failing all the tests. According to both, moral theory is in serious trouble.

After briefly defending moral theory against the charge of in principle untestability, I defend it against the charge of contingent but unmitigated failure. The worries about untestability are, I suggest, easily met. Yet the very ease with which they are met belies the significance of meeting them; all manner of unacceptable theories are testable. The interesting question is not whether moral theory is testable but whether moral theory *passes* the relevant tests. Recently, it has become popular to hold that a moral theory passes only if it is explanatorily potent; that is, only if it contributes to our best explanations of our experiences. The problem with moral theory, on this view, is that it apparently contributes not at all to such explanations.[1] Working out a plausible version of the demand for explana-

This essay first appeared in *Midwest Studies* (Morris: University of Minnesota Press, 1988), 12:433–57. Earlier versions were delivered at the 1985 Eastern Division meetings of the American Philosophical Association, at the Research Triangle Ethics Circle, and at the University of Wisconsin–Madison, University of Notre Dame, Virginia Polytechnic Institute, University of California–Irvine, Duke University, and University of California–San Diego. The essay has benefited considerably from exposure to these audiences, and especially from comments made by Kurt Baier, Douglas Butler, Joseph Camp, Jr., David Gauthier, Joan McCord, Warren Quinn, Michael Resnik, Jay Rosenberg, Robert Shaver, and Gregory Trianosky.

1. For example, see Gilbert Harman's *The Nature of Morality* (New York, 1977) [selections reprinted in this volume] and "Moral Explanations of Natural Facts—Can Moral Claims Be Tested

tory potency is surprisingly hard. Even so, once a plausible version is found, I argue, (some) moral theories will in fact satisfy it. Unfortunately, this too is less significant than it might seem, for any argument establishing the *explanatory* potency of moral theory still falls short of establishing its *justificatory* force. (My arguments are no exception.) And, as I will try to make clear, the pressing worries concerning moral theory center on its claim to justificatory force; its explanatory force is largely beside the point. So much the worse for moral theory, one might be inclined to say. If moral theory goes beyond explanation, it goes where the epistemically cautious should fear to tread. Those who demand explanatory potency, however, cannot afford the luxury of dismissing justificatory theory. Indeed, the demand for explanatory potency itself presupposes the legitimacy of justificatory theory, and this presupposition can be turned to the defense of moral theory's justificatory force. Or so I shall argue.

2. Observational insulation

Keeping in mind that observation is theory-laden, one way to put the charge of untestability is to say that moral theory appears not to be appropriately observation-laden; unlike scientific theories, moral theories seem forever insulated from observational implications.

This objection to moral theory emerges naturally from a variation on the empiricist verification principle. Of course, as a criterion of meaning, the verification principle has for good reason been all but abandoned. Still, taken as a criterion of justifiability, rather than as a criterion of meaning, the principle seems to impose a reasonable requirement: if there is no way to verify the claims of a proposed theory observationally, then there is no way to justify the theory (unless all its claims are analytic).[2] Even if moral claims are meaningful, then, they might nonetheless be impossible to justify.

In favor of thinking moral theory untestable is the apparently unbridgeable chasm dividing what is from what ought to be.[3] After all, claims concerning moral obligations cannot be deduced from nonmoral claims ('ought', it is often said, cannot be derived from 'is'); which suggests (to some) that 'ought-claims' are not

---

against Moral Reality?'' *Spindel Conference: Moral Realism, Southern Journal of Philosophy* 24 (1986), Supplement: pp. 57–68; J. L. Mackie, *Ethics: Inventing Right and Wrong* (New York, 1977) [selections reprinted in this volume]; Simon Blackburn, *Spreading the Word* (New York, 1984); Francis Snare, "The Empirical Bases of Moral Scepticism," *American Philsophical Quarterly* 21 (1984), 215–25; and David Zimmerman, "Moral Theory and Explanatory Necessity," in *Morality, Reason and Truth*, ed. David Copp and David Zimmerman (Totowa, N.J., 1985), pp. 79–103.

2. Although this change in emphasis, from meaning to justification, represents a natural development of the verification principle, it constitutes a significant change. With it comes the rejection of the verifiability principle as grounds for noncognitivism.

3. As Reichenbach notes: "Science tells us what is, but not what should be." *The Rise of Scientific Philosophy* (Berkeley, 1951), p. 287.

'is-claims'. Since observation is always of what is, we may have reason to suspect that observation is irrelevant to what ought to be.

This argument for the is/ought distinction is too strong, though. It mistakenly assumes that definitional reducibility is a prerequisite for putting what ought to be on an ontologically equivalent footing with what is. No matter what we know about the nonmoral facts of the case, the argument emphasizes, we cannot uncontroversially infer the moral facts. Moral assertions are not definitionally reducible to nonmoral assertions. Since nonmoral assertions report what is, and since moral claims are not reducible to these others, then moral claims must not report what is. So the argument goes.

Remarkably, by similar lines of reasoning we would be constrained to admit that the claims made in psychology are not claims about facts, for psychology, no less than morality, resists definitional reduction. No matter what we know of the nonpsychological facts of the case, we cannot uncontroversially infer the psychological facts. Psychological assertions are not definitionally reducible to nonpsychological assertions. Since nonpsychological assertions report what is, and since psychological claims are not definitionally reducible to these others, then (the argument would have it) psychological claims must not report what is. Consequently, if the argument offered in support of the is/ought distinction worked, we would find ourselves stuck with an *is/thought* distinction as well. Psychology, we would have to say, reports not what is but merely what is thought—which is silly.[4]

Yet even if we put aside the is/ought distinction, the claim that moral theory is not properly observation-laden still extracts admirable support from common sense. For if people or actions or states of affairs have a worth or a dignity or a rightness about them, this is something we seemingly cannot sense directly. And most moral theories recognize that we cannot by construing moral properties as not directly observable. The unobservability of moral properties cannot pose a special problem for moral theory's testability, however, since in *this* respect, moral theory is no different from those (obviously testable) scientific theories that postulate unobservable entities.

Moreover, on at least one standard construal of what counts as an observation, some moral claims will actually count as observation reports. Specifically, if one takes an observation to be any belief reached noninferentially as a direct result of perceptual experience, there is no reason to deny that there are moral observa-

---

4. Although it is true that what is thought to be is not always so (just as what ought to be is not always so), reports that something is thought to be (or that something ought to be), are still assertions concerning what is the case. Moral theory is as concerned with what is as is psychology. In making claims about what ought to be, moral theory is claiming that what ought to be *is* such and such. Moral theory characteristically makes assertions such as "Killing humans for entertainment *is* wrong"; "An action *is* made worse if it results in excruciating pain for others"; "The Ku Klux Klan *is* a morally corrupt organization."

tions. After all, just as we learn to report noninferentially the presence of chairs in response to sensory stimulation, we also learn to report noninferentially the presence of moral properties in response to sensory stimulation.

On this liberal view of observation, what counts as an observation depends solely on what opinions a person is trained to form immediately in response to sensory stimulation, and not on the the the content of the opinions.[5] Since such opinions are often heavily theory-laden and are often about the external world rather than about our experiences, the account avoids tying the notion of observation to the impossible ideal of theory neutrality or to the solipsistic reporting of the contents of sensory experience.

Of course, we may be too liberal here in allowing *any* opinion to count as an observation simply because it is reached directly as a result of perception. Surely, one is tempted to argue, we cannot observe what is not there, so that some opinions—no matter that they are directly reached as a result of perception—may fail to be observations because they report what does not exist. As a direct result of perception, I may believe I felt a friend's touch; but in the absence of her touch, my report seems most properly treated as an illusion, not an observation. Taking this into account, it is tempting to distinguish what are merely perceptually stimulated judgments from actual observations, thus reserving 'observation' for those perceptually stimulated judgments that are accurate.

If there were some observation-independent way to determine which judgments are accurate, we might legitimately dismiss a given class of purported observations (say, moral observations) on the grounds that they fail to report the facts accurately. Yet once the prospect of divining some set of basic (and indubitable) empirical statements is abandoned, so too must be the hope of establishing what things exist without appeal (at least indirectly) to observations. If some observations are needed to support the theories we then use to discredit other observations, we need some account of observation that allows us to isolate observations as such without assuming their accuracy has already been shown. Observations (in some ontologically noncommittal sense) will be needed to legitimize the theories we use to separate veridical from nonveridical observations. It is this ontologically noncommittal sense of 'observation' that may be characterized simply as any opinion reached as a direct result of perception; and it is in this sense of 'observation' that we must allow that there are moral observations. Once moral observations are allowed, the admission that moral theories can be tested against these moral observations will quickly follow. Just as we test our physical principles against observation, adjusting one or the other in search of a proper fit, so we can test our moral principles against (moral) observation, adjusting one or the

5. Paul Churchland defends this account in *Scientific Realism and the Plasticity of Mind* (Cambridge, Mass., 1979). See also Norwood Russell Hanson, *Patterns of Discovery* (Cambridge, Mass., 1958); and Wilfrid Sellars, *Science, Perception, and Reality* (London, 1963).

other in search of a proper fit. (Many have exploited the availability of this sort of observational testing and—unsatisfyingly—treated it as the sole criterion we have for the acceptability of theories.)[6]

So neither the is/ought distinction nor the unobservability of moral properties seems to support the charge of untestability. In fact, there is reason to think moral theory passes the testability requirement in the same way any respectable scientific theory does—even if moral properties count as unobservable. Of course, how scientific theories manage to pass the testability requirement is a notoriously complicated matter. As Duhem and Quine have emphasized, scientific theories do not pass the testability requirement by having each of their principles pass independently; many of the theoretical principles of science have no observational implications when considered in isolation. Observationally testable predictions may be derived from these scientific principles only when they are combined with appropriate background assumptions.[7]

In the same way, certain moral principles may not be testable in isolation. Nevertheless, when such principles are combined with appropriate background assumptions, they too will allow the derivation of observationally testable predictions. To test the view that an action is wrong if and only if there is some alternative action available that will bring about more happiness, we might combine it with the (plausible) assumption that punishing the innocent is wrong. From these two principles taken together, we get the testable prediction that there will never be a time when punishing the innocent brings more happiness than any other action that is available. Alternatively, consider Plato's contention that 'virtue pays'. If combined with some account of what virtue is and with the (non-Platonic) view that 'payment' is a matter of satisfying preferences, we get as a testable consequence the prediction that those who are virtuous (in whatever sense we settle on) will have more of their preferences satisfied than if they had not been virtuous. Or again, if a moral theory holds that a just state does not allow capital punishment, and if we assume some particular state is just, we get as a testable consequence the claim that this country does not allow capital punishment.

In each case our moral principles have observationally testable consequences when combined with appropriate background assumptions. Experience may show that punishing the innocent does sometimes increase happiness or that misery often accompanies virtue or that the state in question does allow capital punishment. Upon making such discoveries we must abandon (or amend) our moral principles or our background assumptions or the confidence we place in our

6. See, as examples, John Rawls, *A Theory of Justice* (Cambridge, Mass., 1971); Ronald Dworkin, *Taking Rights Seriously* (Cambridge, Mass., 1977); and Philip Pettit, *Judging Justice* (London, 1980).

7. See Pierre Duhem, *The Aim and Structure of Physical Theory* (Princeton, N.J., 1954); and W. V. O. Quine, "Two Dogmas of Empiricism," in *From a Logical Point of View* (Cambridge, Mass., 1964), pp. 20–46.

discoveries. Something has to give way.[8] Of course, we can often make adjustments in our overall theory in order to save particular moral principles, just as we can adjust scientific theories in order to salvage particular scientific principles. In science and ethics, background assumptions serve as protective buffers between particular principles and observation. Yet those same assumptions also provide the crucial link that allows both moral and scientific theories to pass any reasonable testability requirement. If the testability requirement rules out relying on background assumptions it will condemn science as untestable. If it allows such assumptions, and so makes room for the testability of science, it will likewise certify moral theory as testable. Once—but only once—background hypotheses are allowed, both scientific and moral principles will prove testable. Hence, if moral theories are unjustified, it must be for reason other than that moral theories have no testable consequences.[9]

3. Explanatory impotence

Disturbingly, just as moral theory survives any reasonable standard for testability, so too do phlogiston theory, astrology, and even occult theories positing the existence of witches. Like moral theories, each of these theories (when combined with appropriate background assumptions) generates testable consequences, and each makes cognitively packed claims about the world. Yet given what we now know about the world, none of these theories has a claim on our allegiance. Although testable, they fail the test.

Quite reasonably, then, we might wonder whether moral theories likewise fail the empirical tests to which they may admittedly be subjected. Perhaps we ought to think of moral theories as failed theories—as theories betrayed by experience. Perhaps we ought to give up thinking there are moral facts for a moral theory to be about, just as we have abandoned thinking there is such a thing as phlogiston, just as we have abandoned the belief that the heavens control our destiny, and just as we have abandoned the idea that bound women who float are witches.

In our search for an understanding of the world, each of these theories seems to have been left in the dust; every phenomenon we might wish to explain by appeal to these theories can be explained better if they are put aside. Like phlogiston theory, astrology, and theories positing witches, moral theories appear explanatorily impotent.

8. See Morton White, *What Is and What Ought to Be Done* (Oxford, 1981); and Nicholas Sturgeon, "Moral Explanations," in *Morality, Reason, and Truth*, pp. 49–78, reprinted in this volume.

9. The last few paragraphs reiterate points made in my "Logical Positivism and the Demise of 'Moral Science'," in *The Heritage of Logical Positivism*, ed. Nicholas Rescher, University of Pittsburgh Philosophy of Science Series (Lanham, Md., 1985), pp. 83–92.

The problem is that we need suppose neither that our particular moral judgments are accurate nor that our moral principles are true in order to explain why we make the judgments or accept the principles that we do. It seems we make the moral judgments we do because of the theories we happen to embrace, because of the society we live in, because of our individual temperaments, because of our feelings for others, but not because we have some special ability to detect moral facts, not because our moral judgments are accurate, and not because the moral theories we embrace are true. Given our training, temperament, and environment, we would make the moral judgments we do and advance the moral theories we do, regardless of the moral facts (and regardless of whether there are any).

To clarify the challenge facing moral theory, consider two situations (I take these from Gilbert Harman, who has done the most to advance the charge of explanatory impotence.)[10] In one, a person goes around a corner, sees a gang of hoodlums setting a live cat on fire, and exclaims, "There's a bad action!" In the other, a person peers into a cloud chamber, sees a trail, and exclaims, "There's a proton!" In both cases, part of the explanation of why the report was made will appeal to the movements of physical objects and the effects these movements have initially on light and eventually on the observers' retinas. A more complete explanation would also have to make reference to the observers' psychological states as well as the background theories each accepted. Certainly the scientist would not make the report she did if she were asleep or, even if awake and attentive, if she did not accept a theory according to which vapor trails in cloud chambers evidence the presence of protons. Had she thought witches left such trails, she might have reported a witch in the chamber instead of a proton. Similarly, the moral judge would not have made the report he did if he were asleep or, even if awake and attentive, if he did not accept a view according to which burning live animals is wrong.[11] Had he thought cats the embodiments of evil, he might have reported the action as right instead of wrong.

Whatever explanations we give of the reports, one thing is striking: protons will form part of our best explanation of why the proton report was made; in contrast, moral properties seem not to form part of our best explanation of why the moral report was made. We will often explain the scientist's belief that a proton was present by appeal to the fact that one was. But, the argument goes, we will not explain the moral judge's belief that burning the cat is wrong by appeal to the wrongness of the act.

Harman elaborates on the problem with ethics by noting that "facts about protons can affect what you observe, since a proton passing through a cloud chamber can cause a vapor trail that reflects light to your eye in a way that, given

---

10. Harman, *The Nature of Morality*, pp. 6–7 (pp. 121–22 in this volume).

11. Of course, neither the scientist nor the moral judge need have a well-worked-out theory in order to make observations. The ability to form opinions (about protons, witches, or morals) as a direct result of perceptual experience is more a matter of effective training than of the conscious application of theory to experience.

your scientific training and psychological set, leads you to judge that what you see is a proton. But there does not seem to be any way in which the actual rightness or wrongness of a given situation can have any effect on your perceptual apparatus."[12] This emphasis on affecting (or failing to affect) an observer's perceptual apparatus suggests (mistakenly, I will argue) that the following *Causal Criterion* underlies the explanatory impotence attack on moral theory:

> The only entities and properties we are justified in believing in are those that we are justified in believing have a causal impact on our perceptual apparatus.

Unless moral properties are causally efficacious and so figure as causes in the explanation of our making the observations that we do, moral theory will fail to meet this criterion's test. Even though the argument from explanatory impotence turns on a different (and more plausible) principle, the Causal Criterion deserves attention because of its intimate ties to the causal theories of knowledge and reference.

4. The Causal Criterion, knowledge, and reference

Any reasonable view of moral theory, and of the language(s) we use to formulate the theory, must (if moral theory is legitimate) be compatible with some account of how we come to *know* about moral properties and how the terms of the language come to *refer* to these properties. Assuming that the causal theories of knowledge and reference are substantially correct, the Causal Criterion is attractive simply because we could neither know about nor even refer to any class of properties that failed the Causal Criterion's test.[13] Thus, by requiring that we believe in only those entities and properties with which we believe we can causally interact, the Causal Criterion encapsulates the demands of the causal theories of knowledge and reference.

According to the causal theory of knowledge, we can get evidence only about that to which we bear some appropriate causal connection.[14] All our knowledge arises from the causal interaction of the objects of this knowledge with our bodies; anything outside all causal chains will be epistemically inaccessible.[15] So, if

12. Harman, *The Nature of Morality*, p. 8 (p. 122 this volume).

13. Actually, the Causal Criterion will be attractive regardless of whether the causal theories are substantially correct, as long as we assume causal contact is a necessary condition for knowledge.

14. Goldman characterizes the appropriate connection in terms of there being a "reliable belief-forming operation." Alvin Goldman, "What is Justified Belief?" in *Justification and Knowledge*, ed. George Pappas (Dordrecht, 1979), pp. 1–23.

15. Mark Steiner defends mathematical entities from this objection in *Mathematical Knowledge* (Ithaca, N.Y., 1975) p. 10. See also Penelope Maddy, "Perception and Mathematical Intuition," *Philosophical Review* 89 (1980), 163–96; Paul Benacerraf, "Mathematical Truth," *Journal of Philosophy* 70 (1973), 661–79; Crispin Wright, *Frege's Conception of Numbers as Objects* (Aberdeen, 1983); and Philip Kitcher, *The Nature of Mathematical Knowledge* (Oxford, 1983).

moral properties are causally isolated—if they fail to meet the Causal Criterion—they will be unknowable. More important, if moral properties make absolutely no difference to what we experience, we can never even form reasonable beliefs about what they are like.

The causal theory of reference makes moral theory look all the more hopeless because it suggests that we cannot even successfully refer to, let alone know about, moral properties (if they fail the Causal Criterion). As the causal theory of reference would have it, words in our language refer because they stand at the end of a causal chain linking the speaker's use of the word to the thing to which the word refers. No appropriate chain can be established between speakers of a language (in this case a language containing moral terms) and causally isolated properties. Such properties will lie outside all causal chains, and so outside those causal chains which establish reference.[16]

Moral theory's trouble seems to be that the properties it ascribes to actions, people, and states of affairs, reflect no light, have no texture, give off no odor, have no taste, and make no sound. In fact, they do not causally affect our experience in any way. Were they absent, our experiences would be unchanged. Since we cannot interact with moral properties, there is no way for us to establish a causal chain between ourselves, our use of moral language, and moral properties. Consequently, our moral terms fail to refer.

So put, this criticism of moral theory is much too quick. Even the causal theory of reference allows success in establishing a referential tie between word and world by description as well as by ostension.[17] It is true that ostension works in establishing reference only if the properties (or entities) referred to are causally present (since we can succeed in our ostensions only by locating something in space and time).[18] However, we may still use descriptions to establish a referential link even to that from which we are causally isolated—as long as the appropriate terms of the description succeed in referring. If moral properties fail the Causal Criterion, we will not be able to refer to them by ostension, but we will nonetheless be able to refer to them as long as we can describe the moral properties in nonmoral terms. (Of course, if the description's terms were moral, they would

---

16. All this is compatible, of course, with there being a causal story of our use of moral language. Since we do live in a community of moral-language users, we are taught how to use moral words and we stand at the end of a causal (in this case, educational) chain that explains our use of moral terms. Despite there being such a causal story, if moral properties are causally isolated, our language will lack the grounding that would allow it to refer; the linguistic chain would lack an anchor.

17. See Saul Kripke, *Naming and Necessity* (Cambridge, Mass., 1980).

18. Incidentally, one may succeed in referring to an ordinary object, one located in space and time, even if the object does not actually have any causal impact on the referrer. Eyes closed, I may enter a room, point to my left and declare, "I'll bet that chair is brown." I will have referred to the chair (assuming one is there), and made a bet about its color, despite my neither bumping into it, seeing it, nor in any other way being causally affected by it. At most, successful ostension requires causal *presence* and not causal *impact*.

be no help in grounding the requisite referential link.)[19] As a result, the causal theory of reference will serve to undermine ethics only if we cannot refer to moral properties by using nonmoral descriptions, and then only if moral properties are in fact causally isolated.[20] Any argument against moral facts using the causal theory of reference, then, must rely on some independent argument that shows that moral facts (if such there be) are both indescribable in nonmoral terms and causally isolated.

Plainly, our ability to refer to moral properties will be small consolation unless we can also secure evidence about the properties to which we refer. Successful reference may prove epistemically useless. So, even if we can succeed in referring to moral properties, the problems raised by the causal theory of knowledge remain.

Not surprisingly, these problems, too, are less straightforward than suggested so far. To tell against moral theory, the causal theory of knowledge (like the causal theory of reference) must be supplemented by an argument showing that moral properties are causally isolated (or that they do not exist).

Against theories concerned with abstract entities (like mathematics and Plato's Theory of Forms), the Causal Criterion, and the causal theories of knowledge and reference, apparently meet no resistance. The theories under attack grant right off that the entities in question are causally isolated (because outside space and time). That abstract entities fail the Causal Criterion appears to be a forgone conclusion.[21]

Against moral theories, in contrast, the charge of causal isolation meets with resistance. Unlike abstract entities, moral properties are traditionally thought of as firmly ensconced in the causal nexus: a bad character has notorious effects (at least when backed by power), and fair social institutions evidently affect the happiness of those in society. The ontology of moral theory will not be an unwitting accomplice in the causalist critique of ethics. Of course, moral theory's resistance does not establish that moral properties actually do satisfy the Causal Criterion (and so the causal theories of knowledge and reference); rather, the resistance

---

19. In this paper I shall leave unchallenged the (eminently challengable) assumption that there is some way to isolate moral from nonmoral language.

20. Note that we need not have naturalistic *definitions* in order to succeed in referring by *description*. Since the Causal Criterion rules out all causally inert properties, while the causal theory of reference allows reference to causally inert properties as long as they are describable, the strictures of the Causal Criterion actually go beyond those of the causal theory of reference.

21. Actually, even when applied to mathematical entities the game is not quite so easily won. For instance, Kurt Godel maintained that we have a mathematical intuition akin to visual perception that establishes a causal link, of sorts, between numbers and knowers (in "What is Cantor's Continuum Problem?" *American Mathematical Monthly* 54 (1947), 515–25). Penelope Maddy (in "Perception and Mathematical Intuition") has defended this possibility by appeal to recent theories of perception. In the process, she has argued that abstract mathematical entities (e.g., sets) will, contrary to initial appearances, satisfy the Causal Criterion.

imposes a barrier over which the causal critique of moral theory must climb. Some argument must be given for thinking moral properties fail to meet the Causal Criterion.

5. The Explanatory Criterion

Regardless of whether moral properties satisfy the Causal Criterion, there are good reasons for thinking the criterion itself too strong. To hold tight to the Causal Criterion (and the causal theories of knowledge and reference that support it) is to let go of some of our most impressive epistemological accomplishments; the claims of mathematics, as well as both the empirical generalizations and the laws of the physical sciences, all fail the criterion's test.

We never causally interact with numbers, for instance.[22] So, if causal contact were really a prerequisite to knowledge (and reference), mathematical knowledge and discourse would be an impossibility. For similar reasons, empirical generalizations (like "all emeralds are green"), as well as natural laws (like the first law of thermodynamics), would fall victim to the Causal Criterion. Although these generalizations and laws may help *explain* why we experience what we do as we do, they *cause* none of our experiences. That all emeralds are green does not cause a particular emerald to be green, nor does it cause us to see emeralds as green.[23] These casualties of the Causal Criterion make it clear that we need to replace the Causal Criterion even if we wish to salvage its emphasis on the link between knowledge (or at least justified belief) and experience.[24]

In forging a new criterion, we should concentrate on the reasons that might be given for thinking it reasonable to believe (as I assume it is) in the truth of many mathematical claims, empirical generalizations, and laws of physics. According to one standard line in the philosophy of science (one embraced by Harman, J. L. Mackie, Simon Blackburn, and many other critics of moral theory), the key to the legitimacy of these scientific and mathematical claims is the role they play in the explanations of our experiences. "An observation," Harman argues, "is evidence for what best *explains* it, and since mathematics often figures in the explanations of scientific observations, there is indirect observational evidence for mathematics."[25] Empirical generalizations and physical laws will likewise find

22. As Harman notes, "We do not and cannot perceive numbers . . . since we cannot be in causal contact with them." *The Nature of Morality,* p. 10 (p. 124 this volume).

23. As Harman argues in *Thought* (Princeton, N.J., 1973), p. 127. Adolf Grunbaum makes the same point in arguing that one scientific law may explain another, even though the first law does not *cause* the second. See "Science and Ideology," *Scientific Monthly* (July 1954), 13–19.

24. Harman recognizes the shortcomings of the Causal Criterion, and it is in his pointing them out that it becomes clear that he does not accept the criterion. See *Thought,* esp. pp. 126–32.

25. Harman, *The Nature of Morality,* p. 10 (p. 124 this volume), my emphasis. There is room, of course, to agree that if the truth of mathematical claims contribute to our best explanations these claims should be believed, while also holding that their truth does not so contribute. This is Hartry Field's position in *Science without Numbers* (Princeton, N.J., 1980).

their justification by appeal to their role in the best explanations of our experience; ''scientific principles can be justified ultimately by their role in explaining observations.''[26] The legitimacy of a theory seems to ride on its explanatory role and not on the causal impact of its ontology. From these points we can extract the *Explanatory Criterion*.

The only hypotheses we are justified in believing are those that figure in the best explanations we have of our making the observations that we do.

Significantly, the Explanatory Criterion retains, even reinforces, the empiricist's demand that epistemology be tied to experience; not only does justification turn on experiential testability, it now requires an *explanatory* link between the truth of our beliefs and our experiences as well. Accordingly, an acceptable theory must do more than have observational consequences; it must also contribute to our explanations of why we make the observations we do.

Two versions of the Explanatory Criterion should be distinguished: the first sets necessary and sufficient conditions for reasonable belief, the second sets only necessary conditions. In its stronger version, the criterion would say:

A hypothesis should be believed *if and only if* the hypothesis plays a role in the best explanation we have of our making the observations that we do.

In its weaker version the criterion would say instead:

A hypothesis should be believed *only if* the hypothesis plays a role in the best explanation we have of our making the observations that we do.

Or, in its contrapositive (and more intuitively attractive) form:

A hypothesis should not be believed if the hypothesis plays no role in the best explanation we have of our making the observations that we do.

Accepting the stronger version of the criterion involves endorsing what has come to be called 'inference to the best explanation'. In this guise, the criterion licenses inferring the truth of a hypothesis from its playing a role in our best explanations of our experiences. To be even remotely plausible, of course, some bottom limit must be set on the quality of the explanations that would be allowed to countenance inferences to the truth of the hypotheses invoked. Despite their being the best we have, our explanations can be so bad that we may be quite sure they are wrong. It would be a mistake to infer the truth of a hypothesis from its being part of our best—but obviously flawed—explanation. Even with a quality

26. Harman, *The Nature of Morality*, p. 9 (p. 123 this volume).

constraint, though, the strong version of the Explanatory Criterion is hopelessly liberal because we have such good grounds for thinking that the best explanations we can come up with, at any given time, are not right.[27] In light of these difficulties, I shall concentrate on the weaker version of the criterion; it raises all the same difficulties for moral theory without endorsing carte-blanche inferences to the best explanation.

The problem with moral theory is that moral principles and moral properties appear not to play a role in explaining our making the observations we do. All the explanatory work seems to be done by psychology, physiology, and physics.[28] A scientist's observing a proton in a cloud chamber is evidence for her theory because the theory explains the proton's presence and the scientist's observation better than competing theories can. The observation of a proton provides observational evidence for a theory because the truth of that observation is part of the best explanation we have of why the observation was made. A moral 'observation' does not appear to be, in the same sense, observational evidence for or against any moral theory, since (as Harman puts it) the truth or falsity of the ''moral observation seems to be completely irrelevant to any reasonable explanation of why that observation was made.''[29]

Underlying the Explanatory Criterion is the conviction that confirmation mirrors explanation: theories are confirmed by what they explain. Added to this view of confirmation is the stipulation, motivated by an empiricist epistemology, that we should assume to exist only what we need to explain *our experiences*.

Put generally, some fact confirms whatever principles and hypotheses are part of the best explanation of the fact. So, the fact that some observation was made will confirm whatever is part of the best explanation of its having been made. But the making of the observation will not provide *observational* evidence for a theory unless the observation itself is accurate. And we will have grounds for thinking an observation accurate, on this view, only when its being accurate forms a part of our best explanation of the observation having been made. Thus, embedded within this overarching view of confirmation is a more specific account of observational confirmation. An observation will provide confirming observational evidence for a theory, according to this account, only to the extent that it is reasonable to explain the making of the observation by invoking the theory while also treating the observation as true.[30]

27. The one reasonable application of this strong version of the Explanatory Criterion would tie its use to the explanations reached at the ideal limit of inquiry—an explanation *we* will almost surely never get. At this Piercean limit, there is sense to saying we can infer the truth of the hypotheses invoked by the (very) best explanation; for only if there is some such link with epistemology will truth be accessible. So used, though, the principle will never actually countenance any of our inferences.

28. According to Harman, ''Moral hypotheses never help explain why we observe anything. So we have no evidence for our moral opinions.'' *The Nature of Morality*, p. 13; see also p. 8, (p. 123 this volume).

29. Harman, *The Nature of Morality*, p. 7 (p. 122 this volume).

30. Theories find observational confirmation only from accurate observations, and some particular theory will find observational confirmation from an accurate observation only if the theory also plays a

For this reason, moral facts, and moral theory, will be vindicated (in the eyes of the Explanatory Criterion) only if they figure in our best explanations of at least some of the accurate observations we make. Unfortunately, as Harman argues, "you need to make assumptions about certain physical facts to explain the occurrence of observations that support a scientific theory, but you do not seem to need to make assumptions about any moral facts to explain the occurrence of the so-called moral observations."[31] Of course, moral facts would be acceptable, according to the Explanatory Criterion, as long as they were needed to explain the making of some observation or other (regardless of whether it is a moral observation). Just as mathematics is justified by its role in explaining physical (and not mathematical) observations, moral theory might similarly be justified by its role in explaining some nonmoral observations. But the problem with moral theory is that moral facts seem not to help explain the making of *any of our observations*.

Importantly, the problem is not that moral facts explain nothing at all (they may explain other moral facts); the problem is that regardless of whether they explain something, they do not hook up properly with our abilities to detect facts. Even if there are moral facts and even if some of these facts would help to explain others, none will be epistemically accessible unless some help to explain our making some of the observations we do. No matter how perfect the fit between the content of our moral judgments and a moral theory, no matter how stable and satisfying a reflective equilibrium can be established between them, the theory will not gain observational confirmation unless it enters into the best explanation of why some of our observations are made. We will be justified in accepting a moral theory on the basis of our observations only if we have reason to believe our observations are responsive to the moral facts. And we will have reason to believe this only if moral facts enter into the best explanations of why we make the observations we do. To be legitimized, then, moral facts must explain certain nonmoral facts; specifically, moral facts must explain our making observations.[32]

In order to highlight the problem faced by moral theory, it is a good idea to go back to the (dis)analogy between the scientist's making the observation "there's a proton" and the moral judge's making the observation "there's a bad action."

---

role in explaining the making of that observation. Nonetheless, the making of some observation *O* will confirm a theory *T* even if the observation is false as long as *T* explains why the false observation was made. Even supposing an observation inaccurate, then, the making of the observation will be confirming evidence (but not *observational* evidence) for whatever theories contribute to the best explanation of the making of that (false) observation. When the report is false, however, it will be the making of the observation, and not its content, that serves to confirm our explanatory theories; and it will be the accurate observation that the false observation was made (and not the false observation itself) that provides observational support for our explanatory theories.

31. Harman, *The Nature of Morality*, p. 6 (p. 121 this volume).

32. Implicit in the Explanatory Criterion, then, is the conviction that legitimate theories must be linked to an acceptable theory of observation. As Putnam argues, "It is an important and extremely useful constraint on our theory itself that our developing theory of the world taken as a whole should include an account of the very activity and process by which we are able to know that a theory is correct." *Reason, Truth and History* (Cambridge, 1981), p. 132.

Our best explanation of why the scientist made the observation she did will make reference to her psychology, her scientific theory, the fact that a vapor trail appeared in the cloud chamber, and *the fact that a proton left the trail.* Our best explanation of why the moral judge made the observation he did will likewise make reference to his psychology, his moral theory, the fact that a cat was set on fire, and even (when the explanation is more fully elaborated) to the fact that the cat and the kids were partially composed of protons. Yet our explanation will not make reference to the (purported) fact that burning the cat was wrong: "It seems to be completely irrelevant to our explanation whether [the judge's moral] judgment is true or false."[33] That burning the cat was wrong, if it was wrong, appears completely irrelevant to our explanation of the judge thinking it wrong.[34]

6. Explanatory relevance and explanatory potency

The explanatory critique of moral theory seems to rest on the claim that moral facts are *irrelevant* to explanations of our observations. So, to flesh out the problem, we need a test for explanatory irrelevance. Nicholas Sturgeon proposes the following: "If a particular assumption is completely irrelevant to the explanation of a certain fact, then the fact would have obtained, and we could have explained it just as well, even if the assumption had been false."[35] With this in mind, Sturgeon argues that, for those who are not already moral skeptics, moral facts will prove to be explanatorily relevant.

Sturgeon's argument runs as follows. To decide whether the truth of some moral belief is explanatorily relevant to the making of an observation, we must consider a situation in which the belief is false, but which is otherwise as much like the actual situation as possible. Then we must determine whether the observation would still have been made under the new conditions. If so, if the observation would have been made in any case, then the truth of the moral belief is explanatorily irrelevant (the observation would have been made even if it had been false); otherwise its truth is relevant. If a supervenience account of moral properties is right (so that what makes a moral judgment true or false is some combination of physical facts), then for some true judgment to have been false, or some false

33. Harman, *The Nature of Morality*, p. 7 (p. 122 this volume).

34. Of course, that the burning of the cat was wrong might be part of the moral judge's (as opposed to our) best explanation of why he made the observation he did, just as for some people the best explanations they had of their observations made reference to phlogiston.

35. Sturgeon, "Moral Explanations," p. 245 of this volume. As Sturgeon recognizes, the test has its limits. It will not be a reliable indicator of explanatory relevance when dealing with two effects of the same cause; neither effect would have occurred without the other (because each would have occurred only if the cause of the other had), even though neither explains the other. And it will not be reliable when using 'that-would-have-had-to-be-because' counterfactuals; it may be that if Reagan had lost the presidential election that would have had to be because he failed to get enough votes, even though his being elected is not explanatorily relevant to his getting enough votes. For other limitations, see Warren Quinn, "Truth and Explanation in Ethics," *Ethics* 96 (1986), 524–44.

judgment true, the situation would have had to have been different in some physical respect.[36] Consider the hoodlums' cat-burning. According to Sturgeon, "If what they are actually doing is wrong, and if moral properties are, as many writers have held, supervenient on natural ones, then in order to imagine them not doing something wrong we are going to have to suppose their action different from the actual one in some of its natural features as well."[37] Whether we would still judge the action wrong given these changes is a contingent matter that turns on how closely tied our moral judgments are to the morally relevant physcial features of the situation. In the case of a curmudgeon, who thinks badly of kids as a matter of principle (averring that "kids are always up to no good"), changing the moral (and so the nonmoral) features of the situation will probably not change his moral judgment. For him, the truth of his judgment is irrelevant to the explanation of his having made it. Fortunately, though, many people do not share this bias and are therefore more attuned to the evidence. Such people would have different opinions had the hoodlums found their entertainment in more acceptable ways (say, by petting rather than incinerating the cat). For those to whom the difference would make a difference, part of the explanation of their judgment would be that burning the cat is wrong.[38]

Notice that the same contingency attaches to the scientist's sighting of a proton. Had the proton not been there, whether the scientist would have thought it was depends on how closely tied her scientific judgments are to the relevant features of the situation. If she is a poor researcher, she might well have reported the proton's presence had there really been only a passing reflection. Again fortunately, many scientists are well attuned to the difference between proton trails and passing reflections. At least these scientists would have made different reports concerning the presence of a proton had the proton been absent. For those to whom the difference would make a difference, part of the explanation of their judgment will be that a proton passed.[39]

36. At this stage, the relevant feature of a supervenience account of moral properties is that if the moral properties of something are changed, then so must be some nonmoral properties; there is no holding the nonmoral properties fixed while altering the moral properties, as there can be no moral difference without a nonmoral difference.

37. Sturgeon, "Moral Explanations," p. 247 of this volume. Which nonmoral facts will have to be altered to change the moral facts is, obviously, open to dispute.

38. As Sturgeon emphasizes, "Hitler's moral depravity—the fact of his really having been morally depraved—forms part of a reasonable explanation of why we believe he was depraved" ("Moral Explanations," p. 234 of this volume.). Had he not been depraved we very likely would not have thought him depraved; for he would not have done all the despicable things he did, and it is his having done such things that leads us to our condemnation.

39. Our background theories will clearly play a central role in determining explanatory relevance. In the cat-burning case, we will rely on our moral theory in deciding what nonmoral features of the world would have been different had the hoodlums' activities been unobjectionable. Similarly, in the passing proton case, we will rely on our scientific theory in deciding what physical features of the world would have been different had a proton not passed. Such a reliance on background theories will certainly offend a thoroughgoing skeptic. But attacks on moral theory are interesting only if some of our views survive the skeptic's arguments, so I shall assume we may legitimately rely at least sometimes on our background theories.

Sturgeon holds that what separates his own position from Harman's is a differential willingness to rely on a background moral theory in evaluating the question: Would the moral observation have been made even if the observation had been false? Sturgeon assumes the observation to be true (relying as he does on his background moral theory), and believes that for it to be false, some nonmoral features of the situation must be assumed different (because the moral properties supervene upon natural properties). When these nonmoral features are changed, he points out, the moral judgments (along with the explanations of why they were made) will often change as well. Harman, though, does without the background moral theory, so he has no reason to think that if the observation were false, anything else about the situation (including the observer's beliefs) would have been different. Consequently, he holds that the observation would have been made regardless of its truth.

As a result, Sturgeon concludes that Harman's argument is not an independent defense of skepticism concerning moral facts, for its conclusion apparently rests on the assumption that our moral judgments are false (or, more accurately, on the assumption that moral theory cannot be relied on in estimating how the world would have been if it had been morally different).[40]

Unfortunately for moral theory, Sturgeon's argument fails to meet the real challenge. The force of the explanatory attack on moral theory may be reinstated by shifting attention from explanatory *irrelevance* to explanatory *impotence*, where

> a particular assumption is explanatorily impotent with respect to a certain fact if the fact would have obtained and we could have explained it just as well *even if the assumption had not been invoked in the explanation* (as opposed to: "even if the assumption had been false").

By charging explanatory impotence, rather than explanatory irrelevance, the explanatory challenge to moral theory survives the admission that we hold our moral theories dear. It also survives the supervenience account's provision of a necessary link between moral and nonmoral properties. For the question becomes: Do we honestly think appealing to moral facts in our explanations of moral judgments strengthens our explanations one bit? Behind this question is the worry that we have been profligate with our theory building (or, perhaps, that we've been unnecessarily and unwholesomely nostalgic about old, and now outdated, theories). The concern is that acknowledging moral facts adds nothing to our ability to explain our experiences. Everything we might reasonably want to explain can, it seems, be explained equally well without appeal to moral facts.[41]

40. Sturgeon, "Moral Explanations," pp. 249–51 of this volume. Independently, John McDowell has made essentially the same point concerning skeptical attacks on moral explanations in his "Values and Secondary Qualities," in *Morality and Objectivity,* ed. Ted Honderich (London, 1985), pp. 110–29, reprinted in this volume.
41. Which, of course, is not to say that we can explain everything that we might reasonably want to explain.

If these worries are well founded, if moral facts are explanatory 'fifth wheels', and if we accept the Explanatory Criterion for justified belief, then moral facts will become merely unjustified theoretical baggage weighing down our ontology without offering compensation. In the face of this threat, pointing out that we happen to rely on moral facts in explaining people's behavior is not sufficient to justify believing in (even supervenient) moral facts; actual reliance does not establish justified belief.

To see the force of the explanatory challenge, imagine that a belief in witches becomes popular among your friends. Imagine, too, that your friends teach each other, and you, how to give 'witch explanations'. You 'learn' that the reason some bound women float when tossed into ponds, and others do not, is that the floaters are witches, the others not; that the mysterious deaths of newborns should be attributed to the jealous intervention of witches; and so on. No doubt, with enough practice you could become skillful at generating your own witch explanations; so skillful, perhaps, that in your unreflective moments you would find yourself offering such explanations. In order to assuage your philosophical conscience, you might entertain a sort of supervenience account of being a witch. Then you might comfortably maintain that being a witch is explanatorily relevant (in Sturgeon's sense) to your observations. All this might come to pass, and still you would be justified in thinking there are not really any witches—as long as you could explain the floatings, the deaths, and whatever else, just as well without appealing to the existence of witches. Presumably, the availability of such alternative explanations is just the reason we should not now believe in witches.

Certainly, things could turn out otherwise; we might find that witch explanations are actually the best available. We might discover that postulating witches is the only reasonable way to account for all sorts of otherwise inexplicable phenomena. Should that happen, our conversion to a belief in witches would, of course, be quite justified.

The question is, to which witch scenario are our moral explanations more analogous? Are our appeals to moral properties just intellectually sloppy concessions to effective socialization, or do we really strengthen our explanatory abilities by supposing that there are moral properties? This is a substantial challenge, and one not adequately answered by the observation that we often rely on moral properties to explain behavior.

Two points about the explanatory challenge deserve emphasis. First, the challenge recognizes conditions under which a belief in moral facts, or witches, would be legitimate. Specifically, the Explanatory Criterion would take these beliefs to be justified if their truth figured in the best explanation of why we have the experiences we do. Second, the challenge will not be met simply by pointing out that witches, or moral facts, do figure in some of our best explanations of the world. For unless these explanations of the world can be properly linked to *our experiences* of the world, there will be no way for us to justify accepting some of the explanations rather than others. The truth of one or another will make no difference to our experience and will be epistemically inaccessible.

7. Supervenience and lenience

The problem with concentrating on explanatory relevance, rather than on explanatory potency, is that it makes a defense of moral properties (and belief in witches) too easy. It permits as justified the introduction of any properties whatsoever, so long as they are construed as supervenient upon admittedly explanatory properties.

The Explanatory Criterion, when interpreted as demanding explanatory potency (rather than mere relevance), promises a stricter standard, one which might separate those properties we are justified in believing instantiated from mere pretenders. Yet the criterion must be interpreted in a way that acknowledges, as justified, belief in two kinds of properties: those which can be reductively identified with explanatorily potent properties and those we have independent reason to think supervene upon, without being strictly reducible to, explanatorily potent properties. These properties demand special attention because, at least initially, they appear explanatorily expendable—despite our belief in them being justified.

For instance, though all the explanatory work of water may be better accomplished by H_2O; all the explanatory work of color, by the wavelengths of light; and all that of psychological states, by neurophysiological states of the brain, we are nonetheless justified in believing that the oceans are filled with water, that roses are red, and that people feel pain and have beliefs. Any criterion of justified belief that would rule these beliefs out as unjustified is simply too stringent. The difficulty (for those attacking moral theory) is to accommodate these legitimate beliefs without so weakening the Explanatory Criterion as to reintroduce excessive leniency.

Of course, some beliefs may plausibly find justification by relying on reductions: because water is H_2O, the justification of our belief in H_2O (by appeal to its explanatory potency) serves equally well as a justification of our belief in water. In cases where identification reductions are available, explanatory potency might well be transitive.[42]

Where identification reductions are not available, however, things become trickier; and it is here that the Explanatory Criterion runs into problems. It seems straightforwardly true that roses are red, for example, but our best explanations of red-rose reports might well make reference to certain characteristics of roses, facts about light, and facts about the psychological and perceptual apparatus of perceivers, but not to the *redness* of the roses (and not to any particular feature of the roses that can be reductively identified with redness). Despite this, the availability of such explanations expands our understanding of colors; it does not show there

---

42. Yet there is at least some question as to whether the reductive hypothesis itself satisfies the Explanatory Criterion. What, after all, do we explain with the help of the reductive hypothesis that we could not explain just as well by assuming the reduced claims fail to refer? See Quinn, "Truth and Explanation in Ethics," for more on the tension between the Explanatory Criterion and reductive hypotheses.

are no colors. Similar points hold not just for those properties traditionally characterized as secondary qualities, but for *all* nonreducible properties.[43]

Recognizing this, Harman attempts to make room for nonreducible properties. In discussing colors, he maintains that they satisfy the demand for explanatory potency because "we will sometimes refer to the actual colors of objects in explaining color perceptions if only for the sake of simplicity. . . . We will continue to believe that objects have colors because we will continue to refer to the actual colors of objects in the explanations that we will in practice give."[44] Thus, pragmatic tenacity is supposed by Harman to be enough to establish explanatory potency; now the criterion will allow, as explanatorily potent, those properties and entities to which we appeal in our best explanations, plus those that are precisely reducible to properties or entities appealed to in our best explanations, plus those that are pragmatically tenacious.

Relying on such a lenient interpretation of the Explanatory Criterion, Harman is able to treat moral facts as threatened only by resorting to what is patently false: he is forced to argue that moral facts are not "useful even in practice in our explanations of observations."[45] If nothing else, however, moral facts are useful, at least in practice, when explaining our observations. Many very useful, and frequently offered, explanations of events in the world (and so our observations of those events) make reference to moral facts. Mother Teresa's goodness won her a Nobel Prize; Solidarity is popular because of Poland's oppressive political institutions; millions died in Russia as a result of Stalin's inhumanity; people are starving unnecessarily because of the selfishness of others; unrest in Soweto is a response to the injustice of apartheid. Even if such explanations could eventually be replaced by others that appeal only to psychological, social, and physical factors, without mention of moral facts, the moral explanations would still be useful in just the way talk of colors remains useful even in light of theories of light. If mere pragmatic tenacity is enough to legitimize color properties, then it ought to be enough to legitimize moral properties.

If the Explanatory Criterion is to challenge the legitimacy of moral theory, it must require more than pragmatic tenacity for justification. Yet, almost certainly, any stronger requirement that remains plausible will countenance moral properties. For the Explanatory Criterion will be plausible only if it allows belief in those properties needed both to identify and to explain the natural regularities that are otherwise explicable only in a piecemeal fashion as singular events (and not as

43. There are some significant differences between moral properties and secondary qualities, not least of which is that we can learn to ascribe secondary qualities without having any idea as to what properties they supervene upon, whereas learning to ascribe moral properties requires an awareness of the properties upon which they supervene. This difference will stand in the way of treating moral properties as strictly analogous to secondary properties. But I think it won't underwrite any plausible version of the explanatory criterion that is still strong enough to rule out moral properties. See Quinn, "Truth and Explanation in Ethics."

44. Harman, *The Nature of Morality*, p. 23.

45. Ibid.

instances of regularities). Consider Hilary Putnam's example of the peg (square in cross section) that will pass through the square hole in a board, but not through the round hole.[46] To explain a single instance of the peg's going through one hole, we might offer a microstructural description of the peg and the board in terms of the distribution of atoms, and then appeal to particle mechanics. But even if we could eventually work out such an explanation, it would suffer from a serious drawback; it would only explain why the particular peg went through a particular hole at a particular time. The explanation will be of no help when we are faced with another board and peg, or even with the same board and peg a moment later (when the distribution of atoms has changed). The explanation will not extend to new cases. And the properties appealed to in giving an explanation at the level of ultimate constituents will be useless in trying to identify and explain the general fact that pegs of a certain size and shape (whether made of wood, or plastic, or steel) will fail to go through holes of a certain size and shape (whether the holes are in a piece of wood, or plastic, or steel). This general fact will be identifiable and explicable only if we appeal to certain macrostructural features of pegs and holes.

In the same way, certain regularities—for instance, honesty's engendering trust or justice's commanding allegiance, or kindness's encouraging friendship— are real regularities that are unidentifiable and inexplicable except by appeal to moral properties. Indeed, many moral virtues (such as honesty, justice, kindness) and vices (such as greed, lechery, sadism) figure in this way in our best explanations of many natural regularities. Moral explanations allow us to isolate what it is about a person or an action or an institution that leads to its having the effects it does. And these explanations rely on moral concepts that identify characteristics common to people, actions, and institutions that are uncapturable with finer-grained or differently structured categories.

Of course, even if moral properties do have a role in our best explanations of natural regularities, we might still wonder whether these properties are anything more, anything over and above, psychological properties and dispositions of individuals, and we might wonder to what extent these 'vices' and 'virtues' have any normative authority. For all that has been said so far, we might have no good reason to think the 'virtues' worthy of cultivation and the 'vices' worthy of condemnation. So even if moral properties ultimately satisfy the demands of the Explanatory Criterion (once we get a reasonable interpretation of its requirements), we will at most have established that certain people, actions, and institutions have those properties we label 'moral'. We will not yet have shown that there is any reason to care about the properties or that some of the properties are better than others.[47]

46. "Philosophy and Our Mental Life," in *Mind, Language, and Reality* (Cambridge, 1975), pp. 291–303.
47. Just as reductions of the mental to the physical fail to capture intentionality, reductions of the moral to the mental fail to capture justifiability.

As long as we concentrate on which properties satisfy the Explanatory Criterion and which do not, the distinctive value of moral properties will remain elusive. The structure of a compelling defense of their value will emerge only after we turn our attention to the presuppositions of the Explanatory Criterion itself.

8. The evaluation of explanations

As the Explanatory Criterion would have it, which hypotheses are justified and which are not will depend crucially on our standards of explanation, since it is by figuring in the *best* available explanations that a hypothesis finds justification. No argument that depends on the Explanatory Criterion will get off the ground unless some explanations are better than others. This poses a dilemma for those who suppose that the Explanatory Criterion will support the wholesale rejection of evaluative facts. Either there is a fact of the matter about which explanations are best, or there is not. If there is, then there are at least some evaluative facts (as to which explanations are better than others); if not, then the criterion will never find an application and so will support no argument against moral theory.

If we say that astronomers, and not astrologers, make the appropriate inferences from what is seen of the constellations or that evolutionary theorists, and not creationists, have the best explanation of the origin of our species, we will be making value judgments. In trying to legitimize these judgments by appealing to our standards of explanation, a reliance on values becomes inescapable (even if the values appealed to are not themselves mentioned in our explanations).

The obvious response to this point is to embrace some account of explanatory quality in terms, say, of simplicity, generality, elegance, predictive power, and so on. One explanation is better than another, we could then maintain, in virtue of the way it combines these properties.[48]

When offering the list of properties that are taken to be the measures of explanatory quality, however, it is important to avoid the mistake of thinking the list wipes values out of the picture. It is important to avoid thinking of the list as eliminating explanatory quality in favor of some evaluatively neutral properties. If one explanation is better than another in virtue of being simpler, more general, more elegant, and so on, then simplicity, generality, and elegance cannot themselves be evaluatively neutral. Were these properties evaluatively neutral, they could not account for one explanation being better than another.

If we are to use the Explanatory Criterion, we must hold that some explanations really are better than others, and not just that they have some evaluatively neutral properties that others do not. Any attempt to wash evaluative claims out as

---

48. For discussions of the (often conflicting) criteria for explanatory value, see Paul Thagard "The Best Explanation: Criteria for Theory Choice," *Journal of Philosophy* 75 (1978), 76–92; and William Lycan, "Epistemic Value," *Synthèse* 64 (1985), pp. 137–64.

psychological or sociological reports, for instance, will fail—we will not be saying what we want, that one explanation is *better* than another, but only (for example) that we happen to like one explanation more or that our society approves of one more. What the Explanatory Criterion presupposes is that there are evaluative facts, at least concerning which explanations are better than others, regardless of whether these facts explain any of our observations.

Even assuming that the Explanatory Criterion presupposes the existence of some *evaluative* facts, the question remains whether we have any good reason for thinking there are moral facts as well. We might be convinced that some explanations really are better than others but still deny that some actions or characters or institutions are better than others. Significantly, though, once it has been granted that some explanations are better than others, many obstacles to a defense of moral values disappear. In fact, all general objections to the existence of value must be rejected as too strong. Moreover, whatever ontological niche and epistemological credentials we find for explanatory values will presumably serve equally well for moral values.[49]

Without actually making the argument, I shall briefly sketch one of the ways one might defend the view that there are moral values. The aim of such an argument would be to show that some actions, characters, or institutions are better than others—just as some explanations are better than others.

This defense of moral values rests on recognizing and stressing the similarities between the evaluation of actions, and so on, and the evaluation of explanations. The crucial similarity is that in defending our evaluations (whether of actions, institutions, or explanations) we must inevitably rely on a theory that purports to *justify* our standards of evaluation as over against other sets of (moral or explanatory) standards. In both cases, we will be engaged in the process of justifying our judgments, not of explaining our experiences. The analogy to keep in mind here is not that between moral theory and scientific theory but that between moral theory and scientific epistemology.[50]

Since we must regard certain evaluative claims (those concerning which explanations are better than others) as true, we will be justified in believing those parts of value theory that support our standards of explanatory value. Just as we take the explanatory role of certain hypotheses as grounds for believing the hypotheses, we must, I suggest, take the justificatory role of certain evaluative principles as grounds for believing the principles. If the principles are themselves not reasonably believed, they cannot support our particular evaluations of explanations; and

---

49. Of course this leaves open the possibility that more specific attacks may be leveled at moral values; the point is just that once epistemic values are allowed, no general arguments against the existence of values can work.

50. Here I part company with other moral realists (for example, Boyd, Sturgeon, and Railton) who seem to hold that moral theory should be seen as being of a piece with scientific theory. See Richard Boyd, ''How to Be a Moral Realist'' in this volume; Nicholas Sturgeon, ''Moral Explanations'' in this volume; and Peter Railton, ''Moral Realism,'' *Philosophical Review* 95 (1986), 163–207.

if we can have no grounds for thinking one explanation better than another, the Explanatory Criterion will be toothless.

Thus, if evaluative facts are indispensable (because they are presupposed by the Explanatory Criterion), we can invoke what might be called *inference to the best justification* to argue for abstract value claims on the grounds that they justify (and at the same time, explain the truth of) our lower-level epistemological judgments. And these very same abstract evaluative principles might well imply lower-level, distinctly moral principles and particular moral judgments. If so, then in defending moral values, we might begin with evaluations of explanations, move up (in generality and abstraction) to principles justifying these evaluations, then move back down, along a different justificatory path, to recognizably moral evaluations of actions, characters, institutions, and so on. That is, to argue for a given moral judgment (for example, that it is better to be honest than duplicitous), we might show that the judgment is justified by some abstract evaluative principle that is itself justified by its relation to our standards of explanatory quality (which are indispensable to our application of the Explanatory Criterion). In this way, particular moral judgments and more general moral principles might find their legitimacy through their connection with the indispensable part of value theory that serves to justify our judgments of explanatory quality.

To take one (optimistic) example: Imagine that we justify believing in some property by appeal to its role in our best explanation of some observations, and we then justify our belief that some explanation is the best available by appeal to our standards of explanatory quality, and finally, we justify these standards (rather than some others) by appealing to their ultimate contribution to the maximization of expected utility. Imagine, also, that having justified our standards of explanatory value, we turn to the justification for cultivating some moral property (for example, honesty). The justification might plausibly appeal to its contribution to the cohesiveness of one's society, and we might in turn justify cultivating properties conducive to the cohesiveness of society by appeal to the benefits available only within society. Finally, we might justify these as benefits by appeal to their maximizing expected utility. Appeal to the maximization of expected utility would then serve both as the best justification for certain standards of explanatory value and as the best justification for cultivating particular moral properties. It would justify both our belief that some particular explanation is better than another and our belief that some moral properties (for example, honesty) are better than others (for example, duplicity). We could then deny the justifiability of moral judgments only by denying the justification of our evaluative judgments of explanations.

So, in pursuing justifications for our standards of evaluation, we might discover that the justificatory principles we embrace have as consequences not only evaluations of explanations but also recognizably moral evaluations of character or behavior or institutions. Justificatory principles might come most plausibly as a package deal carrying both explanatory and moral evaluations in tow.

[279]

No doubt this picture is overly optimistic. Most likely, the justificatory principles invoked in justifying particular standards of explanatory quality will not be so neatly tied with the justifications available for having or developing certain (recognizably moral) properties or with the justifications available for condemning other (recognizably immoral) properties. In following out the two lines of justification—that is, in justifying particular evaluations by appeal to principles, which we in turn justify by appeal to more general principles—we may never arrive at a single overarching justificatory principle. Indeed, it is highly unlikely that we will ever get such a principle either in epistemology or in moral theory.[51] It is even less likely that we will ever find a single principle that serves for both.

Although inference to the best justification legitimizes both the lower-level standards of explanatory value (simplicity, generality, and so on) and—more important—the very process of justification, the substantive principles the process engenders will probably vary according to what is being justified. When we are justifying a belief that some property is instantiated, one set of justificatory principles will come into play; when we are justifying the having or cultivating of some property, a completely different set of justificatory principles may prove relevant. The two paths of justification might neither coincide nor converge.

Yet a failure of convergence would not undermine moral justifications. The legitimacy of moral theory does not require any special link between explanatory and moral justifications. What it does require is that moral properties figure both as properties we are justified in believing exemplifiable and as properties we are justified in cultivating.

In constructing explanatory and justificatory theories, we may discover any of four things: (1) that moral properties are neither possessable nor worth possessing, in which case (I assume) moral theory loses its point; (2) that moral properties (for example, honesty, kindness) are possessed by some but that there is no justification for thinking some better than others, in which case only an unexciting conclusion will have been established—like atomic weights, virtue and value would exist, and claims involving them would have a truth value, but they would be normatively inert; (3) that we have no reason to believe moral properties are exemplified, but we do have reason to cultivate them, in which case a unique version of the is/ought distinction will have been established—there are no instantiated moral properties, even though there ought to be; or finally (4) that moral properties are acutally possessed and (some) are worth possessing, in which case moral theory will have found its strongest defense. Which of these four positions is right can be settled only against the background of an accepted justificatory theory.

51. In "Coherence and Models for Moral Theorizing," *Pacific Philosophical Quarterly* 66 (1985), 170–90, I argue that we have good reason for rejecting any proposed unifying fundamental principle we might find.

Of course, whether we are justified in believing moral properties are both possessable and worth possessing is an open question. Yet it is a legitimate question, and it is a question that can be answered only by engaging in moral theorizing; that is, only by attempting the justifications and seeing where they lead.

12

Moral Reality

Mark Platts

1. The nature of ethical realism

My concern is to determine quite what would be involved in extending a realistic view of language to include moral discourse and so moral thought. I do not know if this extension can defensibly be made; I am sure that much hangs upon the question.

Two major ethical traditions which can be construed as realistic are the intuitionist and the utilitarian. Some of my reasons for focusing upon the former, as the more plausible tradition, will emerge on the way.

The form of ethical intuitionism I shall consider has three main positive characteristics and two important negative ones. The first positive characteristic is that moral judgments are viewed as *factually cognitive,* as presenting claims about the world which can be assessed (like any other factual belief) as true or false, and whose truth or falsity are as much possible objects of human knowledge as any other factual claims about the world. This amounts in part to the denial of anything in the literal meaning of a moral judgment which compels us to assess those judgments on some dimension other than (or in addition to) that of the true and the false. It thus amounts also to the claim that if a moral judgment is true, it is true in virtue of the (independently existing) real world, and is true in virtue of that alone. This characteristic, or set of characteristics, is shared by utilitarianism.

The second positive characteristic of this form of intuitionism is that, in general, it is *austerely realistic.* Within the theory of sense for the language of which they are a part, particular moral expressions—'sincerity', 'loyalty', 'honesty', 'prudence', 'courage', 'integrity'—will generally be handled in an austere

This essay was originally published as chapter 10 of Mark Platts; *Ways of Meaning* (London: Routledge and Kegan Paul, 1979).

interpretative manner trading upon the structure of designation and use, whether they be treated as nominatives, adjectives, or adverbs. We may be able to give informal glosses upon some of these expressions, glosses which will help both to exhibit their interrelations (like those between prudence and courage) and to connect with some of the procedures sometimes used in determining their application; but such glosses are not acknowledged within the theory of sense.[1] The motivation behind this is not just to avoid the (usually fruitless) task of producing 'informative' decompositional analyses of these expressions producing truly necessary and sufficient conditions for their application, but also to deny any attempt to *reduce* (obviously) moral claims to (supposedly obviously) non-moral claims. The converse of this is the denial of the claim that moral judgments are *inferences* from non-moral judgments. This pair of denials amounts to a strong version of the claim that moral judgments are *autonomous* of non-moral judgments.

The parallel to exploit in understanding this autonomy is the relation, discussed earlier in another context, between there being a certain arrangement of black dots on a white card and there being a face there pictured to be seen. There is only a face there to be seen because the dot-arrangement is as it is; the dot arrangement *fixes* (subject, perhaps, to existing conventions of pictorial representation) whether or not there is a face there to be seen. Still, we do not *see* the face by *attending* to that dot-arrangement, where that arrangement is characterised in terms free of picture and face-vocabulary—say, by a mathematical grid-system. Indeed, the more we so attend, the less likely we are to see the face that is there pictured. Thus we do not *infer* that the face is there from judgments in this non-pictorial, non-facial vocabulary about the arrangement of the black dots.

This is a dangerous parallel to invoke, both because of the role in such perceptual judgments of conventions of pictorial representation, and, more importantly, because the judgments that finally issue—'That's a face'—are not *literally* true. Still, it need not lead us astray. The picture it invites in the moral case is this: once all the non-moral facts about a situation are fixed, so are all the moral facts; but we could know everything about those non-moral facts while being in utter ignorance of the moral facts. If we now go on to make moral judgments about the case, we do not do so by attending to the non-moral facts, the facts described in vocabulary free of moral import; we do not *infer* the moral facts from the non-moral facts. The more we attend to the non-moral descriptions of the case, the less likely we are to see the moral aspects of that case.[2] But, unlike the picture case, when we make moral judgments about the situation, what we say can be literally

---

1. In these first two characteristics of moral realism, and in the ensuing discussion of moral relativism in the light of those characteristics, the area in which bridges can be built between austerity and full-blooded realism is approached. In my specification of this second characteristic as *austere realism,* I assume these bridges to have been built.

2. The plausibility of this element of the picture presented is difficult to assess without a much surer grasp than we yet have upon the contrast between moral and non-moral facts. See also footnote 5.

true or false; and, again unlike the picture case, there is no question of that truth or falsity being the result of conventions. It is the result of the (independent) world.

This motivation for this conception of autonomy is as coherent (or incoherent) as the contrast between moral and non-moral descriptions, between moral and non-moral facts. (Not that the import, in specific cases, of what has just been said requires adherence to such a general contrast.) Does, for example, the utilitarian deny this autonomy in his attempts to define moral terms using the vocabulary of pleasure, utility, or whatever? If his definitions worked, would this not show merely that the vocabulary of pleasure, utility, or whatever is moral vocabulary? Such definition is as compatible with *elevation* as with reduction. Still, for our purposes it will suffice to construe the (usual) austerity of the treatment of moral vocabulary within the theory of sense as being simply the denial of the possibility of *any* 'informative' definition of the terms so treated. If, subsequently, sense is given to the contrast between moral and non-moral vocabulary; and if, subsequently, sense is given to the claim that a definition is *reductive* rather than *elevatory;* then the motivation just sketched will emerge as a (coherent) consequence of the austerity accorded to moral vocabulary within a realistic, interpretative theory of sense.

Competence in understanding the sayings of others, and of themselves, implies in realistic semantics that speakers can have knowledge of the truth-conditions of sentences that transcend at least their present capacities for determining whether those sentences are true or false. This general consequence of realistic semantics must be accepted in the area of moral vocabulary by anyone wishing to treat such vocabulary realistically. A speaker can know, have a grasp of, the truth-conditions of a moral sentence even if those truth-conditions are beyond his (present) recognitional abilities. I shall return to this consequence of realism in morals a little later.

The third positive aspect of this form of ethical intuitionism is designed to admit, what the utilitarian must deny, the possibility of genuine moral dilemmas, of genuine moral conflicts. In one standard version of ethical intuitionism, the direct object of intuition is The Good. If that is the direct object of intuition, it is difficult to see how genuine dilemmas can arise: just as for the utilitarian the only problem-case is that in which the utilitarian consequences of, say, two possible actions are more or less indistinguishable, so, for this kind of *monistic* intuitionist, the only problem-case is that in which each of two possible actions shares to the same extent the property of being good. Such cases are not true dilemmas. The version of intuitionism I want to consider does admit of such dilemmas by being *pluralistic*. For this version, there are *many* distinct ethical properties whose occurrence can be detected—sincerity, loyalty, honesty, and so on—and there is no reason *a priori* to assume that they cannot conflict, even, perhaps, in tortuous ways.

Such a pluralism seems to me a desirable feature in any plausible form of intuitionism (or of any other moral theory). First, there seems the brute fact that moral dilemmas do occur, and with rather painful frequency. Only moral laziness or moral blindness could hide this from us. Second, we only think of an action (or attitude or person) in terms of whether it is good or not *either* when we are being lazy *or* as a consequence of having thought about it in terms of other, more specific, terms of moral appraisal, so that *The Good* becomes an *indirect* object of moral judgment. Wittgenstein says that we only call a picture 'beautiful' when we cannot be bothered to think of anything more specific (or interesting) to say about it. The same is true of calling something 'good'. The interesting, basic terms of moral description are things like: 'sincere', 'loyal', 'compassionate', and so on. We have a grasp of each of these ideas independent of (indeed, determining) our grasp of 'good'. There is therefore no *a priori* reason to believe that, in their instantiations, these ideas cannot conflict in deep ways, nor any reason to believe that subsuming them under the (derivative) expression 'good' will dissolve such conflicts. Third, acknowledgement of pluralism with the consequent acknowledgment of genuine moral dilemmas may serve to free us from a naive conception of morality as a decision-procedure. Doubtless, our moral thought about the world is geared, in part, to helping us decide what to do. But there seems no reason to believe that that thought does, or acceptably could, yield a printout as to what we should do in every situation; there is certainly no reason to believe that there is one kind of decision-procedure, some one golden rule (Do the best!), that will determine, in any given state of affairs, what we should do. Pluralistic intuitionism, unlike utilitarianism, requires the abandonment of the false hope for such a procedure.

The ethical intuitionism that I am considering, then, is cognitively factual, austerely realistic, and pluralistic. Now for what it is not. First, it is no part of this intuitionism to suggest that we detect the moral aspects of a situation by means of some *special faculty* of the mind, the intuition. We detect moral aspects in the same way we detect (nearly all) other aspects: by looking and seeing. Any further claim, like that positing a distinctive faculty of ethical intuition, is a contribution to the unintelligible pseudo-psychology of the faculties of the mind. Second, contrary to a persistent strand in classical intuitionist thought, certainty plays no rcle in this form of intuitionism. This is a consequence of taking realism seriously. By the process of careful attention to the world, we can improve our moral beliefs about the world, make them more approximately true; by the same process, we can improve our practical understanding, our sensitivity to the presence of instances of the moral concepts that figure in these beliefs. But this process of attention to improve beliefs and understanding will go on without end; there is no reason to believe that we shall ever be justified in being certain that most of our moral beliefs are true, and no reason to believe that we shall ever be justified in being certain that we have now completely understood, any of the moral concepts occurring in these beliefs. Our moral language, like all the realistic parts of that language, transcends our present practical comprehensions in trying to grapple with an independent, indefinitely complex reality; only ignorance of that realism could prompt the hope for certainty.

2. Relativism and reason

We have, now, a picture of a realistic ethical intuitionism, together with some of the considerations that prompt its particular form. But is it defensible? More generally, what difficulties does this attempt to embrace moral discourse within semantic realism encounter?

One familiar set of difficulties, which merits a much more detailed discussion than I can give it here, is raised by the moral relativist. He begins by pointing to the 'observed' fact that moral judgments about some particular action, or about some particular kind of action, vary drastically even within a given community, let alone between different communities; he further points to the 'observed' fact that categories of moral assessment can vary drastically between communities. The utilitarian will often give a different moral assessment from that which (we) old-fashioned moralists give. The act of vengeance by the Sicilian may be deemed by us to be wrong, by his native community to be right. It may even be that the Sicilian has moral concepts that we lack, and vice versa, and claims to be able to 'detect' instances of these concepts that we cannot 'detect', and vice versa. Even if we can *roughly* translate the terms he uses in morally describing his action, that translation cannot be a good one, our grasp upon the relevant concept must be lacking, if we cannot see how that characterisation makes that action even *prima facie* morally desirable, let alone morally praiseworthy *tout court*. But there is no way of resolving such moral 'disputes', no way of ensuring that we can first be led to see things as the Sicilian sees them (or that he be led to see them as we see them), and no way of then deciding, *externally* from any such system of moral perception, which is the correct one. Each system is coherent in its own terms, and there is no external standpoint for viewing, for understanding, and deciding between such systems.

In such a situation, the relativist claims, there is no sense to the claim of objective, cognisable truth in moral judgments, and our intuitionist picture of them must be wrong. The nature of those judgments is, perhaps, closer than we realised to the nature of pictorial judgments discussed earlier. Moral judgments, too, are partly the result of conventions which could have been, and in some places are, otherwise. Moral judgments, too, cannot be called literally true or false. We do indeed, from our particular standpoint, see moral aspects of situations; but what we see is the result of sets of (perhaps unconscious) conventions, and has no independent objective reality.

The substantial, quite general worries the relativist raises are these: when do differences in judgments imply differences in concepts? And when do differences in judgments (failures of intersubjective agreement) compel us to abandon the view of the subject-matter of the judgments as objective? Both are familiar questions from the writings of the later Wittgenstein. I wish I had detailed answers to those questions; but even while lacking such answers, I think we can see that the relativist is prone to answer these questions in too simple and hasty a manner. His case is thus not proven.

First, the 'simple fact' of differences of moral judgment does not yet imply the falsity of moral realism. In moral judgments, as in others, people can, and do, make mistakes. What realism requires is that their errors be *explicable*—in realistic terms. It is not, for example, difficult to explain the erroneous moral judgments of many white (and some black) South Africans. Their perceptions are clouded by their desires and fears in just the way that many of our own factual judgments are clouded: *of course* my wife is completely faithful to me, *of course* my son is quite exceptionally academically talented! The popularity of Lamarck- ian genetic theory in Stalinist Russia, even though that theory is probably false, does not make us doubt the realistic reading of genetic claims, for we can see quite easily what blinded its proponents to its deficiencies; they fell victim to the perpetual human tendency of allowing something other than the evidence as to how things are to affect their beliefs about how things are. For the intuitionistic realist, this same pattern of explanation can be applied to moral errors.

Second, such errors can sometimes be corrected by rational consideration. Many proponents of Lamarckian theory came to see the error of their ways with the passage of time and discussion. Given reflection upon our own experiences and psychological characteristics, we can come to see the error of our moral views; we can come to change our moral views in an intelligibly nonarbitrary manner. We can come to see how that dreadful experience in the potting-shed, or that charismatic personality of the priest, has quite *irrelevantly* influenced our views upon the good and the right. If the errors of others are explicable, our own are discoverable.

Third, moral differences are exactly what a realist should expect. Moral con- cepts exhibit the characteristic of *semantic depth*. Starting from our grasp upon them through our knowledge of the austere truth-conditions of sentences contain- ing them, we have to struggle to improve our sensitivity to particular instantiations of them. This process proceeds without limit; at no point, for the realist, can we rest content with our present sensitivity in the application of these concepts. So at no point can we rest *secure* in all our present judgments involving these concepts. For the realist, moral language, like all other realistic areas of language, is trying to grapple, in ways which transcend our present practical capacities, with a recalcitrant world. Moral differences are no more surprising (or perturbing) to the kind of realism under discussion than are scientific, or historical, differences.

This leads to a fourth point against too hasty a relativism. The fact that people use different criteria in detecting the presence of an abstractly described moral feature does not yet show that they have literally different concepts. Their central

[287]

grasp upon that concept is grounded in the austere theory of sense for the term expressing that concept. That austere concept can be the same even if they use different procedures, yielding different results, in detecting instances of that feature. This must be so if genuine moral conflicts are to arise between people, genuine disputes, for example, as to what is a manifestation of generosity. Many of the differences the relativist points to are not differences in the concept of a particular vice or virtue, but differences of quasi-empirical view as to what counts as a manifestation of the virtue, as to how that virtue is to be detected.[3] Perhaps there is no substantial, non-verbal issue here; or perhaps there is a point where differences about the manifestations of a virtue require us to say that the disputants hold different concepts; but that point, if it exists, is neither as clear nor as quickly reached as the relativist suggests. The *semantic* depth of moral notions is reflected in our austere, realistic treatment of terms expressing them; it leaves quite unclear the (important) question as to *when* different concepts are to be attributed to different speakers. But for the realist, that is a desirable initial lack, not to be remedied in too simplistic a manner.

Fifth, the fact that others apparently lack moral concepts that we possess (or vice versa) no more shows realism to be false in moral matters than the fact that others lack scientific concepts we possess (or vice versa) shows realism in science to be false. What the realist requires is an explanation, in realistic terms, of that lack; and there is usually no shortage of these.

The relativist can, up to a point, concede the foregoing. He is likely to insist upon two points, one incidental, the other central. The incidental point he will insist upon is that no account has yet been given of *how* we can come to understand the (different) moral concepts of others, of how we can come to see things morally as they see them. We have no account, that is, of how we can truly understand different moral schemes of thought in the way that we can understand different scientific, or historical, systems of thought. And the central point he will insist upon is this: that process of comprehension, on any account, will issue in a free choice. Having once come to see things morally as an alien sees them, there is no procedure open to us for deciding which is the *true* way of seeing things. At such a reflective point, we have a *free* choice of adopting (or rejecting) the alien way rather than living with the (usually unthinking) choice originally made for us by those who taught us our initial moral scheme. Incidental as the first point is, it bears closely upon the second: for the anti-realist conceives of the process of understanding others' moral views as comparable to the (perhaps usually *practical*) process of being educated into a different set of *conventional* procedures for seeing things, and so views the state so achieved as itself merely conventional.

Once more, this position is less simple than it seems. Consider, first, the parallel with pictured faces. Having once seen the face, we are convinced that there is (non-literally) a face there to be seen. The failure of others to see that face

3. That is, many of the facts which motivate relativism do indeed point to different *conceptions* of a virtue; but this is not yet to show that different *concepts* are involved. 'Different conceptions *of* one thing' implies realism, not relativism.

does not make us doubt its 'presence' as a possible object of detection. And there are, of course, many ways in which we might try to draw another's *attention* to that 'presence'. In so doing, we demonstrate our conviction that there is something there to be seen; we do not attempt merely to encourage hallucinations. But the closer parallel for the realist is, perhaps, that with the detection of an ambiguity in a spatial figure. Having once seen that ambiguity, we are convinced that there is, literally, an ambiguity there to be seen. The failures of others to see that ambiguity does not make us doubt its presence, and there are many ways in which we might attempt to draw somebody's *attention* to that ambiguity. If we succeed, we succeed in making them see what was all along there to be seen.

This parallel with ambiguous figures is useful since the existence of ambiguities in spatial figures need not be in any clear sense a matter of convention, nor is the process of drawing an ambiguity to someone's attentions always correctly represented as a process of educating them into conventions of seeing. Its limitation is, however, clear: there is no sense to the thought that there is *one* correct way of seeing such a figure. Indeed, the best is to be capable of seeing all its ambiguities.

Still, the idea is left available to the realistic intuitionist of accounting for the process of understanding the ways others see things morally, and of explaining to them one's own way of seeing things, as a process of attention being drawn to features previously overlooked or misperceived. There is no guarantee that this will work; but, then, there is no guarantee that any realistic vocabulary can be taught. What the realist has to deny in the moral case to produce the requisite disanalogy to the example of perceiving spatial ambiguities is that in *radical* cases of moral difference the alien way of seeing things can be fully (perceptually) understood alongside one's own. Cases of misperception—of false claims that an action instantiates some virtue we recognise—are no problem for the realist; the important case is that where an alien claims to have a grasp upon a distinctive virtue that we cannot grasp. But, for the realist, if we cannot see it, then (tentatively) it is not there to be seen. People can have empty moral predicates (which they believe to be non-empty), just as they can have empty proper names (which they believe to be non-empty); trivially, no realistic account can be given of such empty moral predicates. Explanations have to be given of their error, as in the case of proper names; but there is no reason to believe that this cannot be realistically done.

The procedure for understanding another's moral view is that of leaving oneself open to his efforts to draw our attention to the (distinctive) features he claims to detect, perhaps by his engaging us in the practices he engages in. Usually, we shall come to see the difference as non-radical, as a difference of conceptions, not of concepts; but if that difference is radical—if we just cannot *see* what he is talking *about*—then the tentative conclusion is that he is in radical error—or that we are.[4] While if his efforts are successful, if we can indeed see things as he sees them, then our original way of seeing them must be rejected. It is an unargued-for

---

4. Thus, in aesthetics we may suppose that we are blind to the point of some alien art form.

dogma that we shall find ourselves in a position comparable to that in the case of spatial ambiguities, an unargued-for dogma that we shall find at least two radically different ways of seeing things with equal (un)reality.

At least two of the worries raised by the relativist can be raised quite independently of that position. The first is the thought that realism leaves no scope for moral discussion and reasoning; the second is that it neglects moral choice and moral responsibility.

The deficiencies in the first thought—that for the intuitionist you either see it or you don't, and that's the end of the matter—have already been partly exhibited. First, for the realistic intuitionist of the kind described, there need be nothing *obvious* about particular moral truths or about specific categories of moral assessment. Discussion with others, like self-reflection, may prompt the *attention* that is needed, both to focus upon particular moral aspects of a given case that would otherwise have been overlooked and to see instantiations of novel moral concepts of which we previously had no grasp. Likewise, discussion and critical reflection may force us to examine our views upon what constitutes a manifestation of a particular virtue, it may force us to think about why we consider various distinctive manifestations to be manifestations of the same (austerely characterised) virtue. Such discussion may induce reflection upon the ways in which manifestations of distinct virtues related to each other, and so make us see, indirectly, the relations between these virtues. And more concretely, moral discussion may lead us to see the ways in which our moral perceptions have been clouded by the inclusion of irrelevant attitudes. Examples of such considerations would concretise these abstract points; but it is anyway clear that the value of moral exchange is great for the species of intuitionism under consideration.

The second of the worrying thoughts, that about moral choice and moral responsibility, is less clear, and less worrying than it seems, partly because it is often muddled with a (mistaken) objection to moral realism to the effect that it cannot account for the (supposed) desirability of a *plurality* of moral views.

It is indeed a consequence of moral realism of the kind considered here that, in some sense, everybody *should* have the same moral beliefs, just as they should all have the same scientific, or historical, beliefs. But that *should* relates to a world in which we are cognitively more secure than is the actual world. The same considerations in this world favour a plurality of moral views as favour a plurality of scientific views: we are tawdry, inadequate epistemic creatures struggling with an indefinitely complex world, and the dialogue between competing—but *competing*—views may make us attend to features of that world which we would otherwise have overlooked. But what we might overlook are features *of the world*, not fictions of our own imagination. Moral pluralism is desirable, not for liberal-cum-aesthetic reasons, but for *epistemic* ones. Only the mistaken identification of the cognitive element in realism with the indefensible hope of moral *certainty* could blind us to this *consequence* of realism.

The point as regards moral responsibility is simple, and, in the context of this defence of pluralism, forceful: my moral beliefs, like all my other beliefs, are *mine*. When I act, or prompt others to act, upon any of my beliefs—my belief as to the best thing to do, my belief about the last train to London—I have to live with the consequences, and with the consequences for others. When expressing a belief, I present a claim as to *how the world is;* I do that as a tawdry epistemic being, and should do it in the light of that fact; if I have been negligent in attending to evidence, or have allowed considerations other than evidence to intrude, then I am indeed culpable. But that remains true of me for all my realistic beliefs and their expressions; it remains unclear why there should be any further role for a distinctive notion of responsibility for my *moral* beliefs.

Once these points about pluralism and responsibility are seen, it remains unclear what force there is to the demand that moral choice be further acknowledged. Only the assumption that moral judgments reflect desires, or some comparable attitude, together with the assumption that desires, or whatever attitude, are a matter of choice, seems to give any force to that demand. But, then, the realist simply denies the first of these assumptions. Perhaps there are further arguments for that assumption to be met; but the claim for moral choice seems to have no independent ground.

The earlier consideration of the scope for discussion and reason within moral intuitionism leaves one deep problem unresolved, a problem arising from the autonomy *vis-à-vis* non-moral facts accorded by intuitionism to moral judgments. The realistic intutionist holds that, while non-moral facts fix moral facts such that two circumstances cannot differ in a moral respect while being alike in all non-moral respects, still moral judgments are not analysable (or translatable) into non-moral terms; the making of a moral judgment is not an *inference* from non-moral facts. The problem now is that that picture appears to be in tension with the role usually accorded to non-moral differences in *accounting* for differences of moral judgment, accounting in a *reason*-giving way. If I make different moral judgments about situations that appear indistinguishable to you, then, the thought is, I have to *justify* that difference by pointing to a non-moral difference, I have to give a non-moral *reason* for the difference in moral judgment. Indeed, this non-moral reason giving is the foundation of moral *consistency:* such consistency precisely requires (because is constituted by) the principle that if two situations are non-morally indistinguishable, we have to give the same moral judgment in each case, together with the principle that if a difference in moral judgment is given it has to be justified by a non-moral reason.[5]

As said, the first of these principles is, as stated, quite compatible with realism: it is a simple consequence of the claim that non-moral facts *fix* moral facts. The problem is with the second principle, with accounting for the role of non-moral reasons in justifying differences of moral judgment; for this role seems simply

5. Simon Blackburn, ''Moral Realism,'' in J. Casey, ed., *Morality and Moral Reasoning* (London, 1971), pp. 101–24.

[291]

incompatible with the autonomy claimed for moral judgments, seems simply incompatible, for example, with the denial of the claim that moral judgments are *inferences* from non-moral judgments.[6]

Still, two points suggest that this objection is less straightforwardly successful than it seems. First, the proper statement of the thesis of autonomy is in terms of the austere, realistic non-decompositional treatment given to moral terms in our theory of sense for moral language. Further expressions of that thesis, or of theses claimed equivalent to it, in terms of the relations between moral and non-moral facts require that content be given to the *contrast* between moral and non-moral facts. Until that contrast is, independently and defensibly, explicated, it is open to the realist to reply to the argument from moral reasons in a simple way: if the giving of a reason intelligibly accounts for a difference in moral judgment, it is itself a *moral* reason. How could it play that role if it were not? Nor, for the realist, is there anything puzzling about the role of those, as it were, lower level, more concrete moral considerations in justifying moral judgments involving austerely characterised moral concepts. Those austerely characterised moral concepts, like any other austerely characterised concepts, have, in judgments about particular cases, to be *applied* to these cases; it is always a reasonable request to ask what criteria of application we are using. Those are indicated by the reasons we give for our judgments. But the considerations so adduced, if reasons for the application of a moral notion, must themselves be *moral* considerations; they need not, however, be suited for incorporation in non-austere analyses of the abstract moral concepts involved.

Second, even if an acceptable content is given to the distinction between moral and non-moral considerations, and even if the thesis of autonomy is then re-expressed in such terms, the force of the objection from moral consistency and moral reasoning remains unclear: for it remains unclear that that notion of consistency and that notion of reasoning have acceptable roles to play. The ideas behind the conceived roles of these notions appear to be these: first, in deciding what moral judgment about a particular situation is correct, we need to consider our judgments about situations encountered in the past; and, second, that what should *ground* our moral judgments about a specific situation is some rule, or set of rules, in which moral and non-moral terms are connected in such a way that the applicability of the latter determines the applicability of the former. But both these ideas are counsels of moral laziness, are counsels of the neglect of the *particular* situation, in all its (non-obvious) complexity, at present before us. In ordinary moral life, the problem is not that of squaring our present judgments with our previous judgments, but that of *attending* to the full, unobvious moral complexity of the present case. In ordinary moral life, determining our moral judgment about

6. Although this appearance of incompatibility may only arise from too crude a view of the relations between *reasons* and *inference*. Why cannot we acknowledge the role of non-moral statements as reasons for moral judgments while denying the inferred character of moral judgments?

a particular case by means of some rule seizing upon non-moral aspects of that case will simply mean that we neglect the full complexity of that particular case.

Doubtless, for the pressing purposes of everyday life, considerations of consistency with previous judgments on the basis of crudely observed similarities and of rules for making specified kinds of judgments in crudely described states of affairs are useful, perhaps indispensable, *rules of thumb;* but for the realist, they are no more than that. In ordinary life, moral situations do not repeat themselves; only insensitivity can suggest that they do. The objection to moral realism starts, not just from the practical exigencies of ordinary moral life, but from a supposed feature of that moral life; but that feature does not exist.

3. Morality and action

The discussion so far of moral judgments has one curious lack: any consideration of the relation between moral judgments and *action*. Reflection upon that lack prompts a further series of objections to the kind of intuitionistic realism here expounded.

Moral judgments apparently connect with action in two ways: first, a moral judgment always purports to give a *reason for action;* and, second, assessment of the *sincerity* of one who makes a moral judgment is determined by seeing how he *acts*. Both of these points might be thought quite inexplicable upon the intuitionistic view of moral judgments.

The worry about reasons for actions, connecting with the earlier worry about moral choice, is this: any full specification of a reason for action, if it is to be a reason *for the potential agent* for action, must make reference to that agent's *desires;* moral judgments always purport to be (sufficiently) full specifications of at least *prima facie* reasons for the agent to act; so moral judgments must include (perhaps, within the antecedent of a conditional) reference to the potential agent's desires. It is thus misleading to assimilate them to straightforward factual descriptions of the world which make no such references. Whether a moral 'description' of a case is 'true' or not depends upon the desires of the person considering it; those desires are, or admissibly can be, a matter of choice; so, therefore, are the moral 'descriptions' given, there being in consequence no question about those 'descriptions' being objectively true or false.

The crucial premiss in this argument is the claim that any full specification of a reason for an action, if it is to be a reason for the potential agent for action, must make reference to that agent's desires. At first sight, it seems a painful feature of the moral life that this premiss is false. We perform many intentional actions in that life that we apparently do not desire to perform. A better description of such cases appears to be that we perform them because we think them desirable. The difficulty of much of moral life then emerges as a consequence of the apparent fact

[293]

that desiring something and thinking it desirable are both distinct and independent.

The premiss can, of course, be held true by simply claiming that, when acting because we think something desirable, we do indeed desire it. But this is either phenomenologically false, there being nothing in our inner life corresponding to the posited desire, or utterly vacuous, neither content nor motivation being given to the positing of the desire. Nothing but muddle (and boredom) comes from treating desire as a mental-catch-all.[7]

There is a weaker, more abstract claim, difficult to state in non-metaphorical terms, which perhaps underlies the premiss about desire just discussed, and which is still incompatible with the realist position developed. Miss Anscombe, in her work on intention, has drawn a broad distinction between two *kinds* of mental state, factual belief being the prime exemplar of one kind and desire a prime exemplar of the other.[8] The distinction is in terms of the *direction of fit* of mental states with the world. Beliefs aim at the true, and their being true is their fitting the world; falsity is a decisive failing in a belief, and false beliefs should be discarded; beliefs should be changed to fit with the world, not vice versa. Desires aim at realisation, and their realisation is the world fitting with them; the fact that the indicative content of a desire is not realised in the world is not yet a failing *in the desire,* and not yet any reason to discard the desire; the world, crudely, should be changed to fit with our desires, not vice versa. I wish I could substitute a less picturesque idiom for that of *direction of fit,* but I cannot. I wish also I were clearer as to whether there are *any* mental states for which the direction of fit is purely of the second kind; desires seem not to be such a candidate, since all desires appear to involve elements of belief. Still, the picturesque distinction may, for present purposes, suffice; and all these purposes require is a conception of factual belief which is purely of the first kind *vis-à-vis* direction of fit.

The point now is that the realist treats moral judgments as being pure members of the first cognitive category of mental states, and that the anti-realist claims that any full specification of a reason for action must make reference to the (potential) agent's mental states of the second category. If thinking something desirable is to be a reason for doing it, then that notion cannot, contrary to the realist's view, be assimilated to pure factual beliefs. Such an assimilation divorces moral judgments from reasons for action.

The realist has two, ultimately connected, lines of response to this challenge. One focuses upon the fact that the realist's assimilation of moral judgments to purely factual beliefs *vis-à-vis* direction of fit with the world is an assimilation only in point of literal sense. The other line of response is simply to demand argument for the general point about reasons for actions made by the anti-realist.

---

7. Cf. M. J. Woods, "Reasons for Actions and Desires," *Proceedings of the Aristotelian Society,* supp. vol. 46 (1972), 189–201.

8. G. E. M. Anscombe, *Intention* (Oxford, 1963), §2.

Both lines of response, like the possible anti-realist counter-responses, are indecisive; I cannot yet see how the issue is to be resolved.

When we express mundane factual beliefs, we often, in doing so, also express, as it were indirectly, other mental attitudes of a less cognitive kind; but these other attitudes do not enter into the literal truth-conditions of what is said. It is open to the realist to hold the same to be true of expressions of moral beliefs: such expression is often, perhaps standardly, also an indirect expression of other attitudes or mental states of a less cognitive kind, of moral *sentiments*. But, first, it is not easy to produce a general, plausible, non-vacuous description of these standard accompaniments of moral expression. And, second, even if such a description of the standard role of moral sentiments can be given, they need not enter into the literal truth-conditions of what is said; the realist thought is thus that the appearance of an objectionable element only arises through the neglect of the crucial distinction between the literal sense of an expression and the matter of what is done upon occasions of utterance of it. For the realist, these further elements cannot affect the basic mode of moral utterance, which must remain that of *assertion;* but that does not mean that these further elements have no role to play in a full account of what is *done,* standardly, by the assertoric utterance of a moral sentence.

One trouble with this concessive response is that it incurs an obligation to explain how this *standard* connection with expressions of other non-cognitive attitudes comes about. It does not *seem* very satisfactory to hold that the obtaining of this connection is simply an inexplicable feature of moral discourse, an (otherwise) inexplicable consequence of the *sui generis* importance of moral facts; but nor does it *seem* very satisfactory to withhold the initial concession by claiming that the connection misleadingly labelled 'standard' is simply a chance, contingent fact. The last claim would require, to say the least, a very sophisticated account of the phenomenology of moral language use to dispel the 'illusion' of a critical connection between moral judgments and moral sentiments.

The other line of realist defence is to demand argument for the claim that any full specification of a reason for action must make reference to a mental state of the second, non-cognitive kind *vis-à-vis* direction of fit with the world. Why should it not just be a brute fact about moral facts that, without any such further element entering, their clear perception does provide sufficient grounding for action? Two anti-realist lines of thought lead nowhere. One is that there must be such a further element to *impel* action; but this contribution to a pseudo-psychology grinds, or should grind, to a quick halt. The other is to claim that there must be such a further element because any list of sufficient conditions for an action being intentional must include the attribution to the agent of such a mental state. Here, the halt is much slower, deriving from a simple point: since no one has presented a truly sufficient set of conditions for an action to be intentional, it must remain quite undecided whether any given putatively necessary condition is indeed a necessary condition; it might well be that some *further* condition, anyway required for

[295]

sufficiency, will reveal the putatively necessary condition to be unnecessary. If it is replied that this could only occur by the new, sufficiency-achieving condition incorporating the previous necessary condition, then the realist has two rejoinders. First, this assumes that *any* set of sufficient conditions for intentional action must include *every* element which within any one set of sufficient conditions is necessary. Doubtless, for *many* kinds of action reference to some 'pro-attitude' or whatever is necessary in giving sufficient conditions for the action being intentional; it does not yet follow that it is a necessary condition in *any* set of sufficient conditions. And second, the realist claims to have a counter-example to the claim that it does follow: viz. the fact that clear moral perceptions are sufficient as reasons for actions. It is utterly unclear how the claim that moral perceptions have this *sui generis* feature is to be settled one way or the other.

These realist manoeuvres are also available when the matter of *sincerity* of moral judgment is raised. The anti-realist thought is that there is a distinctive notion of sincerity involved in expression of moral 'beliefs' since such sincerity is assessed by seeing what the person does in a way that is different from the assessment of sincerity in expressions of ordinary factual beliefs.

To this the realist can reply, first, that the *sui generis* feature of moral perceptions just relied upon in meeting the worry about reasons for actions does indeed import the requisite connection between sincerity and action. If a person does indeed see the moral nature of a situation clearly, and does indeed express that perception sincerely, then what he does will, subject to later qualifications, accord with what he says; at least, it will accord with what he says to the degree that such accord is plausibly required in deeming his expression sincere. For that clear moral perception, *ex hypothesi* presents sufficient reason for him to act (which is not yet to say that he *will* so act). Second, the appearance of a distinctive notion of sincerity in expression of moral judgments may only arise upon a mistaken view of the notion of sincerity in expression of ordinary, factual beliefs. We tend to treat this latter notion as being a more derivative one than it is. Starting from the identification of sincerity with *truthfulness,* we focus upon the notion of *truth* occurring there, and interpret it realistically: we think of there being two, quite independent elements, the inner, private mental states of the person, and what he says *about* those states, as his *expression* of those states; we then think of sincerity, of truthfulness, as being the result of the latter *fitting* the former. There is then no question of what the person does as opposed to what he *says* determining his sincerity. But there is another way of seeing the elements so described. Starting from what he says *and* does, we invoke the (non-derivative) notion of sincerity in 'constructing' his inner, private life, we gain access to that life by seeing what he says *sincerely* and what he does *sincerely,* where this notion of *doing sincerely* cannot *ex hypothesi* be grounded in the way the first picture suggests. How, if at all, it can be 'informatively' grounded is another question; and how, if at all, he is in a superior position to us in his access to his inner life is another question too. But what is clear is that the notion of sincerity we have acquired has its core, and must have its core, in the second, not the first, picture. And what is also clear is that in this alternative picture what a person does can affect our view of his sincerity in the expression of his ordinary, factual beliefs. Of course, there may still be differences between the factors that guide our construction of the inner mundanely factual beliefs and those that guide our construction of the inner moral beliefs, as there may be differences in the analysis of an intentional action in which the one plays a part and in the analysis of an intentional action in which the other plays a part. But the appearance of radically different notions of sincerity— radically different in focusing on or shunning non-verbal actions—is not obviously correct.

It is anyway very difficult to state in a plausible form the connection between sincerity of moral expression and action. One idea is that such expression is sincere if, and only if, in a suitable circumstance the moral agent concerned acts accordingly. But this is far too strong. While lying on a beach, we see a young child playing in the sea; suddenly, we notice a swarm of sharks approaching; heroically, you plunge into the sea to save the child, risking death in the process. I describe your action in the most flattering moral terms. The next day, on the beach when alone, I see a similar incident happening; being less heroic, I do not plunge into the sea. Does that show that I was *insincere* the previous day when expressing my moral beliefs about your action? Of course not—and not because (or not just because) the notion of moral consistency phrased in terms of similar circumstances is sufficiently vague to leave me with sufficient logical leeway to escape the charge, but because (or also because) whatever the relation is between moral sincerity and action, it does not imply that I avoid the charge of insincerity only by being *heroic*.[9] It is no part of the moral life that insincerity is avoided only by being saintly or heroic; our notion of sincerity is far more realistically connected with our view of *what people are like*. Here there is a substantial problem for *any* ethical theory in accounting for the relations between sincerity and action; but it is not a problem specific to the ethical realist, and there is no reason to believe that he is less well equipped to solve it than any other theorist.

There are two further, importantly distinct questions about the realist view of the relation between moral judgment and action. One is this: when will an agent act upon his moral judgments? It is a mistake to say: when, and only when, those judgments are sincere expressions of what he believes to be clear moral perceptions. The important mistake is not the neglect of actions based on (knowingly) cloudy perceptions, nor the naivety of the assumptions about the saintly and the heroic built into this answer, but is rather the belief that there is *any* useful answer to this question. Agents act upon their judgments when, and only when, they act upon them; beyond that, nothing useful can be said. Saying that they act upon their judgments only when they decide to act upon them, or when they *will* that

9. J. O. Urmson, "Saints and Heroes," in J. Feinberg, ed., *Moral Concepts* (Oxford, 1969), pp. 60–73.

they act upon them, is either vacuous or false. The question in the philosophy of action should not be: when does an agent act upon his judgments? It should rather be: when he has acted intentionally upon his judgments, what is the correct analysis of his having done so? An answer to the latter question will not give the *kind* of pseudo-predictive answer the first question requires; for there is no such answer.

But there is a second, more substantial question. How can the realist position, as developed here, account for the possibility of weakness of will? That is, how, on this account, is it possible for an agent knowingly to perform an intentional action against his best moral judgment? This is quite different from a question tantamount to the first, namely, how is it possible for an agent not to act in accordance with his best moral judgment?

The puzzle about weakness of the will, so-called, is this. Aside from genuine moral dilemmas, clear perception of the moral character of a situation gives, we have said, sufficient reason for action—which is not yet to say that action will ensue. The distinctive feature of clear moral perception is that it gives us a *compelling* reason to act. There are many reasons why action might not follow; but the puzzle misdescribed (for the realist) as weakness of will is how it can be that intentional action does follow, action done for a reason, which conflicts with the action suggested by that *compelling* moral perception. What could 'compelling' mean here if not that if the agent does *any* intentional action, he will do that action?

Two connected peripheral matters should first be mentioned to be put aside. First, it need not be relevant in this context to ask why moral considerations are overriding, if they are. The case to focus upon is that in which the akratic action performed is one which has no very substantial non-moral reason behind it. Second, cases where strong passions are involved are not the most puzzling ones; for in such cases, it is plausible to say that at the moment of action passion clouds the agent's perception. He simply does not see at the point of action the relevant moral aspects. The really puzzling case would satisfy the following description: in a cool, detached frame of mind the agent performs an action with no great point that does not accord with his clear moral perception of the situation.

One might question whether there are such cases, whether this could ever be a correct description of a case, but I shall not now consider this. Instead, I shall suggest that the *semantic depth* of moral concepts is the key to the dissolution of this puzzle.

Moral concepts have a kind of semantic depth. Starting from our austere grasp upon these concepts, together perhaps with some practical grasp upon the conditions of their application, we can proceed to investigate, to experience, the features of the real world answering to these concepts.[10] Precisely because of the

---

10. These, and the ensuing remarks, are meant as preliminary ruminations about the theory of moral language implicit in Miss Murdoch's brilliantly thought-provoking *The Sovereignty of Good;* the detailed development of that theory of language seems to me a matter of great importance.

realistic account given of these concepts and of our grasp upon them—precisely because they are designed to pick out features of the world of indefinite complexity in ways that transcend our practical understanding—this process of investigation through experience can, and should, proceed without end. Our grasp upon what, say, *courage* is can, and should, improve without limit; we must rest content with the thought that at death *approximate* understanding is all that we can hope for. But all along we have a grasp of what the concept is, as manifested by our grasp upon austere T-sentences involving it; and perhaps, all along we have a grasp upon some *gloss,* some *dictionary definition,* of the term picking out that concept. But for the realist this austere grasp, that knowledge of the dictionary, is the beginning of understanding, not the end; there is, for us, no end, yet that starting point is far indeed from it.[11] Just the same could be said, by the realist, about scientific concepts.

This invites the following picture of the most puzzling kind of akratic action—*only* a picture, but that is a persistent feature in this area. At least at the moment of action, the akratic's perception is not *cloudy* but shallow: the concept he is then employing in his moral perception is the skeleton, austere concept, the shallow, dictionary defined concept, not the concept fleshed out by years of experience. He regresses to an earlier point in his conceptual development—*if* he ever moved beyond that point. He regresses to a *formal, non-experiential* understanding of the moral notions involved. He becomes like a Martian who translates our dictionary but has had no *experience* of our moral world. For a moment, he is morally blind: he has to be, for if he saw the moral reality through an experientially enriched concept, he could not act as he does. He has forgotten all that experience has taught him, all that gives moral concepts life; he is like the man whose perception of beauty has been jaded to the point of mere encyclopaedic knowledge. He sees but he does not feel; and he does not feel because he does not *see* sufficiently. Morality, for him, is a dead language.

Such thoughts prompt another challenge to realism, the challenge of undue optimism. The manner in which akrasia is made intelligible apparently makes genuine evil unintelligible. A genuinely evil action, on this anti-realist view, is one done in full, vivid, deep knowledge of its evil character. The genuinely evil man is not the nihilist, with only dictionary knowledge of moral concepts (if that), nor the man who allows non-moral considerations to *override* the moral, nor the man who makes errors of moral judgment: it is the man who knowingly does the evil precisely *because* it is evil, and whose knowledge of its evil character is as deep as ours. It is just that he *chooses* to be evil. For the realist, this is simply not possible.

It is not, however, obvious that the realist is wrong in denying this possibility. First, most such cases as readily fit some other description—the nihilist, the shallow, the overrider, the erroneous—as they do the anti-realist's description.

11. Experience can enrich our conception of what, say, courage is; our concept can meanwhile remain the same.

And, second, this is supported by the ways in which we make the evil actions of others intelligible to ourselves. We see them as seduced by the trappings of evil (the devil has a monopoly of elegance); we see them as desiring to shock, to present an (excessively) rakish image to the world; we see them as insanely unintelligible. We do not pretend to see how the fact that an action is evil is ever a *prima facie* reason for doing it: we hunt for one, intelligible reason, and when the hunt fails, we plead incomprehension—and invite the agent to plead insanity.

'I did it because it was the loyal thing to do' may seem erroneous both in detail and in principle, but is at least intelligible. 'I did it because it was the disloyal thing to do' is unintelligible until *further* reasons are adduced. This linguistic point reflects a fact of moral reality: that seeing the good compels action (without ensuring it), while seeing the bad repels it (without preventing it). The final substantial challenge to moral realism comes in the form of a demand for an explanation of this posited fact; I am unclear what explanation can be given of it.

'Human language is like a cracked kettle on which we beat out tunes for bears to dance to, when all the time we are longing to move the stars to pity.' Flaubert's thought is not a novel one, but is as profound as any utterance of a philosopher. In this chapter, I have tried to show how the [realist] view of language previously sketched will lead us, finally, from the fixation with kettles to wonder at the stars. I have tried to show one way in which all the technicalities earlier discussed can bring us closer to an understanding of the profound questions that contemporary philosophy of language so often appears to neglect. I do not know whether the way considered, the way of realism, will ultimately lead to a true understanding of those questions; I do think that that way is deeply attractive and of the very greatest importance.

Bibliography

Aiken, Henry David. *Reason and Conduct.* New York: Alfred A. Knopf, 1962.

Alston, William. "Meta-Ethics and Meta-Epistemology." In A. I. Goldman and Jaegwon Kim, eds., *Values and Morals.* Dordrecht: Reidel, 1978, pp. 275–97.

Ayer, A. J. *Language, Truth and Logic.* New York: Dover, 1952.

Baier, Kurt. *The Moral Point of View.* Ithaca: Cornell University Press, 1958.

Baldwin, Thomas. "Ethical Non-Naturalism." In Ian Hacking, ed., *Exercises in Analysis.* Cambridge: Cambridge University Press, 1985, pp. 23–46.

Becker, Lawrence. *On Justifying Moral Judgments.* New York: Humanities Press, 1973.

Blackburn, Simon. "Moral Realism." In John Casey, ed., *Morality and Moral Reasoning.* London: Methuen, 1971, pp. 101–24.

——. "Rule-Following and Moral Realism." In Steven Holtzman and Christopher M. Leich, eds., *Wittgenstein: To Follow a Rule.* London: Routledge and Kegan Paul, 1981, pp. 163–87.

——. *Spreading the Word.* New York: Oxford University Press, 1984.

——. "Errors and the Phenomenology of Value." In Ted Honderich, ed., *Morality and Objectivity.* London: Routledge and Kegan Paul, 1985, pp. 1–22.

Bond, E. J. *Reason and Value.* Cambridge: Cambridge University Press, 1983.

Bradie, Michael. "Rationality and the Objectivity of Values," *Monist* 67 (1984), 467–82.

Brambough, Renford. *Moral Skepticism and Moral Knowledge.* Atlantic Highlands, N.J.: Humanities Press, 1979.

Brandt, R. B. *Ethical Theory.* Englewood Cliffs, N.J.: Prentice-Hall, 1959.

——. *A Theory of the Good and the Right.* New York: Oxford University Press, 1979.

Brink, David O. "Moral Realism and the Sceptical Arguments from Disagreement and Queerness," *Australasian Journal of Philosophy* 62 (1984), 111–25.

——. "Internalism and Moral Realism," *Spindel Conference: Moral Realism, Southern Journal of Philosophy* 24 (1986), Supplement, 23–42.

——. *Moral Realism and the Foundations of Ethics.* Cambridge: Cambridge University Press, forthcoming.

Broad, C. D. "Is 'Goodness' a Name of a Simple Non-Natural Quality?" *Proceedings of the Aristotelian Society* 34 (1933–34), 249–68.

——. "Some Reflections on Moral-Sense Theories in Ethics," *Proceedings of the Aristotelian Society* 45 (1944–45), 131–66.

Brody, Baruch. "Intuitions and Objective Moral Knowledge," *Monist* 62 (1979), 446–56.

Brown, Erik. "Sympathy and Moral Objectivity," *American Philosophical Quarterly* 23 (1986), 179–88.

Brown, S. C., ed. *Objectivity and Cultural Divergence.* Cambridge: Cambridge University Press, 1984.

Carson, Thomas. *The Status of Morality.* Dordrecht: Reidel, 1984.

Coburn, Robert. "Relativism and the Basis of Morality," *Philosophical Review* 85 (1976), 87–93.
——. "Morality, Truth and Relativism," *Ethics* 92 (1982), 661–69.
Cohen, M. F. "Knowledge and Moral Belief," *Australasian Journal of Philosophy* 42 (1965), 168–88.
Cohon, Rachel. "Are External Reasons Impossible?" *Ethics* 96 (1986), 545–56.
Conly, Sarah. "The Objectivity of Morals and the Subjectivity of Agents," *American Philosophical Quarterly* 22 (1985), 275–86.
Cooper, David E. "Moral Relativism," *Midwest Studies*. Morris: University of Minnesota Press, 1978, 3:97–108.
Copp, David, and David Zimmerman, eds. *Morality, Reason and Truth*. Totowa, N.J.: Rowman and Allanheld, 1985.
Dancy, Jonathan. "On Moral Properties," *Mind* 90 (1981), 367–85.
——. "Ethical Particularism and Morally Relevant Properties," *Mind* 92 (1983), 530–47.
Daniels, Norman. "Wide Reflective Equilibrium and Theory Acceptance in Ethics," *Journal of Philosophy* 76 (1979), 256–82.
——. "Reflective Equilibrium and Archimedian Points," *Canadian Journal of Philosophy* 10 (1980), 83–103.
——. "Some Methods of Ethics and Linguistics," *Philosophical Studies* 37 (1980), 21–36.
Darwall, Stephen L. *Impartial Reason*. Ithaca: Cornell University Press, 1983.
Donagan, Alan. *The Theory of Morality*. Chicago: University of Chicago Press, 1977.
Dworkin, Ronald. *Taking Rights Seriously*. London: Duckworth, 1978.
——. *A Matter of Principle*. Cambridge, Mass.: Harvard University Press, 1985.
Ewing, A. C. *The Definition of Good*. New York: Macmillan, 1947.
——. *Ethics*. London: English Universities Press, 1962.
Falk, W. D. *Ought, Reasons, and Morality*. Ithaca: Cornell University Press, 1986.
Findlay, J. N. *Language, Mind and Value*. London: George Allen and Unwin, 1963.
Finnis, John. *Fundamentals of Ethics*. Washington, D.C.: Georgetown University Press, 1983.
Firth, Roderick. "Ethical Absolutism and the Ideal Observer," *Philosophy and Phenomenological Research* 12 (1952), 317–45.
Foot, Philippa. "Moral Arguments," *Mind* 67 (1958), 502–13.
——. "Moral Beliefs," *Proceedings of the Aristotelian Soceity* 59 (1958–59), 83–104.
——. *Virtues and Vices*. Los Angeles: University of California Press, 1978.
——. *Moral Relativism*. The Lindley Lectures. Lawrence: University of Kansas, 1978.
——. "Moral Realism and Moral Dilemma," *Journal of Philosophy* 80 (1983), 379–98.
Forester, Mary Gore. *Moral Language*. Madison: University of Wisconsin Press, 1982.
Frankena, William. "The Naturalistic Fallacy," *Mind* 48 (1939), 464–77.
——. "Obligation and Value in the Ethics of G. E. Moore." In P. A. Schlipp, ed., *The Philosophy of G. E. Moore*. LaSalle, Ill.: Open Court, 1942, pp. 91–110.
——. "Ethical Naturalism Renovated," *Review of Metaphysics* 10 (1957), 459–73.
——. "Obligation and Motivation in Recent Moral Philosophy." In A. I. Melden, ed., *Essays on Moral Philosophy*. Seattle: University of Washington Press, 1958, pp. 40–81.
Gauthier, David. *Morals by Agreement*. Oxford: Oxford University Press, 1985.
Gay, Robert. "Ethical Pluralism: A Reply to Dancy," *Mind* 94 (1985), 250–62.
Gewirth, Alan. "Positive 'Ethics' and Normative 'Science'," *Philosophical Review* 69 (1960), 311–30.
——. *Reason and Morality*. Chicago: University of Chicago Press, 1978.
Gibbard, Allan. "Moral Judgment and the Acceptance of Norms," *Ethics* 96 (1985), 5–21.

——. "Reply to Sturgeon," *Ethics* 96 (1985), 34–41.

Grice, Geoffrey Russell. *The Grounds of Moral Judgement*. Cambridge: Cambridge University Press, 1967.

Guttenplan, Samuel. "Moral Realism and Moral Dilemmas," *Proceedings of the Aristotelian Society* 80 (1979–80), 61–80.

——. "Hume and Contemporary Ethical Naturalism," *Midwest Studies*. Morris: University of Minnesota Press, 1983, 8:309–20.

Hall, Everett. *What Is Value?* New York: Humanities Press, 1952.

Hare, R. M. *The Language of Morals*. New York: Oxford University Press, 1952.

——. *Freedom and Reason*. New York: Oxford University Press, 1963.

——. "Some Questions about Subjectivity." In *Freedom and Morality*. The Lindley Lectures. Lawrence: University of Kansas Press, 1976, pp. 191–208.

——. *Moral Thinking*. New York: Oxford University Press, 1981.

——. "Supervenience," *Aristotelian Society, Supplementary Volume* 58 (1984), pp. 1–16.

——. "Ontology in Ethics." In Ted Honderich, ed., *Morality and Objectivity*. London: Routledge and Kegan Paul, 1985, pp. 39–53.

Harman, Gilbert. "Moral Relativism Defended," *Philosophical Review* 84 (1975), 3–22.

——. *The Nature of Morality*. New York: Oxford University Press, 1977.

——. "Relativistic Ethics: Morality as Politics." *Midwest Studies*. Morris: University of Minnesota Press, 1978, 3:109–21.

——. "What Is Moral Relativism?" In A. I. Goldman and Jaegwon Kim, eds., *Values and Morals*. Dordrecht: Reidel, 1978, pp. 143–61.

——. "Metaphysical Realism and Moral Relativism," *Journal of Philosophy* 79 (1982), 568–75.

——. "Human Flourishing, Ethics, and Liberty," *Philosophy and Public Affairs* 12 (1983), 307–22.

——. "Justice and Moral Bargaining," *Social Philosophy and Policy* 1 (1983), 114–31.

——. "Is There a Single True Morality?" In David Copp and David Zimmerman, eds., *Morality, Reason and Truth*. Totowa, N.J.: Rowman and Allanheld, 1985, pp. 27–48.

Harrison, Jonathan. *Our Knowledge of Right and Wrong*. New York: Humanities Press, 1971.

——. *Hume's Moral Epistemology*. Oxford: Oxford University Press, 1976.

——. "Mackie's Moral 'Scepticism'," *Philosophy* 57 (1982), 173–91.

Honderich, Ted, ed. *Morality and Objectivity*. London: Routledge and Kegan Paul, 1985.

Hudson, W. D., ed. *The Is-Ought Question*. London: Macmillan, 1969.

Hurley, S. L. "Objectivity and Disagreement." In Ted Honderich, ed., *Morality and Objectivity*. London: Routledge and Kegan Paul, 1985, pp. 54–97.

Jensen, Henning. "Gilbert Harman's Defense of Moral Relativism," *Philosophical Studies* 30 (1976), 401–8.

——. "Hume on Moral Agreement," *Mind* 86 (1977), 497–513.

Kim, Jaegwon, and A. I. Goldman, eds. *Values and Morals*. Dordrecht: Reidel, 1978.

Klagge, James. "An Allied Difficulty Concerning Moral Properties," *Mind* 93 (1984), 370–80.

——. "Supervenience: Perspectives vs. Possible Worlds," *Philosophical Quarterly* (forthcoming, 1988).

Kolnai, Aurel. *Ethics, Value and Reality*. Indianapolis: Hackett, 1978.

Krausz, Michael, and Jack Meiland, eds. *Relativism: Cognitive and Moral*. Notre Dame: University of Notre Dame Press, 1982.

Kupperman, J. J. *Ethical Knowledge*. New York: Humanities Press, 1970.

——. "Moral Objectivity." In S. C. Brown, ed., *Objectivity and Cultural Divergence*. Cambridge: Cambridge University Press, 1984.

Bibliography

Lewy, Casimir. "G. E. Moore on the Naturalistic Fallacy," *Proceedings of the British Academy* 50 (1964), 251–62.

Lovibond, Sabina. *Realism and Imagination in Ethics.* Minneapolis: University of Minnesota Press, 1983.

Lyons, David. "Ethical Relativism and the Problem of Incoherence," *Ethics* 86 (1976), 107–21.

McCloskey, H. J. *Meta-Ethics and Normative Ethics.* The Hague: Martinus Nijhoff, 1969.

McConnell, Terrance. "Metaethical Principles, Meta-Prescriptions, and Moral Theories," *American Philosophical Quarterly* 22 (1985), 299–309.

McDowell, John. "Are Moral Requirements Hypothetical Imperatives?" *Proceedings of the Aristotelian Society, Supplementary Volume* 52 (1978), 13–29.

———. "Virtue and Reason," *Monist* 62 (1979), 331–50.

———. "Noncognitivism and Rule Following." In Steven H. Holtzman and Christopher M. Leich, eds., *Wittgenstein: To Follow a Rule.* London: Routledge and Kegan Paul, 1981, pp. 141–62.

Machan, Tibor. "Epistemology and Moral Knowledge," *Review of Metaphysics* 36 (1982), 23–49.

MacIntyre, Alasdair. *After Virtue.* Notre Dame: University of Notre Dame Press, 1981.

MacKenzie, J. C. "Moral Scepticism and Moral Conduct," *Philosophy* 59 (1984), 473–79.

Mackie, J. L. "A Refutation of Morals," *Australasian Journal of Philosophy* 24 (1946), 77–90.

———. *Ethics: Inventing Right and Wrong.* New York: Penguin, 1977.

———. *Hume's Moral Theory.* Boston: Routledge and Kegan Paul, 1980.

———. "Anti-Realisms." In *Logic and Knowledge.* Oxford: Oxford University Press, 1985, pp. 225–45.

———. *Persons and Values.* New York: Oxford University Press, 1985.

Marcus, R. B. "Moral Dilemmas and Consistency," *Journal of Philosophy* 77 (1980), 121–36.

Melden, A. I. "On the Method of Ethics," *Journal of Philosophy* 75 (1948), 169–81.

———. "Reasons for Action and Matters of Fact," *Proceedings of the American Philosophical Association* (Pacific Division) 35 (1962), 46–60.

Miller, Richard. "Ways of Moral Learning," *Philosophical Review* 94 (1985), 507–56.

Montague, Phillip. "On the Relation of Natural Properties to Normative and Evaluative Properties," *Philosophy and Phenomenological Research* 35 (1975), 341–51.

Moore, G. E. *Principia Ethica.* New York: Cambridge University Press, 1903.

———. *Ethics.* Oxford: Oxford University Press, 1912.

———. "A Reply to My Critics." In P. A. Schilpp, ed., *The Philosophy of G. E. Moore.* Evanston, Ill.: Northwestern University Press, 1942, pp. 235–75.

———. "Is Goodness a Quality?" In *Philosophical Papers.* London: George Allen and Unwin, 1959, pp. 89–101.

———. "The Conception of Intrinsic Value." In *Philosophical Studies.* Totowa, N.J.: Littlefield, Adams, 1968.

Munro, D. H. *Empiricism and Ethics.* Cambridge: Cambridge University Press, 1967.

Murdoch, Iris. *Sovereignty of the Good.* Tel Aviv: Schocken, 1971.

Nagel, Thomas. *Mortal Questions.* New York: Cambridge University Press, 1979.

———. "The Limits of Objectivity." In Sterling M. McMurrin, ed., *The Tanner Lectures on Human Values.* Salt Lake City: University of Utah Press, 1980, pp. 77–139.

———. *The View from Nowhere.* Oxford: Oxford University Press, 1986.

Nielsen, Kai. "The 'Good Reasons' Approach and 'Ontological Justification' of Morality," *Philosophical Quarterly* 9 (1959), 116–30.

——. "Conventionalism in Morals and the Appeal to Human Nature," *Philosophy and Phenomenological Research* 23 (1962–63), 217–31.

——. "On Deriving an Ought from an Is: A Retrospective Look," *Review of Metaphysics* 32 (1979), 487–514.

Norton, David. *"David Hume": Common Sense Moralist, Sceptical Metaphysician.* Princeton, N.J.: Princeton University Press, 1982.

Nozick, Robert. *Philosophical Explanations.* Cambridge, Mass.: Harvard University Press, 1981.

Nowell-Smith, Patrick. *Ethics.* Baltimore: Penguin, 1954.

Perry, R. B. *General Theory of Value.* New York: Longmans, Green, 1926.

Perry, Thomas D. *Moral Reasoning and Truth.* Oxford: Oxford University Press, 1976.

Pettit, Phillip. "Evaluative 'Realism' and Interpretation." In Steven H. Holtzman and Christopher M. Leich, eds., *Wittgenstein: To Follow a Rule.* London: Routledge and Kegan Paul, 1981, pp. 211–45.

Phillips, D. Z., and H. O. Mounce. *Moral Practices.* Tel Aviv: Schocken, 1970.

Piper, Adrian M. S. "Instrumentalism, Objectivity, and Moral Justification," *American Philosophical Quarterly* (forthcoming).

Platts, Mark. *Ways of Meaning.* London: Routledge and Kegan Paul, 1979.

——. "Moral Reality and the End of Desire." In Mark Platts, ed., *Reference, Truth and Meaning.* London: Routledge and Kegan Paul, 1981, pp. 69–82.

Post, John F. *Faces of Existence.* Ithaca: Cornell University Press, 1987.

Postow, B. C. "Moral Relativism Avoided," *Personalist* 60 (1979), 95–100.

——. "Werner's Ethical Realism," *Ethics* 95 (1985), 285–91.

Price, A. W. "Varieties of Objectivity and Values," *Procedures of the Aristotelian Society,* 82 (1982/83), 103–19.

Prichard, H. A. "Does Moral Philosophy Rest on A Mistake?" *Mind* 21 (1912), 21–37.

Prior, A. N. *Logic and the Basis of Ethics.* New York: Oxford University Press, 1949.

——. "The Autonomy of Ethics," *Australasian Journal of Philosophy* 38 (1960), 199–206.

Putnam, Hilary. *Reason, Truth, and History.* New York: Cambridge University Press, 1981.

——. *How Not to Solve Ethical Problems.* The Lindley Lectures. Lawrence: University of Kansas, 1983.

——. *The Many Faces of Realism.* LaSalle, Ill.: Open Court, 1987.

Quinn, Warren. "Moral and Other Realisms: Some Initial Difficulties." In A. I. Goldman and Jaegwon Kim, eds., *Values and Morals.* Dordrecht: Reidel, 1978, pp. 257–63.

——. "Truth and Explanation in Ethics," *Ethics* 96 (1986), 524–44.

Railton, Peter. "Moral Realism," *Philosophical Review* 95 (1986) 163–207.

Rawls, John. "Outline of a Decision Procedure for Ethics," *Philosophical Review* 60 (1951), 177–97.

——. *A Theory of Justice.* Cambridge, Mass: Harvard University Press, 1971.

——. "The Independence of Moral Theory," *Proceedings of the American Philosophical Association* (1975), 5–22.

——. "Kantian Constructivism in Moral Theory," *Journal of Philosophy* 77 (1980), 515–72.

——. "Justice as Fairness: Political Not Metaphysical," *Philosophy and Public Affairs* 14 (1985), 223–51.

Rice, Phillip. *On the Knowledge of Good and Evil.* New York: Random House, 1955.

Ross, W. D. *The Right and the Good.* New York: Oxford University Press, 1930.

——. *Foundations of Ethics.* Oxford: Clarendon Press, 1939.

Russell, Bruce. "Moral Relativism and Moral Realism," *Monist* 67, (1984), 435–51.

Bibliography

Sayre-McCord, Geoffrey. "Coherence and Models for Moral Theorizing," *Pacific Philosophical Quarterly* 66 (1986), 170–90.
Scheffler, Samuel. "Moral Scepticism and Ideals of the Person," *Monist* 62 (1979), 288–303.
Schlick, Moritz. *Problems of Ethics*. New York: Dover, 1962.
Scott, Robert B., Jr. "Five Types of Ethical Naturalism," *American Philosophical Quarterly* 17 (1980), 261–70.
Sidgwick, Henry. "The Establishment of Ethical First Principles," *Mind* 4 (1879), 106–11.
———. *The Methods of Ethics*. New York: Macmillan, 1907.
Singer, Peter. "Sidgwick and Reflective Equilibrium," *Monist* 57 (1974), 490–517.
Smart, J. J. C. *Ethics, Persuasion and Truth*. London: Routledge and Kegan Paul, 1984.
Snare, Frank. "Externalism in Ethics," *Philosophical Quarterly* 24 (1974), 362–65.
———. "The Argument from Motivation," *Mind* 84 (1975), 1–9.
———. "Three Sceptical Theses in Ethics," *American Philosophical Quarterly* 14 (1977), 129–36.
———. "The Diversity of Morals," *Mind* 89 (1980), 353–69.
———. "The Empirical Bases of Moral Scepticism," *American Philosophical Quarterly* 21 (1984), 215–25.
Stace, W. T. *The Concept of Morals*. New York: Macmillan, 1962.
Stevenson, C. L. *Ethics and Language*. New Haven, Conn.: Yale University Press, 1944.
———. *Facts and Values*. New Haven, Conn.: Yale University Press, 1963.
———. "Ethical Fallability." In Richard DeGeorge, ed., *Ethics and Society*. New York: Doubleday, 1966, pp. 197–217.
Sturgeon, Nicholas. "Brandt's Moral Empiricism," *Philosophical Review* 91 (1982), 389–422.
———. "Gibbard on Moral Judgment and Norms," *Ethics* 96 (1985), 22–33.
Toulmin, Stephen. *The Place of Reason in Ethics*. Cambridge: Cambridge University Press, 1961.
Trigg, Roger. *Reason and Commitment*. Cambridge: Cambridge University Press, 1973.
Urmson, J. O. "On Grading," *Mind* 59 (1950), 145–69.
———. *The Emotive Theory of Ethics*. Oxford: Oxford University Press, 1968.
Von Wright, G. H. *The Varieties of Goodness*. London: Routledge and Kegan Paul, 1963.
Veatch, Henry. *For an Ontology of Morals: A Critique of Contemporary Ethical Theory*. Evanston, Ill.: Northwestern University Press, 1971.
Warnock, G. J. *The Object of Morality*. London: Methuen, 1971.
Wellman, Carl. "Emotivism and Ethical Objectivity," *American Philosophical Quarterly* 5 (1968), 90–99.
———. "Ethical Disagreement and Objective Truth," *American Philosophical Quarterly* 12 (1975), 211–21.
Werner, Richard. "Ethical Realism," *Ethics* 93 (1983), 653–79.
———. "Ethical Realism Defended," *Ethics* 95 (1985), 292–96.
Westermarck, Edvard Alexander. *Ethical Relativity*. New York: Humanities Press, 1932.
White, Morton. *What Is and What Ought to Be Done*. New York: Oxford University Press, 1981.
Williams, Bernard. *Problems of the Self*. Cambridge: Cambridge University Press, 1973.
———. *Moral Luck*. New York: Cambridge University Press, 1981.
———. "The Scientific and the Ethical." In S. C. Brown, ed., *Objectivity and Cultural Divergence*. Cambridge: Cambridge University Press, 1984, pp. 209–28.
———. *Ethics and the Limits of Philosophy*. Cambridge, Mass.: Harvard University Press, 1985.

Wong, David. *Moral Relativity*. Berkeley: University of California Press, 1984.
Zimmerman, David. "Meta-Ethics Naturalized," *Canadian Journal of Philosophy* 10 (1980), 637–62.
——. "Moral Realism and Explanatory Necessity." In David Copp and David Zimmerman, eds., *Morality, Reason and Truth*. Totowa, N.J.: Rowman and Allanheld, 1985, pp. 79–103.

Index

Abrams, M. H., 157n, 239n
absolutism, 30, 138n
action: intentional, 293–98; reason for, 81, 104–5, 214–15, 293–300
Adams, R. M., 239n
Aeschylus, 47
aesthetic concepts, 36; convictions, 180n; criticism, 35, 89, 102, 180n, 289n; experience, 35; explanation, 256n; feeling, 35; judgment, 35, 92, 114; responses, 179; statements, 27; terms, 35; values, 95, 114
Agamemnon, 47
agglomeration principle, 53–56
Anscombe, G. E. M., 116, 294
anthropocentrism, 138, 142, 155; and objectivity, 164
anti-realism, 2–5, 6n, 9–11, 13, 22, 59, 61, 67, 70, 72–73, 199, 205, 208, 212, 215–16, 219–21, 223, 288–89, 294–96, 299; constructivist, 219; moral, 217
Aquinas, Thomas, 116
argument from queerness, 109, 111–14, 117, 175, 230
argument from relativity, 12, 109–11, 117, 229
argument from religious experience, 39–40
Aristophanes, 147
Aristotle, 106, 116, 141, 157, 161–62, 178
arithmetic, 82, 84, 89; transfinite, 85–91, 94
Armstrong, D. M., 188, 191
assertibility, 6n, 144–50, 157–59
assertion, 8, 32, 37, 40, 258; conditions, 144–46; ethical, 28; factual, 28; legal, 16; metaphysical, 38; moral, 231, 253, 258, 295; religious, 37–38
attitudes, 4, 16, 35, 97, 99, 106–7, 114, 141, 166–67, 175; moral, 9, 33, 43, 114, 290–91, 295–96
Austin, J. L., 55, 162n
Ayer, A. J, 4n, 7, 8n, 27n, 155n

Baier, Kurt, 4n, 19
Beanblossom, R. E., 4n
Beckerman, Wilfrid, 139n

behavior, 76, 83, 92–93, 114–15, 194, 273; mathematical, 78; moral, 35, 78, 80
beliefs, 2n, 41–46, 50–51, 73–74, 80, 82–83, 92–93, 112, 120–22, 129, 150, 154n, 191–92, 195, 206, 209–10, 233, 236n, 237–38, 241, 248, 252n, 258, 267, 272–73, 275, 287, 290–91, 294; arithmetical, 83; cognitive, 77; commonsense, 233; conflict of, 41–42, 46, 49–52; conflicting, 42, 44, 50; consistent, 42; ethical, 185; factual, 42, 46, 282, 294–97; foundational, 191; justified, 191, 233, 235, 249, 253, 263, 266–67; 273–74, 278, 280–81; mathematical, 78, 80, 233, 235; moral, 8–9, 12–13, 78, 80, 83, 109, 183–87, 201, 206–8, 211, 224–25, 233–36, 238–39, 243, 245–46, 249, 250n, 253–54, 270, 285, 290–91, 295–97; perceptual, 191; reasonable, 264, 267; reflective, 92–93; religious, 38; scientific, 73, 183–85, 233, 288, 299; system of, 121, 149; theoretical, 152; true, 42, 44, 50–51, 156, 191, 195, 201, 208–10, 267, 294
Bellah, Robert, 2n
Belnap, Nuel, Jr., 21n
Benacerraf, Paul, 263
Benedict, Ruth, 18
Bentham, Jeremy, 140
Berkeley, George, 101
Berlin, Isaiah, 157n
Blackburn, Simon, 2n, 9n, 12n, 59, 75, 83n, 167n, 175n, 176n, 177n, 178, 179n, 180n, 257n, 266, 291
Bostock, David, 72n
Boyd, Richard, 13n, 21n, 181–82, 184, 187–92, 194–95, 198–99, 209, 216–17, 240, 242n, 252n, 278n
Boyle, Robert, 98
Bradley, F. H., 163, 165
Brandt, Richard, 17n
Brentano, Franz, 157n
Brink, David, 12n, 14n, 183
Burnyeat, M. F., 178n
Byerly, Henry, 188

[309]

Index

Index

homeostatic mechanism, 197–98, 203, 204n, 214, 218
Homer, 88
Hume, David, 10–11, 13, 99, 106, 111–12, 114, 151, 163, 176, 215
Hutchenson, Francis, 106
Huxley, Aldous, 158n

idealism, 7–9, 14, 70; subjective, 101
Ideal Observer theories, 17
imperatives, 103–7, 115, 117
instrumentalism, 7–14, 182, 236n; scientific, 8n
intersubjectism, 14–22
intersubjectist, 14–15, 18; truth-conditions, 15
intersubjectivity, 100
intersubjective, 105
intuition, 30, 39, 111, 113, 167, 180, 185, 193, 199–200, 206–9, 284–85; ethical, 111, 200, 285; mathematical, 265n; moral, 13, 99, 181n, 184–85, 206–8, 223; physical, 193, 207–8; scientific, 192–93, 200, 206–7
intuitionism, 111, 160, 164, 282, 285, 287, 290–91, 293; ethical, 282, 284–85; mathematical, 158–59; moral, 291; pluralistic, 285; realistic, 289–91; realistic ethical, 286
invention, 81, 89, 133, 155, 158–60, 164, 180n

James, William, 154n
Jeffrey, R. C., 153
Jennings, Paul, 141
Johnson, Samuel, 135
judgment, 60–61, 99–100, 102, 121, 133–34, 144–45, 148–51, 155, 179, 221–22, 233, 244, 246, 248, 259, 270–72, 278–79, 287, 293, 298; aesthetic, 35, 92, 114; conflicting, 53; deliberative, 133, 142n, 144, 148–49, 150n, 159n, 162n, 166n; directive, 133, 166n; empirical, 158; ethical, 28–34; evaluative, 103, 148, 150, 162n, 279; evidential, 190; factual, 134, 159n, 185–86, 221, 287; geographical, 148; historical, 148; inferential, 192; intuitive, 193; moral, 9n, 12–14, 18, 28–30, 32, 41, 45, 57, 60, 65–66, 72, 77, 92, 96–97, 101–2, 105, 107, 108–10, 111–15, 122, 144, 182, 185–86, 205–6, 208, 214–15, 221–22, 251–53, 262, 269–72, 279, 282–83, 285–87, 291–99; natural, 60; nonmoral, 221–22, 283, 292; perceptual, 121, 192–93, 283; practical, 133, 148, 150n, 158–60, 166n; prescriptive, 100–1; psychological, 28; scientific, 271n; sociological, 28; trained, 192–93, 200, 206–7; validity of ethical, 29–33; valuational, 134n; value, 16–18, 34, 105, 144, 149, 150n, 277

justification, 19, 77, 80, 83, 123–24, 191, 229, 233, 235, 238, 248–49, 252n, 257, 267, 269, 273–81, 291; moral, 280

Kamin, L. J., 211
Kant, Immanuel, 35–36, 88n, 101–5, 115
Kenny, Anthony, 43
Kim, Jaegwon, 63, 75, 250
kind, 71, 74, 196, 198, 201; homeostatic cluster, 198–99, 201; natural, 68, 73, 184, 195–98
Kitcher, Philip, 263n
Klagge, James, 13n
Kneale, William, 133n
knowledge, 4n, 5n, 6n, 10n, 27, 34–36, 78, 101–2, 106–7, 111–12, 118, 157, 167, 181, 184, 187–91, 194, 199, 205–6, 208–9, 216–17, 249, 263–66, 282, 284; causal theories of, 188, 191, 235, 252n, 263, 265–66; causal theory of moral, 215; empirical, 183, 192; experimental, 205; foundational, 191; framework, 72–73; mathematical, 266; moral, 40, 111, 182–84, 200–2, 205, 207–10, 233–35; naturalistic conception of, 184, 200, 208, 213; naturalistic conception of moral, 215; naturalistic conception of scientific, 216; naturalistic theories of, 188, 195; objective, 189–91; perceptual, 171n, 191, 208–9; realistic conception of, 184, 208, 213; realistic conception of moral, 200–2; realistic conception of scientific, 216; religious, 36, 38, 40; scientific, 72, 183–84, 190–93, 201, 206, 208–9, 216; theories of, 188, 195; theoretical, 188–90; of truth-conditions, 284
Kolakowski, Leszek, 157n
Kolnai, A. T., 147, 154n, 157n
Kripke, Saul, 188, 195, 252n, 264n
Kuhn, Thomas, 182, 184, 187, 190, 193

language: descriptive, 4; evaluative, 166; meta-, 21, 143–46; moral, 7–9, 19, 41, 107, 113, 116, 118, 143, 184, 199, 212, 226, 250n, 264, 265n, 285, 287, 292, 295, 298n; object, 21n, 143–47; philosophy of, 182, 184, 187–88; prescriptive, 4; scientific, 216; theoretical, 190
law, 16, 70, 116, 193, 266; divine, 116, 208; of excluded middle, 159–60; moral, 106, 208; natural, 16, 266; physical, 266; scientific, 7, 266
Lazara, Vincent, 188
Lear, Jonathan, 84n, 85n, 89n
Lehrer, Keith, 4n
Leich, Christopher, 9n, 77n, 175n, 177n

Library of Congress Cataloging-in-Publication Data

Essays on moral realism.
  Bibliography: p.
  Includes index.
1. Ethics. I. Sayre-McCord, Geoffrey, 1956–   .
BJ1012.E85   1988   170'.42   88–47753
ISBN 0–8014–2240–X (alk. paper)
ISBN 0–8014–9541–5 (pbk. : alk. paper)